Current Perspectives on Sexual Selection

History, Philosophy and Theory of the Life Sciences

Volume 9

For further volumes:
http://www.springer.com/series/8916

Thierry Hoquet
Editor

Current Perspectives on Sexual Selection

What's left after Darwin?

Springer

Editor
Thierry Hoquet
Faculty of Philosophy
Université Jean Moulin Lyon 3
Lyon
France

ISSN 2211-1948 ISSN 2211-1956 (electronic)
History, Philosophy and Theory of the Life Sciences
ISBN 978-94-017-7990-6 ISBN 978-94-017-9585-2 (eBook)
DOI 10.1007/978-94-017-9585-2

Springer Dordrecht Heidelberg New York London

Softcover reprint of the hardcover 1st edition 2015

Printed on acid-free paper

Springer is part of Springer Science+Business Media (www.springer.com)

Introduction

Opening Pandora's Boxes in Sexual Selection Research

Today, about one hundred and fifty years after the publication of Darwin's *Origin of Species*, how do biologists, historians, and philosophers reassess the strengths and weaknesses of sexual selection? Answering this is the overall aim of this book.

At least two major concepts of contemporary biology originate in the works of Charles Darwin: natural selection and sexual selection. Originally, sexual selection dealt with the competition for mates, while natural selection was more concerned with individual survival. As Darwin says in the first edition of his *Origin of Species* (1859, p. 88): "This depends, not on a struggle for existence, but on a struggle between the males for possession of the females; the result is not death to the unsuccessful competitor, but few or no offspring. Sexual selection is, therefore, less rigorous than natural selection."

The concept of sexual selection was amply developed and refined in Darwin's *Descent of Man, and Selection in Relation to Sex*, along with several other issues like the "proportions of the sexes" (Darwin 1871; see Veuille, this volume). For Darwin, sexual selection "has led to the development of secondary sexual characters." (1871, vol. 1, p. 271) This is fully consistent with Darwin's commitment to an individualistic stance (Ruse, this volume).

More specifically, sexual selection consists of two different processes, "the power to charm the female" and "the power to conquer other males in battle." (1871, vol. 1, p. 279) Thus, there are two selecting forces within the process of sexual selection. They are usually termed "female choice" and "male-male competition." The first leads to ornaments, the second to armaments. On first inspection, sexual selection is just one kind of selection, with a different kind of selector: just as artificial selection is operated by breeders, so sexual selection (at least, its intrasexual component, female choice) is operated by mates, so natural selection is operated by a metaphorically personified agent called "nature". While Darwin's contemporaries readily accepted male-male competition, several issues were raised on the question of female choice. How could female animals develop the ability to discriminate between males or to consciously weigh up their differences? And, more importantly, was there not a contradiction between natural and sexual selection? As one of Darwin's fiercest critics put it: "A crest of topaz is no better in the struggle for existence than a crest of sapphire. A frill ending in spangles of the emerald is no better in the battle of life than a frill ending in the spangles of the ruby. A tail is not affected for the purposes of flight, whether its marginal or its central feathers are decorated with white." (Argyll 1867, pp. 247–248)

Sexual selection claims to account for beauty in animal features: however, it raises the question of whether it can be consistent with the action of natural selection. While natural selection focuses on useful features, sexual selection opens up the possibility of an aesthetic sensibility and suggests that beauty somewhat determines animal anatomy and behaviour. How can evolutionary theory take into account what is obviously beautiful as well as the preference animal individuals seem to show towards beauty?

This question puzzled several naturalists who expressed doubts on sexual selection, faulting it for its anthropomorphical overtones. The most prominent of them is probably Alfred Russel Wallace, who scornfully asserted:

> A young man, when courting, brushes or curls his hair, and has his moustache, beard or whiskers in perfect order and no doubt his sweetheart admires them; but this does not prove that she marries him on account of these ornaments, still less that hair, beard, whiskers and moustache were developed by the continued preferences of the female sex. So, a girl likes to see her lover well and fashionably dressed, and he always dresses as well as he can when he visits her; but we cannot conclude from this that the whole series of male costumes, from the brilliantly coloured, puffed, and slashed doublet and hose of the Elizabethan period, through the gorgeous coats, long waistcoats, and pigtails of the early Georgian era, down to the funereal dress-suit of the present day, are the direct result of female preference. In like manner, female birds may be charmed or excited by the fine display of plumage by the males; but there is no proof whatever that slight differences in that display have any effect in determining their choice of a partner. (Wallace 1889, pp. 286–7)

Wallace readily accepted male-male competition but rejected female choice. His name now stands for utility while Darwin's stands for beauty (Prum 2012, and Hoquet & Levandowsky, this volume). Following Wallace's criticisms, sexual selection became the laughing point of Darwinian theory. Biologists like Thomas Hunt Morgan considered it useless and dedicated several books to showing its irrelevance (1903, 1919). In his *The Scientific Basis of Evolution*, Morgan refers to secondary sexual characters and states that "Darwin attempted to explain their origin historically in his theory of sexual selection, while the modern attitude is to accept their presence as given, and to devote attention to the physiology of their development in the individual." (1932, p. 152) Here we understand an underlying opposition between ultimate and proximate causes. Are evolutionary or ultimate factors the most fundamental causes that account for animal traits—like courtship and display, what Darwin called "strange antics"— or should proximate factors (like hormones or immediate rewards) also be taken into account (see the contributions of Watanabe and Kreutzer & Aebischer in this volume)?

Due to these puzzles and misconceptions, it is often claimed that sexual selection underwent a long century of absence, from its original wording in Darwin's *Descent* (1871) to its "rediscovery" by Trivers (1972) in the wake of sexual liberation. Examples of what I would call the "century of neglect" narrative—allegedly lasting from 1871–1972—can be found in Cronin (1991) or Zuk (2002). However, it requires several important caveats. In this narrative, Ronald A. Fisher (1915, 1930) is singled out as having played an important role and is often quoted as a notable and singular exception among his peers for his idea of a "runaway process". In fact, the story is much more complex, especially regarding female choice (Milam 2011, Gayon, this volume). Indeed, throughout the 20th century one can list many names and projects that contributed to the development of sexual selection as both a theory and a field for experimental studies: in the 1910s, the entomologist Frank E. Lutz (1879–1943), at the Cold Spring Harbor Biological Laboratory, was doing experimental work on sexual selection with drosophila strains (Lutz 1909); from the 1910s to the 1930s, Julian Huxley made field studies of great crested grebes (1914, 1938), while Theodosius Dobzhansky (1944), Ernst Mayr (1946) and their colleagues were actively researching the question of mate choice using drosophila as a model organism and making several observations in the field.

Their work prompted a British plant geneticist, A.J. Bateman, to study intrasexual selection in drosophila. His paper, published in *Heredity* in 1948, was, at first, rather overlooked until, in 1972, it was quoted by Robert Trivers, triggering an exponential rise in citations as well as the exceptional fortune Bateman's paper has enjoyed in the last quarter of a century. From Darwin to Bateman to Trivers, a common thread is the depiction of two stereotypical sex roles: coy females versus eager males. This view became a motto for many sociobiologists and, conversely, it was targeted by many critics of sexual selection, including feminist scholars. Trivers' concept of "parental investment" explains how natural selection may be reconciled with sexual selection. If parental investment increases the progeny's chances of surviving, then each individual offspring can be considered as an investment. This, in turn, reduces the parent's ability to produce further offspring. The basic principles of the theory are that females are limited by the number of offspring they can successfully rear and that males are limited by the number of eggs they can fertilize. But its implications are subtler than that. Due to unequal parental investment, one sex (often, but not necessarily, the female) plays the role of "limiting resource" while the other (often, but not necessarily, the male) is mainly a contributor of genes to the making of the offspring, with little or no further contribution to parenting beyond this, and it is he who is competing for access to mates. Sex cells are, of course, an element of parental investment, though certainly not its only component. In this framework, due to their limited stock of costly eggs, females are the sex-limiting factor, with males, on the contrary, having a large number of small sperm. Gestation and lactation, in mammals for instance, are also important components of female investment, while in other orders, like sea-horses, females simply deliver the eggs whereas pregnant males provide the more taxing investment of developing the embryos.

Severed from the idea of stereotyped sex roles, sexual selection becomes a "logical imperative" (as is suggested by Parker and Pizzari, this volume). In the summary of the fourth chapter of his *Origin of Species*, Darwin gave what has been called the "recipe" approach to natural selection: if ingredients x, y and z are given, then evolutionary change will necessarily follow. The recipe is applicable, in principle, to any entity capable of some form of reproduction (Godfrey-Smith 2009). Similarly, the recipe version of sexual selection starts with gamete competition and anisogamy, thus emphasizing pre-copulatory factors rather than post-copulatory ones. In this version, anisogamy is the primary sexual difference, playing a crucial but not exclusive role in sexual selection. Other factors may also be included in sexual selection in order to construct it as a testable hypothesis (Gowaty, this volume).

Another important set of questions relates to the well-worn issue of genetic determinism (Waage and Gowaty 1997): what difference exists between the "decisions" of the animal individuals (humans included) and their evolutionary roots or causes? If selection operates only on genes, then what relation is there between genetic sequences and behaviour? Questions bearing on the nature of homosexuality, of maternal instinct, and of altruism have been particularly debated, raising again and again the classical opposition between nature and nurture, but also between ultimate and proximate causes. The issue here is twofold. First, as several authors emphasize (especially Ah-King and Gowaty, this volume), new understandings of the mechanisms of gene expression and epigenetics lead to a revised account of individual development and of the way genes are said to determine phenotypes. Another question is whether the depiction of standard sex-roles (coy vs eager, choosy vs promiscuous, passive vs active) is flawed: does this mean that sexual selection is condemned to doom and oblivion (Roughgarden, this volume)?

The question of sex roles tightly links sexual selection to social issues such as women's liberation. In a foreword to Cronin (1991, p. ix), John Maynard-Smith contrasted current enthusiasm for sexual selection with its previous and long-lasting disrepute, accounting for how "neglect of sexual selection turned to enthusiasm during the 1970s and 1980s." Maynard-Smith supported the hypothesis that "this change of attitude" should be ascribed "to the influence of the women's movement," and affirmed that, "It is certainly not the case that the new research, theoretical and empirical, has been carried out by ardent feminists, but I think it may have been influenced, even if unconsciously, by attitudes towards female choice in our own species." The role that feminism played in the history of sexual selection is a contested area, one that is broached in several chapters of this volume (e.g., Ah-King and Gowaty). One of the earliest and probably most important contributions to the feminist critique of Darwin's ideas is the work published in 1875 by a Unitarian minister, Antoinette Blackwell. Blackwell criticized Darwin for failing to see how females actually developed new features. After all, both Darwinian mechanisms, male-male competition and female choice, account for evolution of male traits: "With great wealth of detail, he [Darwin] has illustrated his theory of how the male has probably acquired additional masculine characters; but he seems never to have thought of looking to see whether or not the females had developed equivalent feminine characters." (1875, p. 16) Primatologist Sarah Blaffer Hrdy has brought out the role played by Blackwell: "Blackwell's informed dissent was drowned out in the wake of popular acceptance of social Darwinism. Her contribution to evolutionary biology can be summed up with one phrase: the road not taken." (1981, p. 13)

The notion of stereotyped sex-roles has been highly criticized. Women scientists like S.B. Hrdy (1981), Linda Fedigan (1982) and Barbara Smuts (1985) have shown how female primates in particular are far from being passive participants in the reproductive process. They challenge the assumption that females only mate for fertilization and show that copulation can occur under a variety of circumstances: primate females engage in non-cyclical and non-reproductive sexual activity with various males. Patricia Gowaty (1992) has argued that many points in evolutionary theory could be used by feminists thinkers, especially those regarding variation of female behaviour or female resistance to male control of female sexuality. She also argued that evolutionary biologists might learn from feminists as well.

Both Gowaty and Zuk warned their colleagues about possible male chauvinism blinding evolutionary biology to important insights and preventing it from making further progress. Zuk, for instance, stressed the fact that "an awareness of such bias can help us recognize the need to listen to a variety of voices, which will ultimately result in less biased and more productive science. [...] I address biologists as a feminist, to explore ways that feminism can affect and hopefully improve the study of the evolution of behavior." (1993, p. 774) Referring to new insights in evolutionary biology, Zuk suggested, following Trivers, that the first creature was actually Eve and that she had not been fashioned out of one of Adam's ribs as the traditional account would have it. As Trivers put it in an exam question quoted by Zuk: "And God made the first creature and called her "Eve," and spoke to her, saying, "You and you alone are capable of nourishing an offspring out of your own body. In my own image I have made you. Out of your body all life here on earth shall flow. Go and reproduce and populate the earth. Natural selection will take care of the rest. For thou art Eve."" As to Adam, he is just an additional creation, in order to allow Eve to fight efficiently against parasites: "And God caused a deep sleep to fall over Eve and She took one of Eve's ribs and fashioned a male out of it.

To him She said, "You I will call Adam or man, for you are not fully woman, but were made from her to help her fight parasites, and, if possible, give her a little pleasure along the way." (1993, p. 775) Such change of perspective, Zuk commented, "illustrates how use of feminist principles can expand conceptual possibilities for biology students." (1993, p. 775) However, Robert Trivers abruptly replied that he "would rather derive [his] feminist principles from evolutionary biology than [his] evolutionary biology from feminist principles."

> Evolutionary biology teaches us that females are primary in evolution, males being secondarily derived to serve female needs; that once both sexes are present, female interests carry at least as much evolutionary weight as male interests; that mating systems with a strong component of female choice naturally evolve to give greater weight to female than male interests; and so on. (Trivers 1994)

Trivers also claims that "too-close adherence to political principles as a guide to reality tempts us to merely replace the self-deceptions of the past with a brand new set"—suggesting that, only if we forget about politics, may we achieve an objective, value-free science, emancipated from the dialectical trap of the feminist Scylla and the masculinist Charybdis. However, he does not explain how he controls his own bias, nor does he make the effort to offer any suggestions on how to do so.

No doubt that, partly due to these entanglements with highly debated topics like gender roles and stereotypes, partly due to its status as sound and testable evolutionary hypothesis, sexual selection today is probably "one of the most dynamic areas" of evolutionary research (Parker and Pizzari, this volume). However, it still has an incredible ability to trigger wild controversies. While some acknowledged the Darwinian mechanism as almost self-evident, others criticized it as fatally flawed. In 2006, Joan Roughgarden and her colleagues vigorously challenged the validity of sexual selection as a central component of modern evolutionary theory (Roughgarden et al. 2006). Their paper boldly stated that sexual selection was 'wrong'. This prompted an abundance of responses from evolutionary biologists (Kavanagh 2006), and it was, no doubt, an instigating factor in my own resolution to clarify the epistemological status of sexual selection through two conferences I organized in 2011 (see below).

From an epistemological point of view, the situation of sexual selection debates is much more complex and interesting than just an opposition between supporters and critics. Gowaty, for instance, actively defends Darwinian sexual selection against Roughgarden's claims, although she has repeated Bateman's experiments and concluded that his results were fatally flawed (Gowaty et al. 2012). Conversely, Cézilly (this volume) acknowledges the overall validity of the "Darwin-Bateman Paradigm", but he agrees with Roughgarden that sexual selection theory has now moved far beyond coy females and eager males, and he takes issue with standardized sex preferences and stereotyped sex roles.

To the epistemologist's eye, sexual selection is a somewhat fuzzy concept and new evidence coming from the field of science in action is a regular occurrence. I had the privilege of being invited to participate at a recent NESCent Catalyst Meeting that was held in Durham (NC) from July 15–17 2013 and organized by Joan Roughgarden. The meeting gathered 34 participants who actively reviewed the state of what Roughgarden calls "sexual selection studies". One of our main goals was to come up with a consensus definition of sexual selection. David Shuker (2010) suggested starting out with the following: "Sexual selection describes the selection of traits associated with competition for mates." Continuing, "More formally, sexual selection is the relationship between a trait and its effect on fitness through sexual competition." (2010, p. e12) Several aspects of this definition were shared by many other participants: that sexual selection is not dependent on traditional sex roles like eager males and coy females; that sexual selection is not dependent on cheap sperm and costly eggs; that sexual selection differs from sexual conflict; and, that sexual selection is different from female choice. However, other aspects were also hotly debated; these are treated in later chapters of this volume. The NESCent meeting showed that Shuker's tentative consensus statement is a sort of Pandora's box, or rather a complex of intricate Pandora's boxes that the participants meticulously and methodically opened, one after the other. I draw extensively on the final report of the conference in the following (Roughgarden et al. 2013).

The main Pandora's box is probably contained in wording like "competition for mates", which spawned several issues: is sexual selection about access to *mates*, or about access to *fertilizable gametes*? Maybe the whole question of "access to mates" or numbers of matings is secondary to number of actual fertilizations. What does "competition for mates" encompass? Not only the Darwinian mechanisms of male-male competition and female choice of males, but also competition to attract better mates. Several participants suggested that behavioural aspects should be precluded from the definition and that evolutionary biologists should only be concerned with an operational protocol for *measuring* sexual selection with a concept like "opportunity for selection" (a claim supported by Shuster and Wade 2003). Post-copulatory versions of these different themes were also suggested. Potentially, even if no consensus definition could actually be achieved, some major components could certainly be kept in mind: one of the major aspects of sexual selection is differential fertilization success, or the variance in the number of fertilizations, owing to the variance within one sex in, for example, the number of mates and/or the number of fertilizations per mating.

But even a very old and somewhat basic question such as "Is sexual selection a component of natural selection?" (Mayr 1972) can receive no clear answer; many biologists happily agree that it is, but some claim that sexual and natural selections are actually quite distinct processes. Is this question merely pedantic? For the historian of science, it is clear that many of Darwin's readers thought at first that sexual and natural selections might contradict each other, that, for instance, ornaments like the cumbersome peacock's tail could be detrimental to individual survival. But, strikingly enough, this important question is still raised today, as it is related to central theoretical concerns: what features of the individual organism can be said to be, in Darwin's terms, "advantageous" or "useful", or, in more recent parlance, "functional" or increasing of "fitness"? What is selected; viability of the individual, or transmission of genes?

Moreover, the concept of sexual selection has been profoundly updated since Darwin. The sharp discrepancy between Darwin's (1871) original insights and the contemporary notion of sexual selection throws some doubts on the unity of the concept (see Roughgarden, this volume). If one takes the historical dimension of sexual selection seriously, then one sees how changing definitions make it somewhat an umbrella-term. While biologists constantly refer to Darwin's views and depict themselves as the true heirs of Darwin's mantle, there is little consistency in these claims of legacy. Darwin had no idea of mathematization of sexual selection; he strongly suggested that sexual selection was linked to typical sex roles. One aspect of sexual selection that Darwin was completely unaware of is the role of selfish genetic elements (SGEs): genes, organelles or micro-organisms might manipulate patterns of inheritance in order to increase their representation in the next generation (Weddell & Price, this volume). The impact of SGEs on sexual selection and especially on male and female reproduction is difficult to measure. Another issue at stake is whether SGEs may play a role in shaping mating systems.

Another Pandora's box in Shuker's initial definition is that it requires genetic covariance between variation in a sexually-selected trait and variation in fitness. This raises the question of whether sexual selection is for "good genes" or, if not, then at least for signals of genetic quality. This even led to the paradoxical hypothesis of the "handicap principle". According to Zahavi, odd or costly traits such as the peacock's tail are in fact *adaptive*: they signal the ability of the individual to survive and function while encumbered with the cost (Zahavi 1975). The perspective that selection is ultimately for quality is called "Wallacean", because of Wallace's emphasis on vigour of males, while the opposite view, purporting that selection is for truly aesthetic, and possibly detrimental, features is called "Fisherian". Fisher's runaway-process (1915) develops a case wherein a female preference for a male trait leads to an increase in the male trait which in turn leads to an increase in the female preference for that trait, and so on—with no necessary positive impact on fitness. One could say that the viability impact of natural selection (how a trait increases individual fitness) may be in opposition to the sexual preference aspect: this hypothesis of indirect benefit of sexually attractive offspring is called the "sexy son" hypothesis.

A recent review suggested that, through sexual selection, females get "sexy sons" rather than "good genes" (Prokop et al. 2012). However, Joan Roughgarden (2009) suggested that an ornament might be "sexy" in a given generation, and the same ornament might also be heritable, but this does not necessarily entail that it will be "sexy" in the next generation. Female preferences may not be heritable, even if male traits are. Lack of cross-generational conservation of preferences is clearly evinced by studies on the collared flycatcher case, which show that the female preference for ornament was not heritable (Qvarnström et al. 2006).

Based on 24 years of study with 8500 birds on the Swedish island of Gotland, heritabilities were measured for several factors: heritability of the white badge (a male ornament, considered a sexually selected character), heritability of fitness, heritability of female preference for the badge, genetic correlation between preference and badge. Results of this study showed that ornament is moderately heritable but that male fitness is weakly heritable. This seems to suggest that the badge does not function in sexual selection, the low correlation of the badge size and male fitness showing that the badge size is not an indicator of good genes. Furthermore, daughters do not inherit the preference of their mother—which also disparages the sexy sons hypothesis. The question here is: does this one species study really carry enough weight to make us cast doubt on the validity of the sexual selection model? Or should it be discarded as irrelevant, dealing only with one species, or with the work of only one group of scientists? Cézilly (this volume) highlights several methodological issues in female choice experiments: repeatability of experiments, bias due to the fact that one team publishing on one topic would work always with the same animal population, etc.

Another theoretical issue at stake here is whether heritability is an important aspect of sexual selection or not. If sexual selection is a behavioural process, then what matters is whether, within one given generation, the bearers of a defined trait leave more progeny than their same-sex conspecifics; it does not necessarily lead to evolutionary consequences, meaning that, although there are several heritable features (badge size, female preference, etc.), heritability may well not be a necessary component of sexual selection. However, many biologists tend to accept that sexual selection (be it intra-sexual competition or inter-sexual preferences or choice) involves heritability since it leads to evolutionary consequences.

On the long debated issue of the evolutionary causes and consequences of female choice, some (like Cézilly, this volume) suggest that we should focus more on female choice *per se* and on female preferences. Are these arbitrary? Rational? Useful? What is the meaning of rationality? Comparison and deliberation may be good criteria, but transitivity may well be the fundamental axiom of rational choice. Others (like Prum, this volume) emphasize the importance of sexual autonomy, defined as the opposite of sexual coercion from the opposite sex.

The question of the null-models is also a cause of strong disagreement in the biological community (see Prum 2010, and this volume): under the search for a null-model, the basic question at stake is: "How can we understand whether sexual selection is, or is not, currently occurring in a given population?" This important issue (which hypotheses should define the "null") is also a Pandora's box about which it seems equally difficult to achieve a definite consensus.

As a philosopher of science, I was very much interested, both on the epistemological and historical levels, in seeing various biologists debate on these issues. I invited a large panel of biologists, historians and philosophers to gather in Paris to try to understand how each of them was variously reassessing the strengths and weaknesses of sexual selection. Two conferences were held in Paris in 2011: "*The aesthetic sense in animals (Le sens esthétique des animaux)*", held at Université Paris-Ouest Nanterre on 13–14 January 2011; "*What's left of sexual selection? (Que reste-t-il de la sélection sexuelle?)*", held at the Musée National d'Histoire Naturelle and at Université Paris-Ouest Nanterre on 19–20 May 2011. Both meetings were funded by the French National Research Agency (Agence Nationale de la Recherche), the Biosex Research Project (ANR-07-JCJC-0073-01) and Université Paris-Ouest (Pôle 5 "L'humain en devenir", EA 373, ED 139). The Institut Universitaire de France and the UMR 7205 OSEB (Origine structure et évolution de la biodiversité, MNHN-CNRS-EPHE) also made these meetings possible, through their logistic support. I thank Malek Bouyahia, Melanie Petremont and Eva Rodriguez at the ANR BIOSEX Project for their support in putting these two conferences together. The Biosex Project (http://biosex.univ-paris1.fr/) was a 4 year research group (2007–2011) under the direction of Elsa Dorlin (then at the university Paris 1 Panthéon-Sorbonne). Our aim at Biosex was to bring together scholars from the humanities and the biological sciences, as well as to build an interdisciplinary interface for reflecting on the ways in which biology has conceptualized sex, especially since Darwin. Assessing the question of sexual selection was a core issue in our project, given the theoretical and political intricacies of this topic and also the way feminism and sociobiology have interacted in order to bring out "a better science."

The two conferences brought together specialists of behavioural ecology, historians of science, and philosophers. The participants were asked to reassess the importance of the concept of sexual selection from historical, epistemological and theoretical perspectives. The title of the second conference in particular, "What's left of sexual selection?", could be understood in two different ways: (a) from a technical biological perspective: is the sexual selection concept still useful 151 years after Darwin introduced it in his *Origin of Species* and 140 years after the publication of his *Descent of Man*? (b) As a general social concern, playing with the "left/right" dichotomy: is "sexual selection" a progressive or a reactionary notion?

The present book is divided into three sections. Section 1 assesses the character of the "second" Darwinian revolution, the one dealing not with natural but with sexual selection. Section 2 studies how current sexual selection theory goes further than Darwin in many respects. Section 3 deals with animal aesthetics, as a developing area in sexual selection studies.

References

Argyll G, Douglas Campbell, Duke of (1867) The Reign of law. Alexander Strahan, London

Bateman AJ (1948) Intrasexual selection in Drosophila. Heredity 2:349–368

Blackwell A (1875) The sexes throughout nature. G.P. Putnam's sons, New York

Cronin H (1991) The Ant and the Peacock: altruism and sexual selection from Darwin to today. Cambridge University Press, Cambridge

Darwin C (1859) The origin of species. J. Murray, London

Darwin C (1871) The descent of man, and selection in relation to sex. J. Murray, London

Dobzhansky T (1944) Experiments on sexual isolation in Drosophila. III. Geographic Strains of Drosophila sturtevanti. PNAS 30–11:335–339

Fedigan LM (1982) Primate paradigms: sex roles and social bonds. Eden Press, Montréal

Fisher RA (1915) The evolution of sexual preference. Eugen Res 7:184–191

Fisher RA (1930) The genetical theory of natural selection. The Clarendon, Oxford Press. (Reprinted 1958, New York: Dover Publications)

Godfrey Smith P (2009) Darwinian populations and Natural selection. Oxford University, Oxford

Gowaty PA (1992) Evolutionary biology and feminism. Hum Nat 3–3:217–249

Hrdy SB (1981) The Woman that never evolved. Harvard University Press, Cambridge

Huxley J (1914) The Courtship-habits of the Great Crested Grebe (Podiceps cristatus); with an addition to the theory of sexual selection. Proc Zool Soc Lond 84–3:491–562

Huxleuy J (1938) The present standing of the theory of sexual selection. In Gavin de Beer (ed) Evolution. Oxford University Press, Oxford, pp. 11–41

Kavanagh E (2006) Debating sexual selection and mating strategies. Science 312:689–697

Lutz FE (1909) The effect of sexual selection. Paper read before the American Society of Naturalists, Boston meeting

Mayr E (1946) Experiments on sexual isolation in Drosophila. VII. The nature of the isolating mechanisms between D. pseudo-obscura and D. persimilis. PNAS 32–5:128–137

Mayr E (1972) Sexual selection and natural selection. In Campbell B (ed) Sexual selection and the descent of man 1871–1971. Heinemann, London, pp. 87–104

Milam E (2010) Looking for a Few Good Males: female Choice in Evolutionary Biology. Johns Hopkins University Press, Baltimore

Morgan TH (1903) Evolution and adaptation. MacMillan, New York

Morgan TH 1919. The genetic and the operative evidence related to secondary sexual characters. Carnegie Institution, Washington

Morgan TH (1932) The scientific basis of evolution. W.W. Norton, New York

Prum R (2010) The Lande-Kirkpatrick mechanism is the null model of evolution by intersexualselection: implications for meaning, honesty, and design in intersexual signals. Evolution 64:3085–3100

Prum R (2012) Aesthetic evolution by male choice: Darwin's really dangerous idea. Phil Trans Roy Soc B 367:2253–2265

Prokop Z, Michalczyk L, Drobniak S, Herdegen M, and Radwan J (2012) Meta-analysis suggests choosy females get sexy sons more than "good genes". Evolution 66:2665–2673

Qvarnström A, Brommer J, Gustafsson L (2006) Testing the genetics underlying the co-evolution of mate choice and ornament in the wild. Nature 441:84–86

Roughgarden J (2009) The genial gene: deconstructing darwinian selfishness. University of California Press, Berkeley

Roughgarden J, Oishi M, Akçay E (2006) Reproductive social behavior: cooperative games to replace sexual selection. Science 311:965–969

Roughgarden J et al (2013 December 11) Sexual Selection Studies: progress, Challenges, and Future Directions. Final Report from a NESCent Catalyst Meeting. Durham NC, July 15-17, 2013. Revision 1.1

Shuker DM (2010) Sexual selection: endless forms or tangled bank? Anim Behav 79:e11–e17

Shuster S, Wade M (2003) Mating systems and strategies. Princeton University, Princeton

Smuts BB (1985) Sex and friendship in baboons. Aldine, New York

Trivers RL (1972) Parental investment and sexual selection. In Campbell B (ed) Sexual selection and the descent of man 1871–1971. Heinemann, London, pp. 136–179

Trivers R (1994) Deriving females and feminism. Bioscience 44–4:210

Waage JK, Gowaty PA (1997) Myths of genetic determinism. In: Gowaty PA (ed) Feminism and evolutionary biology: boundaries, intersections, and frontiers. Chapman & Hall- International Thomson Pub., New York, pp 585–613

Wallace AR (1889) Darwinism. An exposition of the theory of natural selection with some of its applications. MacMillan, London

Zahavi A (1975) Mate selection: a selection for a handicap. J Theor Biol 53:205–214

Zuk M (1993) Feminism and the study of animal behavior. Bioscience 43–11:774–778

Zuk M (2002) Sexual selections: what we can and can't learn about sex from animals. University of California, Berkeley

Contents

List of Contributors

Verena Aebischer Laboratoire Parisien de Psychologie Sociale (EA 4386), University Paris Ouest, Nanterre—La Défense, Nanterre, France

Malin Ah-King Centre for Gender Research, Uppsala University, Uppsala, Sweden

Frank Cézilly UMR CNRS 6282 Biogéosciences, Université de Bourgogne, Dijon, France

Jean Gayon Institut d'histoire et de philosophie des sciences et des techniques, Université Paris 1 Panthéon-Sorbonne, Paris, France

Patricia Adair Gowaty Department of Ecology and Evolutionary Biology and Institute of Environment and Sustainability, UCLA, Los Angeles, CA, USA

Smithsonian Tropical Research Institute, Washington, DC, USA

Polistes Foundation, Belmont, MA, USA

Department of Ecology and Evolutionary Biology and Institute of Environment and Sustainability, Los Angeles, CA, USA

Smithsonian Tropical Research Institute, Panama, USA

Thierry Hoquet Universite Jean Moulin Lyon 3, Lyon, France

Michel Kreutzer Laboratoire d'Ethologie, Cognition et Développement (EA 3456), University Paris Ouest, Nanterre—La Défense, Nanterre, France

Michael Levandowsky Haskins Laboratories, Pace University, New York, NY, USA

Geoff A. Parker Institute of Integrative Biology, University of Liverpool, Liverpool, UK

Tommaso Pizzari Edward Grey Institute, Department of Zoology, University of Oxford, Oxford, UK

Tom A. R. Price Institute of Integrative Biology, University of Liverpool, Liverpool, UK

Richard O. Prum Department of Ecology and Evolutionary Biology, and Peabody Museum of Natural History, Yale University, New Haven, Connecticut, USA

Joan Roughgarden Hawaii Institute of Marine Biology, University of Hawaii, Kapaa, HI, USA

Michael Ruse Department of Philosophy, Florida State University, Tallahassee, USA

Michel Veuille Ecole Pratique des Hautes Etudes, Paris, France

Shigeru Watanabe Department of Psychology, Keio University, Tokyo, Japan

Nina Wedell Biosciences, University of Exeter, Penryn, UK

About the Authors

Verena Aebischer is an associate professor in social psychology at West Paris University. Her research and teaching focus on intergroup relations and span a range of areas such as gender stereotypes, cultural diversity, and have lately evolved into a particular interest on how and why relationships may encourage toxic behaviour in people.

Malin Ah-King is an evolutionary biologist and gender researcher at the Centre for Gender Research at Uppsala University, Sweden. Her research aims at problematizing the portrayal of biological sex as stable, making visible stereotypic gender and heteronormative notions in theory and research as well as developing non-normative models of variation in nature's sexes and sexualities.

Frank Cezilly is a population biologist and behavioural ecologist at the University of Burgundy, in Dijon, and a senior member of the Institut Universitaire de France. His main research interests are the study of mate choice and mating patterns, and the influence of parasites on the behaviour of their hosts.

Jean Gayon is Professor of Philosophy at University Paris 1-Panthéon Sorbonne, senior member of the Institut Universitaire de France, and Director of of the Institute for History and Philosophy of Science and Technique (Paris). Author of *Darwinism's Struggle for Survival*, his work is in history and philosophy of biology, with special emphasis on genetics and biometry, general philosophy of science, social and ethical problems raised by the life sciences.

Patricia Adair Gowaty is a Distinguished Professor in the Department of Ecology & Evolutionary Biology, UCLA, Los Angeles. She studies the evolutionary dynamics of social behaviour in wild-living eastern bluebirds and in several species of captive Drosophila, as well as the evolutionary dynamics of reproductive decision-making using theory and numerical experiments.

Thierry Hoquet is Professor of Philosophy of Science in the Faculty of Philosophy, University of Lyon 3, France, and a junior member of the Institut Universitaire de France. His research focuses on the reception of the Darwinian theory: especially how translations have impacted on the general understanding of "natural selection" (see "Translating natural selection: true concept, but false term?", *Bionomina*, 3 (2011), pp. 1–23; and "Translating "natural selection" in Japanese: from "shizen tōta" to "shizen sentaku", and back?" (with KIJIMA Taizo), *Bionomina*, 6(2013), pp. 26–48.

Michel Kreutzer Ethologist, is Professor in the Department of Psychology and researcher in the Laboratory of comparative ethology and cognitive processes at West Paris University, France. He studies bird song productions and receptions using the theory of sexual selection. He investigates both the competition between males and the female song preferences.

Michael Levandowsky is a population ecologist at the Haskins Labs, Pace University, NY, and a visiting scientist at the American Museum of Natural ahistory. His interest in the history of Darwin's theory of sexual selection goes back to conversations he had with the evolutionary biologist Stephen Jay Gould, when they were fellow students, at Antioch College and Columbia University.

Geoff Parker is an evolutionary biologist and behavioural ecologist at the Department of Evolution, Ecology and Behaviour in the Institute of Integrative Biology, University of Liverpool. His research interests focus mainly on theoretical concepts in several areas in evolution, and particularly in sexual selection and sexual conflict.

Tommaso Pizzari is an evolutionary biologist at the Edward Grey Institute in the Department of Zoology of the University of Oxford. His work focuses largely on experimental investigations of sexual selection and reproductive strategies.

Tom Price is an evolutionary biologist at the Institute of Integrative Biology at the University of Liverpool. He studies sexual selection and sexual conflict in a variety of insects, mainly fruit flies. He is particularly interested in the impact of selfish genes on conflict and cooperation at other levels of selection.

Richard Prum is the William Robertson Coe Professor of Ornithology in the Department of Ecology and Evolutionary Biology at Yale University. Prum studied at Harvard University, and the University of Michigan, and taught at the University of Kansas for twelve years before he moved to Yale in 2004. He has worked on phylogenetics, ethology, sexual selection and sexual conflict, evolutionary developmental biology of feathers, syringeal anatomy and evolution, the physics and chemistry of avian coloration, and avian color perception.

Joan Roughgarden received a Bachelor of Science in Biology and a Bachelor of Arts in Philosophy from University of Rochester in 1968 and a Ph.D. in Biology from Harvard University in 1971. She taught at Stanford University since 1972 as Professor of Biology and Geophysics. In 2011 she retired from Stanford and moved to Hawaii. In 2012 she resumed active duty as Adjunct Professor, Hawaii Institute of Marine Biology, University of Hawaii with research and teaching responsibilities. Joan's current research focuses on the evolution of social behavior and mutualistic interactions emphasizing the role of cooperation using cooperative game theory together with the economic theory of the firm. She has written about this subject in her recent books, "The Genial Gene" (2009) and "Evolution's Rainbow" (2004).

Her previous work includes investigations of the community ecology of Caribbean Anolis lizards, the coupling of oceanic and benthic processes in the population dynamics of intertidal invertebrates, and the theory of ecological economics. Joan has published more than 180 papers and 8 books.

Michael Ruse is the Lucyle T Werkmeister Professor of Philosophy and the Director of the Program in the History and Philosophy of Science at Florida State University. A historian and philosopher of science, especially evolutionary theory, he is the author of many books including *The Darwinian Revolution: Science Red in Tooth and Claw* (Chicago, 1979); *Monad to Man: The Concept of Progress in Evolutionary Biology* (Harvard, 1996); and *Can a Darwinian be a Christian? The Relationship between Science and Religion* (Cambridge, 2001). His next book is *The Gaia Hypothesis: Science on a Pagan Planet* (Chicago, 2013).

Michel Veuille is a professor of population genetics at the Ecole Pratique des Hautes Etudes (EPHE), Paris. He is interested the study of selection processes, and in the evolution of ideas in evolutionary genetics. As a geneticist he developed methods to detect natural selection, including the study of sex-ratio segregation distorsion from DNA variation in Drosophila. He also studied sexual selection in fruit flies using individual replica of the same genotype.

Shigeru Watanabe is a Professor emeritus in Keio University in Tokyo. His research interest is comparative cognition and comparative neuroscience. He published several researches on painting discrimination and music discrimination in animals.

Nina Wedell is a Professor of Evolutionary Biology and a Royal Society Wolfson Award holder at the College of Life & Environmental Sciences at the University of Exeter. She is interested in the evolutionary ecology of sex, and in particular the impact of selfish genetic elements in reproduction and how they shape species' ecology and evolution.

Part I
In Darwin's Footsteps: Historical Issues

Chapter 1
Sexual Selection: Why Does it Play Such a Large Role in *The Descent of Man*?

Michael Ruse

Abstract *The Descent of Man* is a strange book. About three-fifths is not on human beings at all, but an extended discussion of Charles Darwin's secondary mechanism of sexual selection. I argue that although the book is surely a little unbalanced, overall the discussion fits into Darwin's strategy of explaining and extending his thinking on evolution, and that truly it marks no significant theoretical shift from his earliest thinking about the nature and causes of evolution. Darwin saw the evolution of human beings at one with the evolution of all organisms.

Keywords Alfred Russel Wallace · Group selection · Individual selection · Hymenopteran sterility · Hybridism · Morality · Human evolution

1.1 Introduction

Charles Darwin's two-volume *The Descent of Man* is in major respects a very odd book. Published in 1871, 12 years after Darwin published the *Origin of Species*, it is devoted to the evolution of our own species, humankind. Except it isn't really. Fully three-fifths is devoted to Darwin's secondary evolutionary mechanism of sexual selection. The first of the two volumes opens in a conventional way with a discussion of human evolution, incuding our social evolution. Then it switches to an extended discussion and review of sexual selection, taking the reader over into the second volume. Finally, towards the end, Darwin returns to our own species and discusses further aspects of our evolution. There is no question but that in respects the discussion is out of kilter. Darwin would have done better to have lifted the discussion of sexual selection and made of it a separate book, perhaps using the subtitle to the *Descent*, "Selection in Relation to Sex." Then he could have published a much-slimmed-down book on our own species, using the results of the book on sexual selection. As it is, Darwin did not have room in the *Descent* for everything he wanted to say about our species and so the next year (1872) published what is in effect a supplementary volume on our species, *The Expression of the Emotions in Man and*

M. Ruse (✉)
Department of Philosophy, Florida State University, Tallahassee, USA
email: mruse@fsu.edu

T. Hoquet (ed.), *Current Perspectives on Sexual Selection,* History, Philosophy and Theory of the Life Sciences 9, DOI 10.1007/978-94-017-9585-2_1

in Animals (Darwin 1872). He might have considered running together the *Descent* and the *Expression* and making one large work, devoted exclusively to human evolution, or at least to human evolution with some comparisons with other animals. But the question that concerns me here is whether, for all that we agree that the *Descent* is ungainly, Darwin takes some radically new tack in the *Descent*. When it comes to human evolution does Darwin switch tactics somewhat dramatically, as for example one might argue that a Marxist or a Christian is bound to do—seeing organic evolution as one sort of thing and human evolution (which one might well not want to call human evolution) as something radically different. I shall argue that this is not the case and that in fact Darwin is pretty consistent throughout.[1]

1.2 Natural Selection

Charles Darwin's grandfather, Erasmus Darwin, was an evolutionist and Charles Darwin grew up knowing about this (Ruse 2008, 2012). However it was not until Charles Darwin was almost thirty that he himself made the move to transmutation (the word "evolution" applying to life's history did not become popular for another 20 years). He did this in part because he had by then jettisoned Christianity and was looking for a law-bound (that is, non-miraculous) picture of the world; in part because of the fossil evidence he had seen in the past few years as he spent time in South America thanks to his status as naturalist aboard HMS *Beagle*; and above all in part because of the peculiar distribution of birds and reptiles on the Galapagos Archipelago, the group of volcanic islands in the Pacific. How could there be different forms from island to island, else they had come from the mainland and changed as they moved from isolated retreat to isolated retreat?

As a graduate of the University of Cambridge, that institution whose greatest scientific alumnus was Isaac Newton, he who had given a causal underpinning to the Copernican Revolution, Darwin just knew he had to find an equivalent causal underpinning to evolutionary change. After some 18 months of frenzied thinking he found this in the fall of 1838, in the mechanism of natural selection. More organisms are born than can survive and reproduce. This leads to what the political economist Thomas Robert Malthus (1826) had labeled a "struggle for existence," and (as Darwin saw) even more a struggle for reproduction. There is constantly appearing natural variation and those that succeed in the struggle will tend to be different from those that do not. Moreover it will be the differences that help in the struggle and so there will be a constant winnowing or selecting of one kind of feature over another. Adding up, this leads to full-blown change.

[1] The argument of this paper strikes me as so obvious that I worry that truly no discussion is needed. I can only say that a recent, good book on the history of sexual selection—*Looking for a Few Good Males: Female Choice in Evolutionary Biology* by Erika Lorraine Milam—neither promotes nor disputes the argument I am about to make (Milam 2010).

But note that it is change of a particular kind. Organic features will be those that help their possessors to survive and reproduce. These features, "adaptations," will have the design-like nature that previous thinkers had taken to be evidence of the good intentions of an extra-natural intelligence, better known as God. Although there is some controversy about this, it seems clear that a major factor behind all of Darwin's thinking at this point was the analogy he was drawing between what happens in the world of animal and plant breeders and what happens thanks to blind law in the world of nature. Breeders select the forms that they want and start from there. Fatter pigs, shaggier sheep, fleshier turnips, and—Darwin made much of this—ever-yet-more fanciful pigeons. It is this vision of change that Darwin transferred to the wild world of animals and plants. From artificial selection to natural selection.

Darwin wrote out his ideas in a short piece (known now as the "Sketch") in 1842 and in a much longer version (the "Essay") in 1844 (Darwin 1909). But he did not publish and the years went by until in 1858 a young naturalist, Alfred Russel Wallace, sent Darwin a short piece with (quite independently discovered) just about all of Darwin's ideas. Darwin then wrote things up quickly and towards the end of 1859 *On the Origin of Species by Means of Natural Selection, or the Preservation of Favoured Races in the Struggle for Life* finally appeared in print. It is worth quoting the two pertinent passages. First to the struggle for existence.

> A struggle for existence inevitably follows from the high rate at which all organic beings tend to increase. Every being, which during its natural lifetime produces several eggs or seeds, must suffer destruction during some period of its life, and during some season or occasional year, otherwise, on the principle of geometrical increase, its numbers would quickly become so inordinately great that no country could support the product. Hence, as more individuals are produced than can possibly survive, there must in every case be a struggle for existence, either one individual with another of the same species, or with the individuals of distinct species, or with the physical conditions of life. It is the doctrine of Malthus applied with manifold force to the whole animal and vegetable kingdoms; for in this case there can be no artificial increase of food, and no prudential restraint from marriage. Although some species may be now increasing, more or less rapidly, in numbers, all cannot do so, for the world would not hold them. (Darwin 1859, p. 63–64)

Then to natural selection:

> Can the principle of selection, which we have seen is so potent in the hands of man, apply in nature? I think we shall see that it can act most effectually. Let it be borne in mind in what an endless number of strange peculiarities our domestic productions, and, in a lesser degree, those under nature, vary; and how strong the hereditary tendency is. Under domestication, it may be truly said that the whole organisation becomes in some degree plastic. Let it be borne in mind how infinitely complex and close-fitting are the mutual relations of all organic beings to each other and to their physical conditions of life. Can it, then, be thought improbable, seeing that variations useful to man have undoubtedly occurred, that other variations useful in some way to each being in the great and complex battle of life, should sometimes occur in the course of thousands of generations? If such do occur, can we doubt (remembering that many more individuals are born than can possibly survive) that individuals having any advantage, however slight, over others, would have the best chance of surviving and of procreating their kind? On the other hand, we may feel sure that any variation in the least degree injurious would be rigidly destroyed. This preservation of favourable variations and the rejection of injurious variations, I call Natural Selection. (Darwin 1859, p. 80–81)

1.3 Individual vs. Group Selection

I want to dig a little more into the nature of natural selection, and to do I shall employ the terms "individual selection" for the view that selection works for and through the individual and "group selection" for the view that selection can work for and through the group—population, race, species—and against the interests of the individual.[2] To see how the ideas behind the terms come into play in Charles Darwin's thinking, let us for a moment go back in history to the years before he was born. Charles Darwin's paternal grandfather, Erasmus Darwin, was not only an early evolutionist, but also a physician, inventor, poet, and above all a great friend and booster of British Industrialism (Ruse 1996). His maternal grandfather was Josiah Wedgwood, the potter, and one of the great drivers of that fundamental revolution at the end of the eighteenth century that changed the face of Britain forever—before it was a rural society, after it was an urban society. The Darwin and Wedgwood families were intertwined as one, Darwin himself following his father's example and marrying one of the Wedgwood girls. They benefited from the science and technology of the day and very naturally they took as gospel the political economy of their caste. Adam Smith was the ultimate authority and this canny Scot stressed and stressed again that no one does or should do anything out of ill-conceived sentiments of charity or fellow feeling. He was not as crude as Gordon Gekko but greed is good. "It is not from the benevolence of the butcher, the brewer, or the baker that we expect our dinner, but from their regard to their own interest" (Smith 1976, pp. 2A, 26–27). Everything comes from pursuit of our own needs and desires. Charles Darwin learnt that message literally at his parents' knees, never forgetting it as he translated it into biology. For Smith this was probably all something to be set in a natural theological context. God—the "Invisible Hand"—had so ordered human nature to get the desired results. Darwin for most of his life probably believed in some attenuated form of this. But, ultimately, he cared little about metaphysical concepts or settings. He was a scientist not a theologian.

Charles Darwin had never heard of "genes," and even less had he heard of "selfish genes" (Dawkins 1976). But I would claim that, thanks to the influence of Smith and his fellow economists, it is the sentiment behind this latter term that infuses and makes comprehensible the very crux of Darwin's thinking about the causes of evolutionary change. For him, the major processes of life's history had to be understood utterly and completely in terms of self-interest. There was no place for disinterested altruism towards others, within or without the species. In other words, Darwin was the archetypal individual selectionist, and it is against this that we should and we can understand Darwin's thinking about natural selection. Certainly this is the case in the *Origin* when he introduces natural selection, or rather when he introduces the struggle that lies behind selection. "Hence, as more individuals are produced than can possibly survive, there must in every case be a struggle for existence, either one individual with another of the same species, or with the individuals of distinct spe-

[2] These are not terms that Darwin uses himself, but as we shall see he grasped clearly the ideas behind the terms so I do not think I am unduly anachronistic in introducing them here.

cies, or with the physical conditions of life." You cannot be much clearer than this. The struggle occurs at the individual level and so any adaptations produced must be for the benefit of and only of the individual. The hand and the eye are paradigms. I grasp for my benefit not for yours, and I see for my benefit not for yours.

Yet although Darwin takes this "selfish" perspective, he would have been no less horrified than Adam Smith to think that cooperation never occurs or that organisms never strive to cooperate. Smith was explicit: "Political economy, considered as a branch of the science of a statesman or legislator, proposes two distinct objects: first, to provide a plentiful revenue or subsistence for the people, or more properly to enable them to provide such a revenue or subsistence for themselves; and secondly, to supply the state or commonwealth with a revenue sufficient for the public services" (Smith 1976, *Wealth of Nations*, Book 4, Introduction). This is not about one person or one small group taking everything with nothing for others. The whole point is that through self-regarding interests and actions, we all benefit. Certainly there are many times when we must fight and push. But cooperation and working together is part of the game. We all individually benefit from such collaboration.

Darwin took this as axiomatic. There is often bloody conflict. But he stressed again and again that the struggle was often metaphorical, and working together pays dividends. It is just that it is never basic or the ultimate end. This comes through clearly in many parts of his discussion, perhaps no more thoroughly than in his treatment of the old notion of a "balance of nature." This supposes that nature has within it a kind of stability, benefiting all. There is a kind of homeostasis to life, because cooperation is in the end the fundamental thing. Darwin has nothing against a balance of nature, but only as long as it benefits the individual. In an earlier version of the *Origin*, a big book on evolution that he was preparing in the middle years of the 1850s, he was explicit. Having made the point (as he makes in the *Origin*) that the struggle does not necessarily mean actual physical fighting, he added: "In many of these cases, the term used by Sir C. Lyell of "equilibrium in the number of species" is the more correct but to my mind it expresses far too much quiescence. Hence I shall employ the word struggle" (Darwin 1975, p. 187). Then in the *Origin* itself, Darwin followed up on this insight: "Battle within battle must ever be recurring with varying success; and yet in the long-run the forces are so nicely balanced, that the face of nature remains uniform for long periods of time, though assuredly the merest trifle would often give the victory to one organic being over another" (Darwin 1859, p. 78). Balance has to be of adaptive worth to individuals—all sharing in a harmonious existence. Sometimes this holds; sometimes not. Apart from the fact that as an evolutionist Darwin saw everything tending to change (that is, break from equilibrium), group balance or equilibrium is not in its own right of biological worth. It persists as long as is needed by individuals, and not a moment longer.

Darwin could not just leave things like that. He had to tackle what seem to be cases of cooperation that simply go against individual interest and that can be explained, if explicable at all, only by relying on selection working for the benefit of the group against the individual—group selection. (Darwin would never have denied that sometimes the group might benefit from the individual's benefit. The tricky cases are where the individual seems to suffer for the benefit of the group.)

In the *Origin* the major case was that of the social insects. These pose a problem because the workers in the nest (invariably female) are sterile, devoting all of their labors to the offspring of others, namely the queens. The actual fact of sterility was not particularly troublesome to Darwin. The situation is analogous to that of cattle, where we kill off and eat the steers. We do so because the steers have features that we want, fine marbled flesh and so forth. They never reproduce themselves, but the farmer gets what he wants by going back to the family and to the breeding stock. The bull may never have the features we want but if he is producing such features in his offspring, then selection can get to work. We select the right bull for the features he will produce and not for his own features. But why sterility at all? Surely this is only of benefit to the group and not to the individual. Today, thanks to our knowledge of genetics, we can dig beneath the surface and ask questions simply not open to Darwin. The popular explanation, in the hymenoptera (ants, bees, and wasps) at least, involves what is known as "kin selection"—because of a peculiar breeding system (males have only mothers and so sisters are more closely related than mothers and daughters) the sterile worker is in fact doing herself more biological good by raising siblings than by raising offspring! So the altruism of the worker is selfish-gene theory in full flight (Hamilton 1964; Hughes et al. 2008). Darwin knew nothing of this so his move was to regard the whole social insect nest as one superorganism (not his language). He thought of the workers as being parts of the whole and not individuals in their own right. So let us say he had a kind of family selection. But note it was a family selection and not group selection. The individuals are really parts and related to the whole as the heart and lungs are part of the whole organism and not individuals in their own right.

Was this a move that bordered on the dishonest? Was Darwin trying to have his individualism at the cost of pretending that he was not taking a route that he really was taking? Darwin would not have thought so. First, note that he was taking an explicit Adam Smith line of thought. He was appealing to a division of labor! "The greatest improvement in the productive powers of labour, and the greatest part of skill, dexterity, and judgment with which it is any where directed, or applied, seem to have been the effects of the division of labour." By splitting up the jobs and not attempting to be a jack of all trades, things are done far more efficiently. Darwin saw that hymenopteran sociality was simply a division of labor in the nest. "And nature has, as I believe, effected this admirable division of labour in the communities of ants, by the means of natural selection" (Darwin 1859, p. 242). Second, we must recall the extent to which Charles Darwin himself was embedded in family. Being often sick, staying at home tended by his wife (who was also his first cousin and between whom marriage was not quite dictated but strongly encouraged), surrounded by many children, visiting relatives (his older sister was married to his wife's brother) for holidays and relaxation—even visits to the dentist in London meant staying with older brother Erasmus—he always thought in terms of us, the family, versus them, the others. The Darwin-Wedgwood clan could have given lessons to the Corleones.

Did Darwin just back into this, as it were, or was the opting for individual over group selection done consciously? At first probably it was done instinctively, but increasingly over the years it was done consciously. In the *Origin*, Darwin considered the issue of hybrids and their sterility. Why is the mule, offspring of horse and donkey, sterile? The group-selectionist argument is obvious. Sterility occurs because it is of benefit to the parental species. They do not want to produce organisms that are literally neither fish nor fowl, adapted well to either parental niche. To the individual selectionist, like Darwin, this would never do. It could not be in the interests of the parents that their offspring be sterile. Sterility therefore had to be a byproduct of the fact that the parental reproductive organs did not fuse entirely happily together.

> [W] e see that when organic beings are placed under new and unnatural conditions, and when hybrids are produced by the unnatural crossing of two species, the reproductive system, independently of the general state of health, is affected by sterility in a very similar manner. In the one case, the conditions of life have been disturbed, though often in so slight a degree as to be inappreciable by us; in the other case, or that of hybrids, the external conditions have remained the same, but the organisation has been disturbed by two different structures and constitutions having been blended into one. For it is scarcely possible that two organisations should be compounded into one, without some disturbance occurring in the development, or periodical action, or mutual relation of the different parts and organs one to another, or to the conditions of life. (Darwin 1859, pp. 265–256)

The co-discoverer of natural selection, Wallace—who incidentally was a committed socialist and thus favored group-type explanations—wrote to Darwin protesting that he had abandoned natural selection too quickly and that he should have opted for a group-selection explanation (Ruse 1980). But, for all that the two men debated the issue back and forth in the 1860s, Darwin stayed put. He could not see how the sterility could have come through benefit to the group. It is the individual or no adaptive explanation at all.

1.4 Humans

In the *Origin*, Darwin said little about our own species, *Homo sapiens*. There was just a throwaway comment almost at the end. "In the distant future I see open fields for far more important researches. Psychology will be based on a new foundation, that of the necessary acquirement of each mental power and capacity by gradation. Light will be thrown on the origin of man and his history." (Darwin 1859, p. 488) Darwin's public reticence should not be confused with a lack of belief. He was ever stone cold certain that humankind was part of the evolutionary picture and that natural causes were in play. Indeed, the first intimation that we have—late in 1838—that he truly had grasped the idea of natural selection comes in a private notebook where he not only applies selection to our species but in the context of our intellectual abilities! "An habitual action must some way affect the brain in a manner which can be transmitted. —this is analogous to a blacksmith having children with strong arms. —The other principle of those children, which *chance*? produced with strong arms, outliving the weaker ones, may be applicable to the formation of instincts, independently of habits. —the limits of these two actions either on form or brain very hard to define." (Barrett et al. 1987, N 42 November 27, 1838). (Note that Darwin was, and indeed always was, a believer in Lamarckism, the inheritance of acquired characteristics.)

One suspects that the reason why Darwin said so little in the *Origin* about humans was that he was keen first to get the basic theory out into the light of day. Not that anyone was deceived or deflected. At once the implications for humankind were grasped and debated, often bitterly. Thomas Henry Huxley, Darwin's great supporter, took the theory as a reason to celebrate materialism. Critics, like Samuel Wilberforce, High Church Bishop of Oxford, took the theory as reason to bemoan the attack on faith. Although it is hard to offer any definitive proof, one suspects that had Darwin been left to himself, he probably never would have written a full-length work on humans. Rather he wanted to amuse himself with little projects such as the investigation into the sexual parts of orchids, something which engrossed him in the year or two after the *Origin* (Darwin 1862). Also, he wanted to write volumes dealing in detail with the claims of the *Origin* itself. And he did in fact write a two-volume treatise on variation in nature and in domestication (Darwin 1868).

But the issue of humans would not go away and finally he was pushed into writing something, a work that eventually appeared in 1871 as *The Descent of Man and Selection in Relation to Sex*. The reason why Darwin felt thus compelled to act is simple. When he discovered natural selection, Wallace had no more desire than did Darwin to introduce extra-scientific religious factors or causes. And this continued for a while in the 1860s, when Wallace wrote a paper on human evolution that Darwin much liked (Wallace 1864). Then, however, Wallace—like a good many other Victorians—became enamored with spiritualism, and refused to listen to the voices of skeptics, who doubted the authenticity of the claims of practitioners (Wallace 1905). He became convinced that something or Someone was guiding evolution, a belief that he held to his dying day. Moreover, in support of his position, drawing on his first-hand acquaintance of native people—unlike Darwin, the more humble and less wealthy Wallace really had lived with the local folk—Wallace argued that there are certain human characteristics that simply could not have evolved through natural selection. Human intelligence was one and human hairlessness was another (Wallace 1870).

Darwin was appalled at Wallace's throwing over the traces as it were. He felt compelled to offer a defense of his theory, and so he launched into an investigation that ended with the *Descent* (and with the supplementary volume, *The Expression of the Emotions*). Honor and science were satisfied. As I have said, the *Descent* is an odd book—first humans, then sexual selection, and finally humans again. If you start with the first section, however, although there is much of great interest, there is nothing much to raise eyebrows about methodology. The Darwin of the *Descent* is the Darwin of the *Origin*. There is discussion about how we are clearly animal, there is discussion about our origins (Africa probably), and then on into culture with discussions about the evolution of religion and of morality. It is interesting how relatively brief is the former and how relatively extended is the latter. Like many middle to late Victorians, Darwin had lost his childhood faith and so thought basically that religion was false, but at the same time was keen to show that this belief did not destroy morality and the foundations of society, so felt the need for a careful look at our ethical sense and behavior. What is striking is the extent to which Darwin pursued the same strategy of individual selection as before, even in the light of cases that seemingly cry out for a group-selection perspective. Take the case of the soldier who throws himself on a grenade that the lives of his colleagues be spared. As it happens, general opinion is that Darwin caved in a bit here. It is true that he invoked something that is today known as "reciprocal altruism"—you scratch my back and I will scratch yours (Trivers 1971). "In the first place, as the reasoning powers and foresight of the members became improved, each man would soon learn that if he aided his fellow-men, he would commonly receive aid in return. From this low motive he might acquire the habit of aiding his fellows; and the habit of performing benevolent actions certainly strengthens the feeling of sympathy which gives the first impulse to benevolent actions. Habits, moreover, followed during many generations probably tend to be inherited" (Darwin 1871, 1, pp. 163–164; note the Lamarckism). But then it is thought he accepted group selection.

> It must not be forgotten that although a high standard of morality gives but a slight or no advantage to each individual man and his children over the other men of the same tribe, yet that an advancement in the standard of morality and an increase in the number of well-endowed men will certainly give an immense advantage to one tribe over another. There can be no doubt that a tribe including many members who, from possessing in a high degree the spirit of patriotism, fidelity, obedience, courage, and sympathy, were always ready to give aid to each other and to sacrifice themselves for the common good, would be victorious over most other tribes; and this would be natural selection. At all times throughout the world tribes have supplanted other tribes; and as morality is one element in their success, the standard of morality and the number of well-endowed men will thus everywhere tend to rise and increase. (Darwin 1871, 1, p. 166)

I am not however convinced that this is quite as definitive as it seems. Darwin is talking about the tribe and he makes it clear that there is going to be a lot of intermarriage in such a group. In other words, there will be a lot of blood relationships. And in an unpublished letter to one of his sons (George, a physicist) later in the decade, he affirms what we might expect, namely that he is thinking of the tribe as akin a hymenopteran nest, where we do get such close relationships.

> To G. H. Darwin 27 April [1876]
> Down Beckenham Kent
> Ap. 27th
> My dear George
> I send "Mind"— it seems an excellent periodical— Sidgwicks Article has interested me much.— It is wonderfully clear & makes me feel what a muddle-headed man I am.— I do not agree on one point, however, with him. He speaks of moral men arising in a tribe, accidentally, i.e. by so-called spontaneous variation; but I have endeavoured to show that such men are created by love of glory, approbation &c &c.—
> However they appear the tribe as a tribe will be successful in the battle of life, like a hive of bees or nest of ants.
> We are off to London directly, but I am rather bad.
> Leonard comes home on May 10th!! Plans changed.

Also in response to a young supporter, George John Romanes, who wrote pressing on him a group-selection perspective, Darwin demurred (Romanes 1895, p. 173).

1.5 Sexual Selection

Against this background, let us turn now to sexual selection. First, what is sexual selection and why did Darwin invoke it in the first place? Second, why does it have such a role in the *Descent*? Third, does it lead to a radical reconceptualizing of Darwin's thinking about evolution now that he is applying it to humans? So, to the first question:

> Inasmuch as peculiarities often appear under domestication in one sex and become hereditarily attached to that sex, the same fact probably occurs under nature, and if so, natural selection will be able to modify one sex in its functional relations to the other sex, or in relation to wholly different habits of life in the two sexes, as is sometimes the case with insects. And this leads me to say a few words on what I call Sexual Selection. This depends, not on a struggle for existence, but on a struggle between the males for possession of the females; the result is not death to the unsuccessful competitor, but few or no offspring. (Darwin 1859, pp. 87–88)

I agree with Michael Ghiselin (1969) that Darwin was surely led to sexual selection, and a division he makes within it of male combat and female choice, because of the analogy with artificial selection. There we find selection for profit, for shaggier sheep and fatter porkers, corresponding to natural selection and then selection for pleasure, for fiercer bulldogs and prettier tail feathers, corresponding to sexual selection for male combat and female choice respectively. This is certainly the context in which sexual selection gets announced to the world in the *Sketch* of 1842.

> Besides selection by death, in bisexual animals <illegible> the selection in time of fullest vigour, namely struggle of males; even in animals which pair there seems a surplus <?> and a battle, possibly as in man more males produced than females, struggle of war or charms. Hence that male which at that time is in fullest vigour, or best armed with arms or ornaments of its species, will gain in hundreds of generations some small advantage and transmit such characters to its offspring. So in female rearing its young, the most vigorous and skilful and industrious, <whose> instincts <are> best developed, will rear more young, probably possessing her good qualities, and a greater number will thus <be> prepared for the struggle of nature. Compared to man using a male alone of good breed. (Darwin 1909, pp. 48–49)

In the *Origin*, sexual selection has a similar (although extended) discussion as a kind of younger brother to natural selection. The background links to artificial selection are made more explicit than ever. Thus, starting with sexual selection through male combat.

> Generally, the most vigorous males, those which are best fitted for their places in nature, will leave most progeny. But in many cases, victory will depend not on general vigour, but on having special weapons, confined to the male sex. A hornless stag or spurless cock would have a poor chance of leaving offspring. Sexual selection by always allowing the victor to breed might surely give indomitable courage, length to the spur, and strength to the wing to strike in the spurred leg, as well as the brutal cock-fighter, who knows well that he can improve his breed by careful selection of the best cocks. (Darwin 1859, p. 88)

Then sexual selection through female choice.

> Amongst birds, the contest is often of a more peaceful character. All those who have attended to the subject, believe that there is the severest rivalry between the males of many species to attract by singing the females. The rock-thrush of Guiana, birds of Paradise, and some others, congregate; and successive males display their gorgeous plumage and perform strange antics before the females, which standing by as spectators, at last choose the most attractive partner. Those who have closely attended to birds in confinement well know that they often take individual preferences and dislikes: thus Sir R. Heron has described how one pied peacock was eminently attractive to all his hen birds. (pp. 88–9)

Basically that is it in the *Origin* for sexual selection. It is introduced but not really used.

1.6 Descent of Man

Why does sexual selection get such major exposure in *The Descent of Man*? The reason is simple. Darwin thought that sexual selection could be used to counter Wallace. Darwin was less interested in general features, like hairlessness, and more in specific features like racial differences (although the latter certainly included the former). But Wallace was right in his overall claim. Some human characteristics cannot be explained by natural selection.[3]

> We are …led to inquire whether slight individual differences, to which man is eminently liable, may not have been preserved and augmented during a long series of generations through natural selection. But here we are at once met by the objection that beneficial variations alone can be thus preserved; and as far as we are enabled to judge (although always liable to error on this head) not one of the external differences between the races of man are of any direct or special service to him. The intellectual and moral or social faculties must of course be excepted from this remark; but differences in these faculties can have had little or no influence on external characters. (Darwin 1871, p. 1, 248–249)

Wallace was wrong in thinking that this means we must turn to supernatural causes. Sexual selection can do the job!

> We have thus far been baffled in all our attempts to account for the differences between the races of man; but there remains one important agency, namely Sexual Selection, which appears to have acted as powerfully on man, as on many other animals. I do not intend to assert that sexual selection will account for all the differences between the races. An unexplained residuum is left, about which we can in our ignorance only say, that as individuals

[3] I confess that this is one topic in Darwin scholarship that for many decades I thought was absolutely and completely without question or doubt. I introduced this connection in my overview of the Darwinian Revolution (Ruse 1979), and by the nature of that discussion I am sure I picked it up from others. In the next decade, Joel Schwartz (1984), in painfully greater detail than I, confirmed my belief that it was Wallace's apostasy that led Darwin to greater reliance on sexual selection. Fairly recently however those entertaining revisionists Adrian Desmond and James Moore (Desmond and Moore 2009) have argued that all of Darwin's thinking on our species starts and basically remains with his detestation of slavery. I confess that although I applaud the extent to which Desmond and Moore have shown the Darwin-Wedgwood family, including Charles, as being fervently anti-slavery, with Robert J. Richards (2009) I see absolutely no merit whatsoever in this claim. It simply isn't so.

> are continually born with, for instance, heads a little rounder or narrower, and with noses a little longer or shorter, such slight differences might become fixed and uniform, if the unknown agencies which induced them were to act in a more constant manner, aided by long-continued intercrossing. Such modifications come under the provisional class, alluded to in our fourth chapter, which for the want of a better term have been called spontaneous variations. Nor do I pretend that the effects of sexual selection can be indicated with scientific precision; but it can be shewn that it would be an inexplicable fact if man had not been modified by this agency, which has acted so powerfully on innumerable animals, both high and low in the scale. It can further be shewn that the differences between the races of man, as in colour, hairyness, form of features, &c., are of the nature which it might have been expected would have been acted on by sexual selection. (pp. 249–250)

This of course calls first for a detailed discussion of sexual selection, and this Darwin was happy to provide at very great length. To be candid, one has to be a little bit of a devotee of the subject to read through all of the detailed descriptions without eyes glazing over. Fortunately we can skip over pretty much all of it, for it is basically background to our main interest about the role of sexual selection in the general Darwinian view of the workings of nature. Starting with the lowest forms of life, we go on through insects, fish, amphibians, reptiles, birds and mammals, until finally—the third section of the *Descent*—we reach our own species. A lot of the discussion, candidly, is pretty Victorian, with all sorts of flat statements about the superiority of males over females and that sort of thing.

> Man is more courageous, pugnacious, and energetic than woman, and has a more inventive genius. His brain is absolutely larger, but whether relatively to the larger size of his body, in comparison with that of woman, has not, I believe been fully ascertained. In woman the face is rounder; the jaws and the base of the skull smaller; the outlines of her body rounder, in parts more prominent; and her pelvis is broader than in man; but this latter character may perhaps be considered rather as a primary than a secondary sexual character. She comes to maturity at an earlier age than man. (Darwin 1871, p. 2, 316–17)

The main thing is that a total commitment to individual selection is shown throughout, even when Darwin is modifying the choice aspect of sexual selection so that it is males making the running rather than females.

> It is well known that with many Hottentot women the posterior part of the body projects in a wonderful manner; they are steatopygous; and Sir Andrew Smith is certain that this peculiarity is greatly admired by the men. He once saw a woman who was considered a beauty, and she was so immensely developed behind, that when seated on level ground she could not rise, and had to push herself along until she came to a slope. Some of the women in various negro tribes are similarly characterised; and, according to Burton, the Somal men "are said to choose their wives by ranging them in a line, and by picking her out who projects farthest *a tergo*, Nothing can be more hateful to a negro than the opposite form." (pp. 345–346)

Generally, it is all a matter of male combat, but it does work the other way some times, with the females having quite an active role to play.

> With respect to the other form of sexual selection (which with the lower animals is much the most common), namely, when the females are the selectors, and accept only those males which excite or charm them most, we have reason to believe that it formerly acted on the progenitors of man. Man in all probability owes his beard, and perhaps some other characters, to inheritance from an ancient progenitor who gained in this manner his ornaments.

> But this form of selection may have occasionally acted during later times; for in utterly barbarous tribes the women have more power in choosing, rejecting, and tempting their lovers, or of afterwards changing their husbands, than might have been expected. (p. 372)

1.7 Individual Selection Triumphs

The one thing I would emphasize about sexual selection is just how much it reinforces the commitment to individual selection. It is all taking place within the species and between competing individuals.[4] There are no friends, just rivals. So in the *Origin*, and then later in the *Descent*, Darwin obviously felt no need to excuse or otherwise qualify sexual selection. It just fits in nicely with his perspective on the workings of nature. It is indeed true that through history, starting with Wallace, there have been many critics of sexual selection. Natural selection's co-discoverer did not want to challenge sexual selection through male combat, but he thought that sexual selection through female choice was altogether too anthropomorphic (Wallace 1870). Who is to say that animals have the same standards of taste as we humans? Wallace of course could not deny that we get sexual dimorphism, especially in birds, but he was more inclined to attribute it to natural selection making females drab rather than sexual selection making males bright. Females are often exposed, sitting on the nest and the like, and so there is good reason for adaptations that make them inconspicuous. Darwin was right in drawing attention to the phenomenon but got it backwards when it came to causes! Probably there is no direct connection, but note also that Wallace now (by the mid-1860s) had committed himself to the plausibility of group selection, so he did not have the fierce dedication of Darwin to individual selection. Natural selection working on females to make them drab is of course basically an individual selection process, but it does not have quite the essential connection to individual selection that any sexual selection explanation must have. It does not, as it were, throw it in your face.

It is true also that in the past few decades, sexual selection has come into its own, and there is now widespread agreement that Darwin was right in thinking it a significant factor in evolution (Campbell 1972). It is true also that this recognition has come at a time that—and surely is connected with the fact that—evolutionists have turned strongly towards an individual selection perspective on the workings of natural selection. But in a way, all of this is really neither here nor there. The point is that in turning to sexual selection as a major factor in human evolution, Darwin was reaffirming the position that he had held ever since he first became an evolutionist, namely that humans are part of the picture and should not in that sense be picked out for special causal treatment.

---

[4] Philosopher Elliott Sober, well known for his fervent devotion to group selection (see for example Sober and Wilson 1997), an enthusiasm presumably not unconnected to his Marxism (Wright et al. 1992), argues that when Darwin turns to sex ratios in the *Descent*, he shows his commitment to group selection (Sober 2011). I counter this in Richards and Ruse 2014.

1.8 Conclusion

We have reached a good point to draw the discussion to an end. I make no judgments in this paper on the value of sexual selection. My interests and claims are rather different. First, I argue that Charles Darwin was throughout his work an absolutely committed individual selectionist. Second, I argue (what hardly needs arguing) that sexual selection is the epitome of an individual selectionist approach. Third, I conclude that while it is true that Darwin's thinking evolved about the need for and importance of sexual selection—going from almost an afterthought to an essential explanatory component in his thinking—the mechanism itself snuggled soundly within his overall picture of evolution and thus its increased importance confirmed rather than refuted his world vision. *The Descent of Man* is an oddly shaped book, but conceptually it is at one with the *Origin of Species*.

References

Barrett PH, Gautrey PJ, Herbert S, Kohn D, Smith S (1987) Charles Darwin's notebooks, 1836–1844. Cornell University Press, Ithaca

Campbell B (1972) Sexual selection and the descent of man. Aldine, Chicago

Darwin C (1859) On the origin of species by means of natural selection, or the preservation of favoured races in the struggle for life. John Murray, London

Darwin C (1862) On the various contrivances by which British and foreign orchids are fertilized by insects, and on the good effects of intercrossing. John Murray, London

Darwin C (1868) The variation of animals and plants under domestication. John Murray, London

Darwin C (1871) The descent of man, and selection in relation to sex. John Murray, London

Darwin C (1872) The expression of the emotions in man and animals. John Murray, London

Darwin C (1909) The foundations of the origin of species: two essays written in 1842 and 1844. Darwin F (ed). University of Cambridge Press, Cambridge

Darwin C (1975) Charles Darwin's natural selection, being the second part of his big species book written from 1856 to 1858. Stauffer RC (ed). University of Cambridge, Cambridge

Dawkins R (1976) The selfish gene. Oxford University, Oxford

Desmond A, Moore J (2009) Darwin's sacred cause: how a hatred of slavery shaped Darwin's views on human evolution. Houghton Mifflin Harcourt, New York

Ghiselin MT (1969) The triumph of the Darwinian method. University of California Press, Berkeley

Hughes WO, Oldroyd HB, Beekman M, Ratnieks FW (2008) Ancestral monogamy shows that kin selection is the key to the evolution of eusociality. Science 320:1213–1216

Malthus TR (1826) An essay on the principle of population (sixth edition). Everyman, London (1914)

Milam EL (2010) Looking for a few good males: female choice in evolutionary biology. Johns Hopkins University, Baltimore

Richards RJ (2009). Review of Darwin's sacred cause: Race, slavery and the quest for human origins. Am Sci (September-October):415–17

Richards RJ, Ruse M (2014) Debating Darwin. University of Chicago Press, Chicago

Romanes E (1895) The life and letters of George John Romanes. Longmans, London

Ruse M (1980) Charles Darwin and group selection. Ann Sci 37:615–630

Ruse M (1996) Monad to man: the concept of progress in evolutionary biology. Harvard University Press, Cambridge

Ruse M (2008) Charles Darwin. Blackwell, Oxford

Ruse M (2012) The Cambridge encyclopedia of Darwin and evolution. Cambridge University Press, Cambridge

Schwartz JS (1984) Darwin, Wallace, and the *Descent of Man*. J Hist Biol 17:271–89.

Smith A (1976) The glasgow edition of the works and correspondence of Adam Smith. Campbell RH, Skinner AS (eds). Oxford: Clarendon Press.

Sober E (2011). Did Darwin write the "origin" backwards? philosophical essays on Darwin's theory. Prometheus, Buffalo

Sober E, Wilson DS (1997) Unto others: the evolution of altruism. Harvard University Press, Cambridge

Trivers RL (1971) The evolution of reciprocal altruism. Q Rev Biol 46:35–57

Wallace AR (1864) The origin of human races and the antiquity of man deduced from the theory of natural selection. J Anthropol Soc Lond 2:clvii-clxxxvii

Wallace AR (1870) Contributions to the theory of natural selection. Macmillan, London

Wallace AR (1905) My life: a record of events and opinions. Chapman and Hall, London

Wright EO, Levine A, Sober E (1992) Reconstructing Marxism: essays on explanation and the theory of history. Verso, New York

Chapter 2
Utility vs Beauty: Darwin, Wallace and the Subsequent History of the Debate on Sexual Selection

Thierry Hoquet and Michael Levandowsky

Abstract We examine the origins of the disagreement of Alfred Russel Wallace and Charles Darwin regarding the significance and mechanism of sexual selection and relate this to differences in their views of human evolution, and of cognitive ability and esthetic sensibilities of various human and nonhuman populations. We trace subsequent versions of these differing views into the twentieth century, and the controversy between R. A. Fisher's Darwinian "runaway" model of sexual selection by female choice (the "sexy son" model), and Wallacean models of sexual selection based on signs of greater fitness of males (the "healthy gene" hypothesis). Models derived from the latter, the "honest signal" and "handicap" models, are discussed, and we note that these different models, based on utility or beauty, are not necessarily mutually inconsistent.

Keywords Darwin, Wallace and sexual selection · Fisher's runaway model of sexual selection · "Good genes" versus "sexy son" models in sexual selection · Female choice in sexual selection · Darwin, Wallace and the evolution of the human mind

2.1 Introduction

The ideas of Charles Darwin (1809–1882) on the evolution of secondary sexual characters, noted in *The Origin of Species* (1859) and developed in *The Descent of Man and Selection in Relation to Sex* (1871, 1874), spurred an important controversy which has remained an area of passionate contention to the present: What does "sexual selection" mean? How does it interact with natural selection? Does it

T. Hoquet (✉)
Université Jean Moulin Lyon 3, Lyon, France
e-mail: thierry.hoquet@univ-lyon3.fr

M. Levandowsky
Haskins Laboratories, Pace University, New York, NY 10038, USA

T. Hoquet (ed.), *Current Perspectives on Sexual Selection,* History, Philosophy and Theory of the Life Sciences 9, DOI 10.1007/978-94-017-9585-2_2

apply to humans? Regarding the definition of sexual selection, Darwin suggested that males were struggling with each other for mates and that females were able to choose. But he provided no explanation for the reasons why females as a whole were generally "the choosing sex", and for why a definite female would choose a definite male rather than any of his competitors: instead, he suggested that females were endowed with an "aesthetic sense", a mysterious taste for beauty, which was governing their choices. This idea was largely criticized, first and foremost by Alfred Russel Wallace (1823–1913), who doubted female animals might have the power to choose.

Wallace's views can be viewed as a "utilitarian" approach that resolves sexual selection into natural selection: in this view, aesthetic traits are eventually always *useful* to their bearer, thus subject to being interpreted as advantageous to the general fitness of the individual. During the first half of the twentieth century, R. A. Fisher attempted to salvage Darwinian sexual selection with a run-away model: suggesting a mechanism for mere aesthetic preferences to develop traits, with no direct benefit. On the other hand, Wallacean utilitarian views have been considered the precursor to the good-gene model, for which wooing signals vigor. Later efforts to develop quantitative models of the early verbal suggestions have led to a controversy between the aesthetic and the utilitarian views.

Our paper is twofold. First we focus on the disagreement between Darwin and Wallace. We give an overview of its roots and its scale, showing how it involved not only sexual, but also aspects of natural selection, and we consider some differences in their understandings of how selection works. This analysis suggests that the classical theme of Wallace refusing "sexual selection" on grounds of his rejection of "female choice" is but a part of a larger picture. In particular we consider possible differences in their views on how the evolutionary principles apply to the human species. In the second part of this paper, we view Darwin and Wallace as rival scientists embodying two competing evolutionary principles, namely, Beauty and Utility, and how this has contributed to shaping the evolutionary debate of sexual selection throughout the Twentieth century and until today, as evidenced by several chapters of this book (see papers by Prum and Cézilly, this volume).

2.2 Darwin and Wallace: a Range of Disagreements

Sexual selection is Darwin's second important concept. In 1871 (t. I, p. 256), he defined it in a rather general fashion, as "the advantage which certain individuals have over other individuals of the same sex and species, in exclusive relation to reproduction". Darwin decomposed sexual selection in two classes of phenomena: in some species, male competition for females is evident while, in others, female choice of males is clearly shown (Darwin 1859, pp. 87–90). Males, particularly in polygynous species, might fight over females, leading to selection for physical size and weaponry, as in the case of the male elk's antlers. Females, on the other hand, would choose among males, but on what basis? How does it help the peahen's posterity to choose a peacock with a splendid tail, for example? Darwinian female choice focuses on beauty for beauty's sake and does not emphasize the utility of exaggerated features, like ornaments. In contrast to this view, Wallace noted that Darwin attributed colors or courtship displays in birds and insects to sexual selection, but he thought that the 'greater vigor' and 'higher vitality' of males might somehow be associated with, or perhaps lead to their greater coloration or activity. He also attributed a role to the protective value of drab colors for females. Besides, he emphasized how elaborate male crests and erectile feathers might function as species recognition signals, or as a means to frighten away predators, not to attract females.

It can be said that Wallace emphasized both the protective and the signaling value of color while Darwin stressed its aesthetic value. But beyond the specific issue of sexual selection, Darwin and Wallace entertained different views on several issues, and not only on the mechanisms that account for sexual dimorphism.

2.2.1 Darwin and Wallace as Codiscoverers

When Darwin received Wallace's Ternate manuscript called "On the tendency of species to form varieties; and on the perpetuation of varieties and species by natural means of selection" on 18 June 1858, he immediately wrote to his friend and mentor, the geologist Charles Lyell: "Your words have come true with a vengeance that I should be forestalled. You said this when I explained to you here very briefly my views of "Natural Selection" depending on the Struggle for existence. I never saw a more striking coincidence. If Wallace had my M.S. sketch written out in 1842 he could not have made a better short abstract! Even his terms now stand as heads of my chapters". And, facing the future publication of Wallace's manuscript, Darwin concluded: "all my originality, whatever it may amount to, will be smashed" (Darwin to Lyell, 18 June 1858, *in* Darwin 1991, p. 107).

In fact, Wallace's letter was not the first, but the third that the self-educated collector had sent to the famous author of the *Voyage of the Beagle*: the two men were in touch since 10 October 1856, and Wallace knew that Darwin was preparing a big book "on species and varieties, for which he ha[d] been collecting information for 20 years". The young naturalist (he was only 35 at the time) thought that Darwin's work might save him the trouble of "proving that there is no difference in nature between the origin of species and varieties", as he wrote to his friend the entomologist Henry Walter Bates (Wallace to Bates, 4 Jan. 1858, *in* Marchant 1916, t. I, p. 67).

The convergences in the ideas of these two men were indeed remarkable. Especially both referred to a form of "struggle for existence" (Wallace 1858, p. 54). Nevertheless, as there were a number of points of differences between them, Darwin and Wallace have also been cast as rivals and competitors by others: some claim that Wallace was the true discoverer of the mechanism of evolution and that Darwin usurped the credit; others that Darwin was the true discoverer and Wallace's initial essay was not an adequate statement of the mechanism of evolution. Such claims no doubt generate a certain notice, but they are not well supported by the comments and attitudes of the two men themselves, each of whom referred to the other as co-discoverer on many occasions, both privately and publicly. The two principals were friends and admired each other; and indeed there was much to admire: in addition to their writings on evolution, both made other, substantial contributions to science, and both would be remembered today, had they never written about evolution.

What were the main differences? First and foremost, Darwin viewed the modification of physical and behavioral traits in domestic animals and plants by selective breeding as a kind of metaphor, an indication of what selection by natural forces might be able to accomplish. Wallace, on the other hand, viewed domestic species as essentially abnormal, and considered that they would rapidly return to the ancestral type if released to the wild. In his initial essay of 1858 he sought to refute

> the assumption that varieties occurring in a state of nature are in all respects analogous to or even identical with those of domestic animals, and are governed by the same laws as regards their permanence or further variation". In contrast, Wallace thought that "there is a general principle in nature which will cause many varieties to survive the parent species, and to give rise to successive variations departing further and further from the original type, and which also produces in domesticated animals, the tendency of varieties to return to the parent form. (Wallace 1858, p. 54)

Thus, Wallace never agreed with Darwin's frequent and prominent use of human artificial selection of domestic varieties as an argument supporting the possibility of nature selecting new varieties. He disputed the usefulness of domestication as a sound analogy for understanding the modification of species in the wild: the possibility that domestic breeds would revert to an original "type" when becoming feral was a stumbling block to Wallace (Gayon 1998; Beddall 1968). Consistently with this critique of Darwin's foundational analogy with the world of breeding, Wallace never fully accepted the phrase "natural selection", as it was based on and encapsulated the analogy between Nature and the breeders' "selecting" actively albeit unconsciously some traits over others. Wallace did not like the term at first, and examination of his copy of the *Origin* shows that he cautiously crossed the word (Beddall 1988). In several letters to Darwin, Wallace repeatedly denounced the "agentive" connotations of the word "selection" and he was constantly urging Darwin to state that nature is not a breeder capable of conscious choice (Gayon 1998; Hoquet 2011). Wallace was so concerned with a possible personification of "selection", that he was even responsible for Darwin's introducing Herbert Spencer's phrase "survival of the fittest" in the fifth edition of the *Origin* (1869). It's ironic that, in spite of his early reluctance to accept the term *selection*, Wallace later became a strong and convinced *selectionist*. His *Darwinism* (1889) reshaped Darwin's theory as a pan-utilitarianism, promoting an interpretation of Darwin that was so radical that George Romanes, another disciple of Darwin, accused Wallace of being "ultra-Darwinian" (see below sect. 2.1).

A third difference bears on what is now called "the levels of selection". Both Darwin and Wallace strongly believed in the causal power of natural selection, but they disagreed on the level at which competition occurs: Darwin referred to competition between individuals, while Wallace, though his initial statement referred to competition among individuals, tended to focus on competition between populations. This difference has been often noted, at least since the work of paleontologist

Henry Fairfield Osborn (1894) (for instance by Bowler 1976; Gayon 2009a; Bock 2009, Ruse, this volume). H.F. Osborn called the Darwin-Wallace moment "one of the most striking of all the many coincidences and independent discoveries in the history of the Evolution idea" (1894, p. 243). However, when Osborn compared Darwin and Wallace's contributions to the 1st July 1858 meeting of the Linnaean Society, he concluded: "remarkable as this parallelism is, it is not complete. The line of argument is the same, but the *point d'appui* is different. Darwin dwells upon *variations in single characters*, as taken hold of by Selection; Wallace mentions variations, but dwells upon *full-formed varieties*, as favorably or unfavorably adapted" (1894, p. 245; emphasized by Osborn). The struggle is much more intense in the Darwinian world, so that the slightest difference in organization or instinct can have the most dramatic effect on individual survival; on the other hand, in the Wallacean world, environmental change occurs, and some varieties happen to be adapted to it. In contrast with Darwin's focus on individual *variation*, Wallace's 1858 paper focuses on *varieties*: "the very clear recognition of the importance of individual differences" came only later in his writings and "marked a significant development in his thought" (Bowler 1976, p. 17). Such difference in emphasis is somewhat reminiscent of a dispute that arose in the twentieth century among the 3 founders of the Modern Synthesis. R. A. Fisher and J. B S. Haldane viewed evolution as proceeding through single gene selection, whereas Sewall Wright emphasized gene interaction and saw collections of genes as the unit of selection, giving rise to Haldane's famous paper "A defence of beanbag genetics" (Haldane 1964).

So we have listed three disagreements of varying importance between Darwin and Wallace: on the value of the analogy with domestic breeds; on the appropriateness of the term "natural selection"; and perhaps on the levels of selection and the difference between variations and varieties. We come now to a major difference that forms the theme of this paper.

2.2.2 Disagreement on Sexual Selection

It should be added that, from the very outset, Darwin and Wallace disagreed about sexual selection, its importance in the development of secondary sexual characteristics, and its role in human evolution. The theme of sexual selection is treated in a short section in the chap. 4 of Darwin's *On the Origin of Species* (1859, pp. 87–90), and is also briefly mentioned in an earlier essay published jointly with Wallace's paper in 1858 (see below). Darwin later wrote an entire two-volume book on the subject (1871), revealing the importance he ascribed to this process. Between the publication of his two major works, and especially around 1867–1869, Wallace and Darwin were both working on the issue of sexual characters and had an extensive correspondence on the subject, trying to resolve their differences (collected and analyzed by Kottler 1980, 1985). It seems that once again, Wallace "still anticipated ideas in the most embarrassing manner" (Irvine 1955, p. 184) and Darwin was obviously annoyed by this new coincidence. He wrote to Wallace, 29 April 1867: "It is curious, how we hit on the same ideas" (Marchant 1916, t. I, p. 184). But 2 days later, on May 1st, Wallace replied to Darwin:

> I had thought of a short paper on The Connection between the colors of female birds and their mode of nidification—but had rather leave it for you to treat as part of the really great subject of sexual selection—which combined with protective resemblances and differences will I think when thoroughly worked out explain the whole coloring of the animal kingdom. (1st May 1867, http://www.darwinproject.ac.uk/entry-5522)

As this last quote shows, Wallace constantly showed unabashed deference to Darwin and sent him all his notes on the topic. But beyond his submission to Darwin's priority, there were strong disagreements between the two men—and Darwin was trying his best to bridge the gap between them and have them come to an agreement. On 23 September 1868, Darwin restated the problem of their divergence between protection and sexual selection: "We differ, I think, chiefly from fixing our minds perhaps too closely on different points, on which we agree" (Marchant 1916, t. I, p. 225). Darwin tried to bring by all possible means closer agreement between him and Wallace. However, eventually, it turns out that Wallace did not think sexual selection was a significant evolutionary factor although he seemed, at times, to waver somewhat.

An early discussion of sexual selection appears in the portion of Darwin's 1844 essay that was read to the Linnaean Society in July 1858, along with Wallace's paper "On the Tendency of Varieties to depart indefinitely from the Original Type." Darwin wrote:

> Besides this natural means of selection, by which those individuals are preserved, whether in their egg, or larval, or mature state, which are best adapted to the place they fill in nature, there is a second agency at work in most unisexual animals, tending to produce the same effect, namely, the struggle of the males for the females. These struggles are generally decided by the law of battle, but in the case of birds, apparently, by the charms of their song, by their beauty or their power of courtship, as in the dancing rock-thrush of Guiana. The most vigorous and healthy males, implying perfect adaptation, must generally gain the victory in their contests. This kind of selection, however, is less rigorous than the other; it does not require the death of the less successful, but gives to them fewer descendants. The struggle falls, moreover, at a time of year when food is generally abundant, and perhaps the effect chiefly produced would be the modification of the secondary sexual characters, which are not related to the power of obtaining food, or to defense from enemies, but to fighting with or rivaling other males. (Darwin 1858, p. 50)

In this passage, as later in the *Origin*, sexual selection appears to be an umbrella term for two different kinds of phenomena: male-male rivalry leading to armaments; female preferences leading to ornaments. The first of these mechanisms, rivalry among males, was generally undisputed. Wallace accepted it (1905, t. II, pp. 17–18) and he considered "a very general fact that the males fight together for the possession of the females. This leads … to the stronger or better-armed males becoming the parents of the next generation … From this very general phenomenon there necessarily results a form of natural selection, which increases the vigor and fighting power of the male animal" (1889, p. 282). Vigor was a rationale for including male-male competition as part of natural selection.

What was really at stake was the idea that female animals have the capacity to choose their mates. Darwin had strongly supported the possibility of female aesthetic choice. In the *Origin*, he wrote: "if man can in a short time give elegant carriage and beauty to his bantams, according to his standard of beauty, I can see no good reason to doubt that female birds, by selecting, during thousands of generations, the most melodious or beautiful males, according to their standard of beauty, might produce a marked effect" (1859, p. 89). For Darwin then, it was not unreasonable to invoke a rudimentary aesthetic sense on the part of the peahen as a factor in the selection of the peacock's tail. Note that he invokes here the model of artificial selection by humans—a model that Wallace presumably would have rejected.

Wallace did not reject female choice *in general*, but he thought that female preferences targeted male vigor, not beauty. Accordingly, he endorsed two main objections that had been raised against females' taste for the beautiful:

a/Female choice, if it seeks for sheer beauty (unrelated to the signaling of any quality), undermines the power of natural selection, as this mechanism is only concerned with benefits. Especially, female preferences for this or that trait would have no evolutionary foundation.

b/Assuming a sense of beauty in lower animals raises the broader question of animal faculties. As Gayon (2009b) puts it, female choice amounts to claiming "that many animals, from fishes to primates, have perceptive, emotional and cognitive abilities that make them able to discriminate and choose their sexual partners. This claim raised no more or less than the problem of the gradual evolution of the mind".

Accordingly, Wallace thought that sexual selection (*sensu* female choice) was an unnecessary hypothesis (Gayon 2009b). Wallace could not accept the notion of peahens with an aesthetic sense and was seeking for usefulness of traits.

These points are now well-established in the literature (Cronin 1991; Milam 2010). But other elements should also be brought to the fore. First, as noted earlier, Wallace's environmentalist conception of natural selection (Nicholson 1960) should be differentiated from Darwin's own understanding of natural selection as "a competitive process within the species, which can change the species even under unchanged conditions" (Gayon 2009b). This may ultimately impact on their contrasted views on sexual selection: Wallace's reluctance to accept sexual selection is linked to the fact that sexual selection is "a purely competitive process among the members of one sex within the species"; while, for Darwin, sexual selection "was based exclusively upon differential reproductive success among individuals of one sex" and "did not rely upon an adaptive advantage" (Gayon 2009b). For Wallace, sexual selection "was outside Darwinism", while for Darwin, sexual selection, "because of its primarily competitive and individualistic nature, revealed something important about how selection in general works in nature" (Gayon 2009b).

Darwin's sexual selection aims at explaining certain largely male traits: weapons and beauty, or sex differences "in structure, colour, or ornament", as Darwin himself puts it. Darwin focuses, among other traits, on exuberant coloration in males. Wallace has a different take on this question. He is interested in the relatively plain or drab coloration of the females compared to the males in many bird species; he thinks it is the result of natural selection, to protect them from predation as they nested, while the males in those cases are less subjected to such selection. The focus on coloring shows the paramount importance of protection. Two positions are open: one (Darwin's) claim that colouration results from female preference for beautiful feathers in males; the other (Wallace's) stresses the protective value of coloration in females. Darwin himself was oscillating (Darwin to Wallace, 16 September 1868):

> You will be pleased to hear that I am undergoing severe distress about the protection & sexual selection: this morning I oscillated with joy towards you: this evening I have swung back to old position, out of which I fear I shall never get. (Marchant 1916, t. I, p. 222–223)

Wallace took great pride in having shown the usefulness of phenomena which were previously regarded as non-adaptive. He was an extreme utilitarian and, as a result, a pan-selectionist. Wallace argued we should look at nature assuming that each feature we see is useful:

> … other slight differences which to us are absolutely immaterial and unrecognizable, may be of the highest significance to these humble creatures, and be quite sufficient to require some adjustments of size, form, or color, which natural selection will bring about. (1889, p. 148)

Wallace also rephrased Darwin's "great general principle" as: "all the fixed characters of organic beings have been developed under the action of the law of utility", entailing for instance that "so remarkable and conspicuous a character as color, which so often constitutes the most obvious distinction of species from species or group from group, must […] in most cases have some relation to the wellbeing of its possessors." (1889, p. 187–188).

Another important issue between Darwin and Wallace is sex-linked inheritance. On that matter, it should be noted that the first words of the section on sexual selection in the *Origin* provides us with an important key to understand Darwin's mechanism: peculiarities appear "in one sex and become hereditarily attached to that sex" (1859, p. 87). Darwin called Wallace's attention on their diverging views on inheritance in his letter dated 5 May 1867 (Marchant 1916, I, p. 185). As Kottler put it (1980, p. 204): "At the heart of their disagreement was a basic difference of opinion about the laws of inheritance." At the climax of the controversy (23 September 1868) Darwin wrote to Wallace: "I think we start with different fundamental notions on inheritance." Wallace believed that, as a rule, variations as they first appeared, were inherited equally by both sexes, and that, afterwards, natural selection had to convert equal inheritance into sex-limited inheritance. Whenever one sex is endangered more than the other (for instance by conspicuous coloration), natural selection would convert the equal inheritance of the variations sexually selected, into sex-limited inheritance, so that the sex in greater danger loses conspicuous coloration. Following his belief in the generality of equal inheritance, Wallace attributed the drab coloration of the less conspicuous sex to natural selection for the sake of concealment of the individuals in greater danger. On the other hand, Darwin was in favor of sex-limited inheritance of traits: in his view, female animals never had to "lose" bright coloration or to be modified for protection—as they never acquired gaudy feathers.

We now understand that it is a misconception to regard Wallace as opposed entirely to sexual selection. Besides, while Wallace thought that sexually dimorphic traits were initially the same in both sexes and natural selection *had to make* the sex in greater danger less conspicuous, Darwin claimed that sexually selected traits are only present in one sex, and natural selection *has to keep* the sex in greater danger less conspicuous (Kottler 1980). Such disagreement reflects the fact that the actual (Mendelian) laws of genetics were of course unknown to both of them.

2.2.3 The Riddle of Human Evolution

The question of sexual selection was also closely linked to the question of human evolution. First, sexual selection was deeply tied with anthropomorphic views. The flavor of Wallace's thinking on the topic can be seen in a well-known passage from his book *Darwinism*:

> It will be seen, that female birds have unaccountable likes and dislikes in the matter of their partners, just as we have ourselves, and this may afford us an illustration. A young man, when courting, brushes or curls his hair, and has his moustache, beard or whiskers in perfect order and no doubt his sweetheart admires them; but this does not prove that she marries him on account of these ornaments, still less that hair, beard, whiskers and moustache were developed by the continued preferences of the female sex. So, a girl likes to see her lover well and fashionably dressed, and he always dresses as well as he can when he visits her; but we cannot conclude from this that the whole series of male costumes, from the brilliantly coloured, puffed, and slashed doublet and hose of the Elizabethan period, through the gorgeous coats, long waistcoats, and pigtails of the early Georgian era, down to the funereal dress-suit of the present day, are the direct result of female preference. In like manner, female birds may be charmed or excited by the fine display of plumage by the males; but there is no proof whatever that slight differences in that display have any effect in determining their choice of a partner. (1889, pp. 286–287)

But beyond these considerations, the genesis of Darwin's concept of sexual selection is, on a deeper level, intimately tied to the puzzle of human races. As Desmond and Moore have shown (2009, p. 282), Darwin's earlier notice of sexual selection was found in a manuscript note on Knox' *Races of man*. Darwin wrote to Wallace (March 1867): "…my sole reason for taking it up [i.e. the subject of man] is that I am pretty well convinced that sexual selection has played an important part in the formation of races, and sexual selection has always been a subject which has interested me much." (Marchant 1916, t. I, p. 182). And again, on 29 April 1867: "in my Essay upon Man I intend to discuss the whole subject of sexual selection, explaining as I believe it does much with respect to man." (t. I, p. 183).

Darwin and Wallace disagreed on the importance of sexual selection in the evolution of secondary sexual characteristics, and also on the question of human evolution.

Darwin sought naturalistic explanations for phenomena, including behavioral phenomena, and considered human capacities such as cognition, emotions, aesthetic feelings to be traits that had evolved, and thus could also exist in other species. He wrote a book about the expression of emotions in animals (1872). While there is still debate on the question of animal cognition, it is fair to say that modern neurology and studies of animal behavior have largely vindicated Darwin's basic viewpoint (e.g., Griffin 2001).

Wallace had a rather Cartesian view of human mental abilities. He believed that some perhaps mystical principle was involved in the generation of the human mind and its consciousness. As a result, the intelligent design movement has apparently adopted Wallace in recent years, claiming that his version of evolution anticipated their claims (Flannery 2008, 2011).

Darwin's view: A Gradation of Mental Powers It is particularly interesting that Darwin published his major discussion of sexual selection in the same book as his treatment of human evolution, which was also not treated extensively in the *Origin*. This juxtaposition may relate to a deeper division between him and Wallace, involving their views on human evolution, as well as their views on sexual selection by female choice: the two differences may in fact have been related.

Why should these two great intellects be likely to differ on the subject of human evolution, and also on the possibility of sexual selection by female choice? There is a strong temptation here to indulge in what is sometimes termed *whig history*—the application of contemporary norms to past historical events or figures. Both Darwin and Wallace were of course Victorians, with constant immersion in the racial, ethnic and gender biases of the period. However, let us note at the outset that Wallace was, among other things, a strong feminist, and it is thus difficult to see his opinion as simply the result of male bias. Similarly, Darwin was an abolitionist and a strong opponent of slavery, and indeed it has been argued that this was a major factor in his development of the theory of evolution by natural selection (Desmond and Moore 2009), so it becomes difficult to ascribe his views on human evolution to racism.

We will argue here that their disagreements about sexual selection as well as about human evolution probably did not arise primarily from Victorian biases, but rather had roots in fundamentally different conceptions about the evolutionary relations of humans to other species, and that this difference in turn reflected, at least in part, differences in their experiences as naturalists.

To understand this difference we need to consider the backgrounds of both men. A major difference in the personal backgrounds of Darwin and Wallace was their experience of non-European peoples and cultures. Darwin had traveled, of course, circumnavigating the globe for 5 years in the *Beagle*, with extensive inland excursions in South America and elsewhere, but he was usually either in the company of fellow Englishmen or in any case supported, protected and cushioned, directly or indirectly, by the great authority of the British Navy. According to all accounts, he was a very tolerant person, not given to aggression or autocratic assertion, and a Whig politically, strongly opposed to slavery. Nevertheless he was also a product of mid-Victorian British culture, with a strong belief in progress and little in-depth personal knowledge of non-European cultures (see Browne 1995, especially chap. 10, pp. 234–253; Desmond and Moore 2009). His expressed surprise when encountering the natives of Tierra del Fuego serves to illustrate this and makes a sharp contrast with the "domesticated" figure of the 3 natives transported back to South America on the *Beagle*:

> The Fuegians rank among the lowest barbarians; but I was continually struck with surprise how closely the three natives on board H.M.S. Beagle, who had lived some years

> in England, and could talk a little English, resembled us in disposition and in most of our mental faculties. (Darwin 1871, t. I, p. 34)

Much of his other information regarding other "savage" peoples came from anecdotal accounts by a variety of travelers. Continuing the quotation above, he states

> If no organic being excepting man had possessed any mental power, or if his powers had been of a wholly different nature from those of the lower animals, then we should never have been able to convince ourselves that our high faculties had been gradually developed. But it can be clearly shown that there is no fundamental difference of this kind. … there is a much wider interval in mental power between one of the lowest fishes … and one of the higher apes, than between an ape and man. (Darwin 1871, t. I, pp. 34–35)

On the other hand,

> Nor is the difference slight in … intellect, between a savage who does not use any abstract terms, and a Newton or Shakespeare. Differences of this kind between the highest men of the highest races and the lowest savages are connected by the finest gradations. … there is no fundamental difference between man and the higher mammals in their mental faculties. (Darwin 1871, t. I, p. 35)

Darwin strongly believed in the unity of the human family, but (it is like the other face of the same coin), as a result, he tended to view, possibly unconsciously, other populations as "uncivilised" (for instance, 1859, p. 38, 140). His belief in progress entailed gradual improvement by stages, entailing that the native Fuegians represented earlier evolutionary stages, less advanced than Europeans. However, he would clearly have disagreed with the theory of retrogression or degradation, espoused by, among others, the Archbishop of Dublin Richard Whately (1855), or George Campbell, 8th Duke of Argyll (1869). According to the theory of degradation, the 'savage races' of mankind presented a degradation from a previously more advanced civilized state, and Darwin clearly refuted these theories in his *Descent* (Darwin 1871, t. I, p. 181; see also Gillespie 1977).

In any case, Darwin clearly thought of the human species as being derived evolutionarily from earlier primates, and from these and many other passages, it's clear he saw no unbreachable barrier separating humans from other animal species, and believed there had been a succession of evolutionary stages from earlier primates to humans, with no impermeable boundary. He thus had no difficulty with the idea that the rudimentary beginnings of human intellectual, moral and a esthetic sensibilities could be found in lower animals, an attitude evident in the very title of another of his books, *On the Expression of the Emotions in Animals and Man* (1872). And he also noted the profound unity of all human beings, in sharing the same basic emotions (see Radick 2010).

A rather direct indication of the link between his views on human evolution and his adoption of sexual selection by female choice is found in an addition to chap. 8 in the second edition of *The Descent of Man*. Where the first edition read:

> No doubt this implies powers of discrimination and taste on the part of the female which will at first appear extremely improbable; but I hope hereafter to shew that this is not the case. (1871, t. I, p. 259)

The second edition reads:

> No doubt this implies powers of discrimination and taste on the part of the female which will at first appear extremely improbable; but by the facts to be adduced hereafter, I hope to be able to show that the females actually have these powers. When, however, it is said that the lower animals have a sense of beauty, it must not be supposed that such sense is comparable with that of a cultivated man, with his multiform and complex associated ideas. A more just comparison would be between the taste for the beautiful in animals, and that in the lowest savages, who admire and deck themselves with any brilliant, glittering, or curious object. (1874, p. 211)

As noted by others (Prum 2012), Darwin was quite serious in ascribing an aesthetic sense to other species. The difference in mentality, emotions and sensibility between humans and other creatures was one of degree, and not fundamental: these qualities were also evolving.

Wallace's View: Beyond the Scope of Natural Selection Wallace presents a contrast. Having spent many years largely on his own, first in South America and then in Southeast Asia, in intimate contact with native populations, he could appreciate from personal experience the competence and intelligence of the peoples he encountered, and was convinced that the mentality and reasoning power of 'savages' was quite comparable to that of 'civilized' Europeans. Thus he had a unitary view of the various kinds of humanity, as a single species, with great variety, but all at the same intellectual level. Like Darwin, he was a mid-Victorian, but his personal history was different, and he was also perhaps more of a maverick than Darwin, far more active politically, espousing a variety of political and social causes.

Reviewing Darwin's *The Descent of Man and Selection in Relation to Sex,* he wrote that the "vast amount of the superiority of man to his nearest primate relatives is what is so difficult to account for" (Wallace 1871, p. 183). Darwin, as we have seen, did not share this view and considered that the differences between other animals and humans, though indeed large, were essentially of degree rather than fundamental.

Further, Wallace continues, "It must be admitted that there are many difficulties in the detailed application of [Darwin's] views, and it seems probable that these can only be overcome by giving more weight to those unknown laws whose existence he admits but to which he assigns an altogether subordinate part in determining the development of organic form" (Wallace 1871). These "unknown laws" were mentioned by Darwin in the *Origin*: the laws of growth, development, inheritance, correlation, the "direct action of the environment," and the laws of habit and instinct—many of these have become major areas of twentieth and twenty-first century biological research but were largely unexplored in Darwin's day. Wallace tended to view some of these as evidences of a controlling Mind or Supreme Intelligence, and his involvement in spiritualism and related subjects in his later years no doubt reflects this view (Slotten 2004).

In a joint review of the tenth edition of Charles Lyell's *Principle of geology* (1867) and of the sixth edition of his *Elements of geology* (1865), Wallace developed the following argument: the brain of the "lowest savages" (Wallace thinks of the Australians or the Andaman islanders) and, probably, those of "the pre-historic races," was an organ barely "inferior in size and complexity to that of the highest types (such as the average European)"; in other terms, two or three thousand years would be sufficient for them to acquire, by a "process of gradual development", the average results of humans of higher civilizations. In contrast to that, the mental requirements of these lowest savages, "are very little above those of many animals": "the higher moral faculties and those of pure intellect and refined emotion are useless to them, are rarely if ever manifested, and have no relation to their wants, desires, or well-being" (Wallace 1869, p. 91–392). Hence the following paradox: "How, then, was an organ developed so far beyond the needs of its possessor? Natural selection could only have endowed the savage with a brain a little superior to that of an ape, whereas he actually possesses one but very little inferior to that of the average members of our learned societies."

In his copy of this text, Darwin wrote in the margin a vehement "No", triple scored and showered with exclamation points (Irvine 1955, p. 187).

A curious and perhaps ironic end-note on Wallace and sexual selection may also be mentioned. He did, in fact, allow for the possibility of sexual selection by female choice in one species, namely in humans. Though he rejected Darwin's speculation that hairlessness and skin color in humans were products of sexual selection, he nevertheless invoked a kind of sexual selection by female choice in another area. He thought violent tendencies in humans would gradually diminish and intelligence increase, as women chose mates and had children preferentially with gentler, more intelligent males (Wallace 1913). He obviously followed the logic of sexual selection by female choice, but was not willing to grant it status as an evolutionary force, except in the (for him, exceptional) case of humans. Darwin, the mid-Victorian, tended somewhat to pessimism, while the younger Wallace, like many Europeans of the Edwardian era, was in some ways an optimist, believing in incremental progress in human society. Wallace died in 1913, too soon to witness the events that began a year later in Sarajevo.

In summary, Darwin's and Wallace's views diverged on two accounts: one related to the mental abilities of animals, one related to the mental capacities of so-called "uncivilized men". On the first point, Darwin thought it was not unreasonable to invoke a rudimentary aesthetic sense on the part of the peahen as a factor in the selection of the peacock's tail, whereas Wallace could not accept the notion of peahens with an aesthetic sense. On the second point, while both men recognized the unity of the human family, Darwin was struck by the lowness of non-European civilizations, to the point that he estimated that the distance between human and animal brains was not so large; while Wallace emphasized the gap between humans and non-human animals and stressed the seeming impotence of natural selection when it came to explaining human higher mental faculties. Darwin had a continuous view of mental powers, from non-human animals to humans; while Wallace, having a sense of a strong discontinuity between humans and non-humans, thought it was a sufficient argument to repel the role of natural selection and call for other (supernatural) agencies.

Darwin saw humans as simply part of an evolutionary continuum, their mental, aesthetic and emotional qualities as differing in degree, but not fundamentally from those of other animals. In contrast, Wallace viewed humans as a special case. In particular, he considered that human mentality, cognition, aesthetic senses, and spirituality could not be a product of natural selection, and it was difficult for him to accept the idea of an aesthetic sensibility in non-human animals. He was not religious in the conventional sense of organized religion, but he viewed humans as more than simply products of natural selection, and this is no doubt connected to his life-long interest in spiritualism and the occult. Eventually, in a review of E.B. Poulton's *Colours of Animals*, Wallace stated (1890, p. 291): "This most interesting question … in all probability, will not be finally settled by the present generation of naturalists."

2.3 Shaping the Darwin/Wallace Debate: What the Positions of Darwin and Wallace Imply.

Now that we have analyzed the complexity of the debate between Darwin and Wallace, we will approach the historical steps towards the rephrasing of their divergence in terms of Utility vs Beauty.

2.3.1 Sexual Selection During the "Eclipse of Darwinism"

The historian Peter J. Bowler and others have noted that, for a period of several decades before and to some extent after 1900, Darwin's version of evolution by natural selection was out of favor with many biologists, who supported other types of evolutionary theory: Lamarckism, theistic evolution, mutationism, orthogenesis (Bowler 1983, 1988). While Darwin's sexual selection is often considered merely an expression of Victorian prejudice, Bowler claims that, "during the eclipse of Darwinism, sexual selection was even less popular among biologists than natural selection was" (Bowler 1984, p. 314). In other terms, in spite of its familiar ring for a Victorian reader, Darwin's concept elicited more criticisms than positive responses. And indeed, although it didn't evoke the great eruption of criticism that greeted the appearance of the *Origin*, Darwin's theory of sexual selection did attract criticism and satire, as in the lampoon by Richard Grant White, *The Fall of Man, or the Loves of Gorillas* (1871), in which gorillas exercise sexual selection by mating with a variety of other species. Such ridicule of sexual selection in the lay press may have served sometimes as proxy for opposition to Darwin's other mechanism, natural selection, and to evolutionary thought in general.

With regard to animal coloration, Darwin's aesthetic hypothesis was viewed as emphasizing *love* and *beauty*, whereas Wallace's adaptationist standpoint emphasized *vigor* and *safety* (mimicry, protection). Ironically, enemies of Darwin's natural selection were quite at ease with the idea of sexual selection as it seemed to resurrect a metaphysical (and non utilitarian) kind of beauty. For instance, the Duke of Argyll's *Reign of Law* (1867) discussed coloration in hummingbirds: he asked why a topaz crest should be selected in preference to a sapphire one. Focusing solely on utility, Darwin's natural selection seemed to be missing the point that is central to Argyll's conception of nature: beauty for its own sake. But sexual selection appeared to restore beauty to nature. It should be noted that early reactions to Darwin's model of sexual selection, including that by Wallace, considered it as separate from, or even contradicting natural selection.

While some recent readers have suggested that Darwin integrated beauty into nature (e.g. Cronin 1991), many of his contemporaries thought the tie between beauty and divine creation was impossible to sever: as soon as Darwin acknowledged the existence of beauty, he had reintroduced a teleological feature in nature. The reluctance of many biologists (including Wallace) to follow Darwin in this model starts with what was considered the special status of beauty. It was necessary to decide whether Darwin's incorporation of an aesthetic quality such as beauty among animals, was consistent with a naturalistic framework.

The rephrasing of the Darwin-Wallace debate in terms of Beauty vs Utility owes a lot to the biologist George J. Romanes (1848–1894), who was described by the *Times* as "the biological investigator upon whom, in England, the mantle of Mr. Darwin has most conspicuously descended" (quoted by Thiselton-Dyer 1888). Romanes was fighting with Wallace over Darwin's legacy. He depicted Wallace as a supporter of a pan-utilitarian stance. For instance, Romanes attributed to Wallace the thought that "natural selection has been the sole means of modification ...Thus the principle of Utility must necessarily be of universal application" (1892, t. II, p. 6). Romanes referred to "two great classes of facts in organic nature: namely, those of Adaptation and those of Beauty. Darwin's theory of descent explains the former by his doctrine of natural selection, and the latter by his theory of sexual selection" (Romanes 1892). Apparently, Romanes committed to both Darwinian mechanisms, but by phrasing the problem this way, with a clear divide between Utility and Beauty, he, willingly or not, confirmed the idea that natural and sexual selection were two rather separate mechanisms, having little to do with each other.

In fact, while Darwin saw at first no contradiction between natural and sexual selection, he saw clearly that sexual selection could lead to the evolution of non-adaptive traits. As a result, his followers asked whether sexual selection challenged what Darwin had called the "paramount power of natural selection" (Darwin 1859, p. 84). In 1877 Eduard von Hartmann claimed that Darwin weakened his case for natural selection by trying to take beauty (and not merely utility) into account (Hartmann 1877; Hoquet 2009).

While Darwin and Wallace debated on the sex-limited character of variation, their followers put forward the topic of the "greater eagerness" of males. Both Darwin and Wallace, but maybe Wallace even more than Darwin, had stressed the idea of the "greater vigor" of males. In 1883, Harvard and Johns Hopkins biologist William Keith Brooks suggested that a more fundamental explanation was required in order to explain why males have stronger passions (Brooks 1883). If it is a general rule that males are more modified than females, then biology needs a theory of heredity that accounts for this fact.

Brooks disagreed with Wallace's hypothesis that females were drab in order to be less visible to predators: he noted that in species where both males and females brood, a colour dimorphism subsists. He supported Darwin's hypothesis that "the excessive exposure of the male to the action of selection, natural and sexual" is the cause of his being modified. Male characters being useful, they have been positively selected for; while there was no such pressure for the evolution of corresponding female traits. "No one can doubt the truth of this statement, but it does not go to the root of the matter. The question is not how peculiarities useful to the male alone have been restricted to that sex, but why the female has not acquired another set of characteristics to fit her for her peculiar needs" (Brooks 1883). Brooks explicitly put forward the problem of sex-limited inheritance. The "provisional hypothesis of pangenesis", described in Darwin's book *Variation under domestication* (1868) was never introduced to later editions of the *Origin of Species*; but it played an important role in W.K. Brooks' understanding of his ideas on sexual selection. Brooks suggested that transmission of gemmules by the mother was more rare than transmission by the father, explaining why males vary more than females.

A different approach to the question of sexual dimorphism was taken by Patrick Geddes and Arthur J. Thomson (1889). For them, "no special theory of heredity is required,—the males transmit the majority of variations, because they have most to transmit". Darwin and Wallace's theory are considered symmetrical: sexual selection is, with Darwin, acknowledged as a minor accelerant, natural selection is, with Wallace, understood as a retarding "brake" on the differentiation of sexual characters, but for these authors, the key to sexual dimorphism is to be found in a constitutional or organismal origin, which they term "the katabolic or anabolic diathesis which preponderates in males and females respectively" (Geddes and Thomson 1889). (It should be noted that these various speculations occurred before the the actual laws of Mendelian heredity were rediscovered).

We see how, with Brooks and Geddes & Thomson, the debate on sexual selection has moved away from the topic of female choice to encompass the question of male vigour and male variability (allegedly superior to that of females).

In a 1903 book dedicated to W.K. Brooks, Thomas Hunt Morgan argued vigorously against sexual selection theory. He gave a comprehensive list of objections, bearing both on natural and sexual selection, mixing cartoonesque and biological remarks. Morgan coarsely caricatured sexual selection in anthropomorphic terms: "It sounds a little strange to suppose that women have caused the beard of man to develop by selecting the best-bearded individuals, and the compliment has been returned by the males selecting the females that have the least amount of beard" (Morgan 1903). His objections also included observations bearing on the laws of sex-limited inheritance: "It is also assumed that the results of the selection are transmitted to one sex only. Unless, in fact, the character in question were from the beginning peculiar to only one sex as to its inheritance, the two sexes might go on forever selecting at cross-purposes, and the result would be nothing"(Morgan 1903).

Thus, Morgan rejected both Darwin and Wallace: the numerous difficulties that the theory of sexual selection had met led to rejecting it as an explanation of the secondary sexual differences amongst animals; but Wallace's explanation of the sex differences as due to the excessive vigour of the male, was equally unsuccessful. In the end, Darwin's theory was only useful as it "served to draw attention to a large number of most interesting differences between the sexes, and, even if it prove to be a fiction, it has done much good in bringing before us an array of important facts in regard to differences in secondary sexual characters". As to the theory itself, it "meets with fatal objections at every turn" (Morgan 1903). For Morgan, the key to sexual differences was to be found in internal (hormonal) factors—a view that was taken up by some major histories of biology of the early twentieth century (Ràdl 1913; Nordenskjöld 1920; quoted by Cronin 1991, p. 50–51).

In the early twentieth century the scientific climate changed rapidly. Mendel's laws of particulate genetics were rediscovered, and this paved the way for development of the science of population genetics. Basic principles and tools such as the principle today known as the Hardy-Weinberg distribution were developed (the relevant papers were published in 1908), and major figures such as R. A. Fisher, J.B.S. Haldane and Sewall Wright began to develop theoretical connections between genetics and evolution by natural selection. Eventually information and insights from genetics, paleontology and ecology would be gathered together into a broad view of evolution that was termed the Modern Synthesis (Mayr and Provine 1980).

There is a common claim that the question of sexual selection was largely ignored during this period by leading evolutionary biologists such as Theodosius Dobzhansky and G.G. Simpson. The great emphasis was always on the effects of natural selection. At most, secondary sexual signals which could not be explained as due to competition among males, such as antlers, were seen as identity signals of species, to prevent hybridization: the peacock's tail would tell the peahen that this male was a conspecific.

In fact, this common claim can be seriously challenged. First there was continuous study of sexual selection in the field of experimental evolution. For instance, the entomologist Frank E. Lutz (1879–1943) published on "the effect of sexual selection" (1911). He worked at the time at the Carnegie Institute's Station for Experimental Evolution, under the head of Charles B. Davenport. Lutz argues strongly in favor of female choice. "The basis upon which these flies discriminate against ultra-veined individuals when choosing a mate is a matter for further study. There is an elaborate 'courtship' in which the flirting of the wings in front of the prospective mate plays a large part. It *seems* as though a choice were made on the basis of sight, but I doubt whether that is the case. However, there is no doubt of the choice. It is a clear case of the undoing of artificial selection by sexual selection." (Lutz 1911, p. 37).

Secondly, as noted by Erika Milam (2010), Theodosius Dobzhansky and his colleagues extensively worked on mate choice (especially male choice) in their studies on reproductive isolation in drosophila.

2.3.2 Beauty for Beauty's Sake? R. A. Fisher and the Runaway Model

Traditionally depicted as the sole major exception to the general neglect surrounding sexual selection, the geneticist and statistician R. A. Fisher (1890–1962) also played a key role in shaping the terms of the Darwin-Wallace debate. Fisher criticized

Wallace's idea that animals do not show any preference for their mates on account of their beauty, and that female birds do not choose the males with the finest plumage. He also revived Darwin's idea of beauty for beauty's sake and developed it into what came to be known as the "runaway" model. Oddly, Fisher, an early pioneer in the field of applied mathematical statistics, did not construct a mathematical model of the process, but his verbal description and discussions became the basis for others to take up that challenge. Fisher provided a more precise verbal statement of an effect perhaps hinted at by Darwin: essentially, a positive feedback mechanism.

In a first paper published in 1915, Fisher rejected Wallace's argument against aesthetic choice as weak: 1/ because of our necessary ignorance of the motives from which wild animals choose between a number of suitors; 2/ because there remains no satisfactory explanation either of the remarkable secondary sexual characters themselves, or of their careful display in love-dances, or of the evident interest aroused by these antics in the female; 3/ because this objection is apparently associated with the doctrine put forward by Wallace that the artistic faculties in man belong to his "spiritual nature" and have come to him independently of his animal nature. But, Fisher acknowledged, the strongest point in Wallace's objections was that Darwin had left unexplained the origin of the aesthetic sense in the lower animals.

In 1930, Fisher gave a succinct summary of the disagreement between Darwin and Wallace in the following terms:

> The theory put forward by Darwin to account for the evolution of secondary sexual characters involves two rather distinct principles. In one group of cases, common among mammals, the males, especially when polygamous, do battle for the possession of females. That the selection of sires so established is competent to account for the evolution, both of special weapons such as antlers, and of great pugnacity in the breeding season, there are, I believe, few who doubt … (Fisher 1930, p. 131)

At first sight, Fisher's account is in full acceptance of the first mechanism identified by Darwin, namely male-male competition. But, at the same time, one can feel the influence of Morgan's hormonal creed in his interpretation of male-male fights: it has become especially clear, according to Fisher, that male-male competition is now beyond doubt, "especially since the investigation of the influence of the sex hormones has shown how genetic modifications of the whole species can be made to manifest themselves in one sex only" (p. 131).

> For the second class of cases, Fisher continued, for which the amazing development of the plumage in male pheasants may be taken as typical, Darwin put forward the bold hypothesis that these extraordinary developments are due to the cumulative action of sexual preference exerted by the females at the time of mating. (p. 131)

Here, Fisher isolated the second factor (intra-sexual selection) from the first one, showing that what Darwin had unified under the general head "sexual selection" should be clearly divided into two different factors. Fisher continued:

> The two classes of cases were grouped together by Darwin as having in common the important element of competition, involving opportunities for mutual interference and obstruction, the competition being confined to members of a single sex. To some other naturalists the distinction between the two types has seemed more important than this common element, especially the fact that the second type of explanation involves the will

> or choice of the female. A. R. Wallace accepted without hesitation the influence of mutual combats of the males in the evolution of sex-limited weapons, but rejected altogether the element of female choice in the evolution of sex-limited ornaments. (Fisher 1930, pp. 131–132)

We see here how Wallace's stance is clearly divided in two parts: acceptance of male-male competition; rejection of female choice. But Fisher raises several objections against Wallace:

a. As Argyll convincingly argued, the hypothesis of protective colouration during brooding is not sufficient;
b. Wallace made errors in assuming that "the effect of selection in the adult is diminished by a large mortality at earlier stages" (Wallace 1889, p. 296). Fisher argues that "if one mature form has an advantage over another, represented by a greater expectation of offspring, this advantage is in no way diminished by the incidence of mortality in the immature stages of development, provided there is no association between mature and immature characters".

In conclusion, Fisher suggested that Wallace's reluctance to accept Darwin's female choice was clearly deriving from his "conviction that the aesthetic faculties were a part of the 'spiritual nature' conferred upon mankind alone by a supernatural act, which supplies an explanation of the looseness of his argument" (p. 134).

Fisher's approach to female choice is original, in that he admits that "with respect to sexual preference, the direct evidence of its existence in animals other than man is, and perhaps always will be, meager" (p. 135). But at the same time, he suggests this should be approached with an evolutionary eye: "the tastes of the organisms, like their organs and faculties, must be regarded as the products of evolutionary change, governed by the relative advantage which such tastes may confer" (p. 136).

This leads Fisher to formulate his idea of a runaway model. The question Fisher posed was: why should a peahen prefer to mate with the peacock with the most splendid tail? His model says: because it's fashionable. Other females also choose the males with the most impressive tails, so if her sons inherit the genes for a splendid tail they will get to mate more frequently, and their genes will spread in the population. But how does the fashion get started? Fisher suggested that, initially, a slightly larger tail may have conferred some minor selective advantage, so that, by natural selection, the genes of females mating with a peacock with a larger tail may have been somewhat favored. With time, though, the main advantage became the fact that more females mated with males with larger tails and the increased number of matings in itself would lead to the greater fitness of large-tailed males. As Fisher noted, this becomes a "runaway", an accelerating process, where both the male trait and female preferences for it increase geometrically (exponentially) until the counter-selective disadvantages of an extreme dimorphism lead to a balance between the opposing forces of natural selection and sexual selection. So the process ends in a dynamic equilibrium.

In Fisher's own words, this 2-step process involves two selective influences:

> (i) an initial advantage not due to sexual preference, which advantage may be quite inconsiderable in magnitude, and

(ii) an additional advantage conferred by female preference, which will be proportional to the intensity of this preference. The intensity of the preference will itself be increased by selection so long as the sons of hens exercising the preference most decidedly have any advantage over the sons of other hens, whether this be due to the first or to the second cause (p. 136).

> The two characteristics affected by such a process, namely plumage development in the male, and sexual preference for such developments in the female, must thus advance together and so long as the process is unchecked by severe counterselection, will advance with ever increasing speed. In the total absence of such checks, it is easy to see that the speed of development will be proportional to the development already attained, which will therefore increase with time exponentially, or in geometric progression (p. 137).

Fisher's runaway process stresses the co-evolution between preferences and traits.

What was Fisher's motivation in developing this insight? It may be related to his early interest in eugenics and his view of sexual selection as a mechanism for 'racial repair' and human progress (Bartley 1994).

Fisher's conjecture was later supported by detailed mathematical analysis (for instance Kirkpatrick et al. 1990). Modeling simulations have shown that the initial selective advantage could be dispensed with, and that the runaway process could begin with an arbitrary signal (O'Donald 1967; Lande 1981; Kirkpatrick 1982). But beyond his formulation of the runaway principle, Fisher's contribution was important in shaping Darwin and Wallace as standing for two rival evolutionary mechanisms.

2.3.3 The Good Gene Model

Meanwhile, a rather different kind of explanation for extreme secondary sexual characteristics such as the peacock's tail was developed. Focusing on the utility of secondary characters, this view is now called the "good gene hypothesis": it claims that beauty has always, eventually, a purpose. This pan-utilitarian stance has come to be viewed as 'Wallacean' although one can find elements of it in texts by Charles Darwin or his grandfather Erasmus Darwin. For instance, in the passage from the 1844 Essay quoted above, Darwin refers to the fact that "the most vigorous and healthy males … must generally gain the victory"—that is, the song, beauty or power of courtship could serve as a signal of a vigorous and healthy male. Similarly, Erasmus Darwin (1794, t. I, p. 503) stated: "The final cause of this contest amongst the males seems to be, that the strongest and most active animal should propagate the species, which should thence become improved."

A striking extension of the good gene hypothesis was proposed by Amotz Zahavi (1975, 1977; Zahavi and Zahavi 1997) and came to be known as "the handicap principle". The basic idea is a counter-intuitive one, and it caused much controversy. According to Zahavi's principle, odd or costly features like the peacock's tail become subject to adaptive explanations. Being able to survive and function while encumbered with the cost, or handicap of an extreme sexual dimorphism, such as the peacock's tail or the heavy antlers of the male elk, or elaborate display behaviour in itself serves as a signal of superior genes in a mate. This proposal aroused immediate negative reactions among many, but with time, it has come to be seen as a real possibility, partly because of the appearances of mathematical models indicating how it might work (Pomiankowski 1987; Grafen 1990), so the handicap mechanism of Zahavi can no longer be dismissed. It has fostered the rise of the new field of signal theory, now become a sub-branch of sexual-selection theory (Maynard Smith and Harper 2003). Zahavi has offered an elegant solution to the riddle of female preference for exuberant traits: a question that Darwin had not asked, "why waste attracts mates and deters rivals" (Zahavi and Zahavi 1997, p. 38). And yet the feeling of exuberance in front of sexually-differentiated traits, still remains a strong argument in favour of the Fisherian runaway process. Interestingly enough, Zahavi, while probably being the most prominent neo-Wallacean today, does not claim this. There is only one reference to Wallace in Zahavi and Zahavi (1997, p. 44): "Wallace, in his argument with Darwin over sexual selection, proposed that the main function of male showing off is species recognition." Accordingly, Zahavi distinguishes two kinds of natural selection: *utilitarian selection*, which favors straightforward efficiency; *signal selection*, which results in costly features and traits that look like "waste". In other words, he recasts Darwin's sexual selection as the difference between *utilitarian* and *signal* selections. The only difference between *signal* and *sexual selection* is that the former is much broader than the latter, including *all* signals, not just those affecting potential mates or sexual rivals.

This distinction clearly raises the question of the meaning of "utility". Darwin had commented in the *Origin* "on the protest lately made by some naturalists, against the utilitarian doctrine that every detail of structure has been produced for the good of its possessor": "They believe, Darwin claimed, that very many structures have been created for beauty in the eyes of man, or for mere variety. This doctrine, if true, would be absolutely fatal to my theory. Yet I fully admit that many structures are of no direct use to their possessors." (1859, p. 199). So Darwin, at least in the *Origin*, did not believe that beauty for beauty's sake was compatible with his natural selection theory. So let's take a difficult textbook case: the famous peacock tail. The good gene model refers to the selection of characteristics such as 'vigor', so its supporters ask if the peacock's tail could be, somehow, *a signal of greater fitness*—the 'honest signal' hypothesis. Perhaps peacocks with longer tails are also healthier or more fecund? The motivation for this hypothesis was no doubt a desire to uncover something that conventional natural selection could work on, and it also has the advantage of immediately suggesting experimental and field studies. In genetic terms, the basic component here would be a linkage disequilibrium between the conspicuous signal (as in the peacock's tail) and other, adaptive physiological features, so that the signal can serve as a proxy for another feature that is in fact the subject of natural selection.

It was found experimentally that, given a choice, peahens evidently preferred peacocks with bigger, more brilliant tails (Petrie et al. 1991), and that the offspring of peacocks with larger, more brilliant tails were healthier in various ways, or in any case tended to survive and produce more offspring (Petrie 1992,1994; Moller and Alatalo 1999). However, there has been a recent debate, initiated by a report of study over several years of feral peacocks in which there appeared to be no preference for mating with males with longer trains (Takahashi et al. 2008; Loyau et al. 2008). This has triggered the claim that the "poster-child" example for sexual selection was actually flawed (Roughgarden 2009). Further examinations of this question have concluded that the situation is complex. Another recent study found that males with smaller, less decorated tails are chosen less often as mates but above a low threshold there appeared to be no advantage to having larger, more decorated tails (Dakin and Montgomery 2011). However, it is possible that part of the explanation for these somewhat disparate results from different groups may be found in a study by Loyau's group, which found a correlation between mating success and the iridescence, or structural colour of the peacock's tail (Loyau et al. 2007). Females may be responding to the quality of the structural colour (which was not measured in the other studies cited), more than to the size or number of eyespots in the tail (On the "peacock tale", see Cézilly, this volume).

Potential complexities of this kind of indirect selection could also arise through the intricacies of pleiotropic pathways and linkage disequilibrium, as suggested by a study by Hale et al. (2009). They found experimentally that (1) male peacocks with longer trains tended to have more diversity in their major histocompatibility complex (MHC), generally considered to signify superior immune response to diseases; and also that (2) females preferentially mate with males with longer more elaborate trains. However, in their multivariate analysis of data from a captive population, they also found that, statistically, peahens lay more and larger eggs for males with a more diverse MHC but not necessarily for males with longer trains. Thus, in this case the linkage disequilibrium, if it exists, may be with some other signaling feature or features beside the train. (Again, though, this study did not attempt to monitor iridescence, or structural colour features of the males' tail feathers.)

2.4 Concluding Remarks

This paper endeavored to give a detailed overview of the debate between Darwin and Wallace: we have differentiated the motivations and views of the founders of the evolutionary paradigm. But beyond the two men Darwin and Wallace, this paper also gave an opportunity to analyze some conceptual issues between natural and sexual selection. It is often argued that sexual selection and natural selection do not contradict each other, but our study of the Darwin/Wallace controversy reveals that from an historical point of view, this seeming harmony between all types of selection is illusory. In any case, sexual selection was clearly seen as different from and even as contradictory to natural selection.

The Darwin-Wallace controversy continues, in a multitude of forms, to fuel new theoretical, experimental and field research. Several papers in this collection still refer to the two major figures, almost ritually, and Darwin and Wallace are energetically hauled over the fuzzy border between good and bad science.

The alleged opposition between the Fisher runaway process and the good-gene hypothesis have been a central concern for many historians of biology. Serving as a grid framing more recent debates, the differences between the Darwinian-Fisherian "sexy son" and the Wallacean-Zahavian "healthy offspring" have been variously rephrased over time, for instance, as "good-taste" vs "good-sense" (Cronin 1991, p. 183). Ridley (1993, p. 143) has likened the conflict to the feud of the Montagues and the Capulets in *Romeo and Juliet*, and suggested that it was rooted more in personality than in objective science: "Those of a theoretical or mathematical bent—the pale, eccentric types umbilically attached to their computers—became Fisherians. Field biologists and naturalists—bearded, besweatered, and booted—gradually found themselves Good-geners". The difference would be more one of "scientific temper": mathematical modeling vs naturalistic fieldwork.

Ultimately, the rhetorical reference to Darwin and Wallace may well serve as an introductory and pedagogical "red herring": readers are lured into this historical battle of the founding fathers, in order to make the pill of highly abstract theoretical modeling easier to swallow. For instance, Grafen (1990) studied a model of the Zahavi mechanism that explicitly excludes the Fisher process and, in his words, "places Zahavi's handicap principle on the same logical footing as the Fisher process". To do this, he required three mathematical appendices, the last of which is 23 pages long and employs relatively advanced tools and concepts (e.g., measures on Banach space) that in general will be familiar only to mathematicians; he then proposes a method of quantifying the relative importance of the Fisher process and the Zahavi principle in both theory and facts (data), and presents a "Fisher index," to indicate the relative importance of the two processes in a given model or situation. Prum (2010) suggests that the pure Fisher process, without linkage disequilibrium between signal and other adaptive genes, should be considered a null hypothesis in a continuum of models. This may prove to be a fruitful way to look at the landscape of theory.

One can question whether these mechanisms are really mutually exclusive. Indeed, both seem plausible and have support. It might well be that they do both occur in separate cases, or perhaps even simultaneously in a single case, and a realization is developing that the two mechanisms may not necessarily be incompatible.

References

Argyll, GD Campbell, 8th Duke of (1867) The reign of law. Alexander Straban, London

Argyll, GD Campbell, 8th Duke of (1869) Primeval man: an examination of some recent speculations. Strahan & co, London

Bartley M (1994) Conflicts in human progress: sexual selection and the Fisherian 'runaway'. Brit J Hist Sci 27:177–196

Beddall BG (1968) Wallace, Darwin, and the theory of natural selection. A study in the development of ideas and attitudes. J Hist Biol 1(2):261–323

Beddall BG (1988) Wallace's annotated copy of Darwin's. Orig Species J Hist Biol 21(2):265–289

Bock WJ (2009) The Darwin-Wallace Myth of 1858. Proc Zool Soc 62(1):1–12

Bowler PJ (1976) Alfred Russel Wallace's concepts of variation. J Hist Med 31:17–29

Bowler PJ (1983) The Eclipse of darwinism: anti-darwinian evolution theories in the decades around 1900. Johns Hopkins University, Baltimore

Bowler PJ (1984) Evolution. The history of an Idea. University of California Press, Berkeley

Bowler PJ (1988) The Non-Darwinian revolution. Reinterpreting a historical myth. The John Hopkins University Press, Baltimore

Brooks WK (1883) The law of heredity. A study of the cause of variation, and the origin of living organisms. John Murphy & Co, Baltimore

Browne J (1995) Charles darwin. voyaging, (A Biography: vol I). Princeton University Press, Princeton

Cronin H (1991) The ant and the peacock. Cambridge University Press, Cambridge

Dakin R, Montgomerie R (2011) Peahens prefer peacocks displaying more eyespots, but rarely. Anim Behav. doi:10:1016/j.anbehav.2011.03.01

Darwin E (1794) Zoonomia, or, the laws of organic life. J. Johnson, London

Darwin CR (1858) On the tendency of species to form varieties; and on the perpetuation of varieties and species by natural means of selection. J Proc Linnaean Soc Zool 3:46–50

Darwin CR (1859) On the origin of Species by means of natural selection, or the preservation of favoured races in the struggle for life. John Murray, London

Darwin CR (1868) The variation of animals and plants under domestication. John Murray, London

Darwin CR (1871) The descent of man and selection in relation to sex. John Murray, 2 vol. London

Darwin CR (1872) The expression of the emotions in man and animals. John Murray, London. (reprinted 2009. Boston: Houghton Mifflin Harcourt)

Darwin CR (1874) The descent of man, and selection in relation to sex, 2nd ed. John Murray, London

Darwin CR (1991) The correspondence of Charles Darwin. In: Burkhardt F, Smith S (eds) (vol 7). Cambridge University Press, Cambridge

Desmond A, Moore J (2009) Darwin's sacred cause: how a hatred of slavery shaped darwin's views on human evolution. Houghton Mifflin Harcourt, Boston

Fisher RA (1915) The evolution of sexual preference. Eugen Res 7:184–191

Fisher RA (1930) The genetical theory of natural selection. The Clarendon Press, Oxford (Reprinted 1958, New York: Dover Publications).

Flannery M (ed) (2008) Alfred Russel Wallace's theory of inteligent evolution. The Discovery Institute, Seattle

Flannery M (2011) Alfred Russel Wallace: a rediscovered life. The Discovery Institute, Seattle

Gayon J (1998) Darwinism's struggle for survival. Heredity and the hypothesis of natural selection. (Translation, Matthew Cobb). Cambridge University Press, Cambridge

Gayon J (2009a) Darwin et Wallace: un débat constitutif pour la théorie de l'évolution par sélection naturelle. In L'évolution aujourdhui à la croisée de la biologie et des sciences humaines. Actes du colloque des 29, 30 et 31 janvier 2009 à l'Académie royale de Belgique. Académie royale de Belgique, Bruxelles, pp 89–122

Gayon J (2009b) Sexual selection: another Darwinian process. C R Biol 333(2010):134–144

Geddes P, Thomson AJ (1889) The evolution of sex. Walter Scott, London

Gillespie NC (1977) The Duke of Argyll, evolutionary anthropology, and the art of scientific controversy. Isis 68(241):40–54

Grafen A (1990) Sexual selection unhandicapped by the Fisher process. J Theor Biol 144:473–516

Griffin D (2001) Animal minds: beyond cognition to consciousness. University of Chicago Press, Chicago

Haldane JBS (1964) A defense of beanbag genetics. Perspect Biol Med 7:343–343

Hale M, Verduijn M, Moller A, Wolff K, Petrie M (2009) Is the peacock's train an honest signal of genetic quality at the major histocompatibility complex? J Evol Biol 22:1284–1294

Hartmann E von (1877) Le Darwinisme. Ce qu'il y a de vrai et de faux dans cette théorie. Baillière, Paris

Hoquet T (2009) Darwin contre Darwin. Comment lire l'Origine des Espèces? Le Seuil, Collection "L'Ordre Philosophique." Paris

Hoquet T (2011) Translating natural selection: true concept but false term? Bionomina 3:1–23
Hrdy SB (1981) The woman that never evolved. Harvard University Press, Cambridge
Irvine W (1955) Apes, angels and victorians. McGraw-Hill, New York
Kirkpatrick M (1982) Sexual selection and the evolution of female choice. Evol Int J org Evol 36:1–12
Kirkpatrick M, Price T, Arnold SJ (1990) The Darwin-Fisher theory of sexual selection in monogamous birds. Evol Int J org Evol 44:180–193
Kottler M (1980) Darwin, Wallace and the origin of sexual dimorphism. Proc Am Phil Soc 124:203–226
Kottler M (1985) Charles Darwin and Alfred Russel Wallace: two decades of debate over natural selection. In: Kohn D (ed) The Darwinian heritage. Princeton University, Princeton, pp 317–432
Lande R (1981) Models of speciation by sexual selection in polygenic traits. PNAS 78:3721–3725
Loyau A, Gomez D, Moureau B, Thery M, Hart N, Saint Jaime M, Bennett A, Sorci G (2007) Iridescent structurally based coloration of eyespots correlates with mating success in the peacock. Behav Ecol 18:1123–1131
Loyau A, Petrie M, Saint Jaime M, Sorci G (2008) Do peahens not prefer peacocks with more elaborate tails? Anim Behav 76:e5–e9
Lutz FE (1911) The effect of sexual selection (1909) In Experiments with Drosophila ampelophila concerning evolution. Carnegie Institution of Washington, Washington, DC
Marchant J (1916) Alfred Russel Wallace, letters and reminiscences. Cassell, London
Maynard Smith J, Harper D (2003) Animal signals. Oxford University Press, Oxford
Mayr E, Provine WB (eds) (1980) The evolutionary synthesis. Perspectives on the unification of biology. Harvard University Press, reprint Cambridge (Mass)-London (England). (1998)
Morgan TH (1903) Evolution and adaptation. MacMillan, New York
Møller A, Alatalo R (1999) Good genes effects in sexual selection. Proc Roy Soc Lond B 266: 85–91
Milam EK (2010) Looking for a few good males. Johns Hopkins University Press, Baltimore
Nicholson AJ 1960. The role of population dynamics in natural selection. In: Sol Tax (ed) Evolution after Darwin, (vol 1). University of Chicago Press, Chicago, pp 477–522
Nordenskjöld E (1920-1924) Biologins Historia; English trans. L. Bucknall Byre, The history of biology. A survey, Tudor Publishing Co., 1928.
O'Donald P (1967) A general model of sexual and natural selection. Heredity 22:499–518
Osborn HF (1894) From the Greeks to Darwin. Columbia University Press, New York
Petrie M (1992) Peacocks with low mating success are more likely to suffer predation. Anim Behav 44:585–586
Petrie M (1994) Improved growth and survival of offspring of peacocks with more elaborate trains. Nature 371:598–599
Petrie M, Halliday T, Sanders C (1991) Peahens prefer peacocks with elaborate trains. Anim Behav 41:323–331
Pomiankowski A (1987) Sexual selection: the handicap principle does work—sometimes. Proc Roy Soc Lond B 231:123–145
Prum R (2010) The Lande-Kirkpatrick mechanism is the null model of evolution by intersexual selection: implications for meaning, honesty, and design in intersexual signals. Evol Int J Org Evol 64:3085–3100. doi:1111/j.1558-5646.2010.01054.x
Prum RO (2012) Aesthetic evolution by mate choice: Darwin's really dangerous idea. Philo Trans Royal Soc B 367:2253–2265
Radick G (2010) Darwin's puzzling. Expression CR Biologie 333:181–187
Ràdl E (1913) Geschichte der biologischen Theorien in der Neuzeit. Engelmann, Leipzig-Berlin
Ridley M (1993) The red queen: sex and the evolution of human nature. Viking, London
Romanes G (1892) Darwin, and after Darwin: an exposition of the Darwinian theory and a discussion of post-Darwinian questions (3 vols). Longmans Green, London
Roughgarden J (2009) The genial gene. University of California Press, Berkeley

Slotten RA (2004) The Heretic in Darwin's court: the life of Alfred Russel Wallace. Columbia University Press, New York

Takahashi M, Arita H, Hiraiwa-Hasegawa M, Hasegawa T (2008) Peahens do not prefer peacocks with more elaborate trains. Anim Behav 75:1209–1219

Thiselton-Dyer WT (1888) Mr. Romanes's paradox. Nature 39:7–9

Wallace AR (1858) On the tendency of varieties to depart indefinitely from the original type. J Proc Linnean Society Zoology 3:53–62

Wallace AR (1869) Sir Charles Lyell on geological climates and the origin of species. The quarterly review (April), 359–394. http://people.wku.edu/charles.smith/wallace/S146.htm

Wallace AR (1871) Review of the descent of man and selection in relation to sex, by Charles Darwin. The academy, March 15, 177–183. http://people.wku.edu/charles.smith/wallace/S186.htm

Wallace AR (1889) Darwinism. An exposition of the theory of natural selection with some of its applications. Macmillan and Co, London

Wallace AR (1890) Review of E.B. Poulton's The colours of animals. Nature 42(1082):289–291. http://wallace-online.org

Wallace AR (1905) My life. Chapman and Hall, UK

Wallace AR (1913) Social environment and moral progress. Cassell & Co, London

Whately R (1855) On the origin of civilization, Young Men's Christian association lectures, (1854–1855). James Nisbet & Co, London

White RG (1871) The fall of man, or the loves of Gorillas. A popular scientific lecture upon the Darwinian theory of development by sexual selection, by a learned Gorilla. Carleton & Co, New York

Zahavi A (1975) Mate selection: a selection for a handicap. J Theor Biol 53:205–214

Zahavi A, Zahavi A (1997) The handicap principle: a missing piece of Darwin's puzzle. Oxford University Press, Oxford

Chapter 3
Darwin on the Proportion of the Sexes and General Fertility: Discovery and Rejection of Sex Ratio Evolution and Density-Dependent Selection

Michel Veuille

Abstract In *The Descent of Man and selection in relation to sex* (1871) Darwin put forward theoretical principles which foreshadowed two breakthroughs of evolutionary thinking in the twentieth century: sex ratio evolution and the principle of a density-dependent trade-off between offspring number and offspring quality. His contribution to the subject long remained unknown since he withdrew them from the second and final edition of the book (1874). This is the biggest change in Darwin's lifelong reflection on evolutionary mechanisms. Why did this scrupulous and prudent scientist once decide to publish these views, then to reject them 3 years later? I will review the reasons he became interested in the issue, the status he gave to the question in his thinking, and the probable reasons he was eventually dissatisfied with the solution he found. I will also consider the status of another of Darwin's hypotheses, with which he was dissatisfied in relation to this question: that of a reciprocal sexual choice between the most vigorous males and the most vigorous females. Darwin's solution to sex ratio evolution was independently discovered in 1930 by Fisher, but it was not until 1984 that Darwin's priority was acknowledged. I will show that Darwin's solution went further than so-called Shaw-Mohler's demonstration in 1953. I will also raise the issue of density-dependent selection, which appeared in contemporary evolutionary biology in 1971 but which was clearly addressed, a century earlier, in the same section of the first edition of *The Descent of Man.*

Keywords Darwin · Sexual selection · Sex ratio · Population density regulation

---

M. Veuille (✉)
Ecole Pratique des Hautes Etudes,
UMR 7205 CNRS, MNHN, EPHE, Paris, France
e-mail: veuille@mnhn.fr

T. Hoquet (ed.), *Current Perspectives on Sexual Selection,* History, Philosophy and Theory of the Life Sciences 9, DOI 10.1007/978-94-017-9585-2_3

3.1 Introduction

A commonly held view is that Darwin's Descent of Man and Selection in relation to Sex was a great synthesis for his time on sexually dimorphic traits in most phyla of animal kingdom, with brilliant insight on a major evolutionary force, sexual selection, including its division into two different mechanisms (sexual struggle and sexual choice), a dual concept still in use today. Another common view is that the book conveys the anthropomorphic explanation that the aesthetic sense of females drives evolutionary change in male ornaments. While it is right that Darwin referred to the aesthetic sense of females as a proximate mechanism (as it would be termed today), he did not make it the ultimate reason that sexual choice existed in animals, and was in search of a deeper and universal selective force.

Until recently it was also commonly believed that sex ratio theory (which explains why males and females are born in nearly equal number in most species) had been first put forward by Fisher in *The Genetical theory of natural selection* (1930), an important book in the history of evolutionary theory, and that this question could be understood only in the age of genetics. It was not until 1984 that Sober mentioned that Darwin had published the same conclusion as Fisher in the—usually non-widely read—first edition of *The Descent of man* (1871) and that nobody could have ever suspected his talented contribution to the subject since he deleted it from the second and last edition (1874) of the book. Sober (2007) presents it as a "retractation". It was also observed that Darwin's first enunciation of this principle had been discussed in the interim meantime, and even put by Carl Düsing (1884) in form of a mathematical model (Edwards 2000) which Fisher could hardly have ignored (Edwards 1998). I will also raise the issue of density-dependent selection, a model predicting that rules of intraspecific competition change with population level. It appears in evolutionary biology in the 1970s but is clearly addressed, a century earlier, in the same section of the first edition of *The Descent of Man.*

This section of the book is strange in many ways. Firstly it consists in a "supplement" inserted in the middle of the book (pp. 300–320 of a work of 828 pages). Second, Darwin there appears to be very different from the prudent and methodical writer we are accustomed to reading. He hardly refers to facts and switches from one subject to another in the order they occur to him in successive reflections. Third, as said before, he there opens a window on a variety of subjects, and will shut it after only 3 years with no real explanation.

These facts are of importance for understanding Darwin's method. They illustrate the solitude of a rigorous thinker in a century during which he was renowned and nevertheless misunderstood. They are also of importance to evolutionary biologists today. Given the paucity of basic knowledge in what was to become population biology, the ability of a formidable thinker to solve, at least at the level of intuition, some of the main issues of twentieth century's evolutionary theory may be a matter for reflection on the minimal requisites for addressing them.

3.2 Sexual Selection, a Question That Seems Simple

After having presented natural selection in *The Origin of Species* (1859), tackling the issue of sexual selection in *The Descent of Man* (1871) allowed Darwin to address a similar question under a different perspective. While avoiding some difficulties encountered with natural selection, he was facing new ones. In natural selection, the selective agent is hard to identify, except in cases like mimicry, where selection is exerted by predators; for instance, the selection effected by birds on mimetic butterflies makes it clear that selection has no purpose and derives from the mere conditions of life. But this can be viewed as a special case. In other cases it may be difficult to show that natural selection is not the personification of nature. While the modification of domestic breeds due to the selection exerted by the agriculturist had given Darwin the experimental demonstration of the possibility of his theory, its translation into a natural phenomenon depended on the assumption that the competitive pressure exerted by crowding upon natural populations had the same constant and long-term effect as the continued choice made by agriculturists.

The two kinds of sexual selection (sexual struggle and sexual choice) offered two different ways to present the comparison between competitive pressure in the wild and the choice exerted by the breeder in a garden. In sexual struggle, it is the direct confrontation between males which generates competition, resulting in the long term evolution in favour of the most successful ones. This process is very similar to that of natural selection, but for its limitation to one sex. In sexual choice (most frequently involving the choice exerted by females among males), the selective agent is a well identified factor—a member of the same species yet acting for the sake of its own, external motivations—which offers a vivid example in which selection in the wild is not merely a metaphor of purpose selection, the "aesthetic sense" of females being the transposition of the constant choice exerted by humans.

Examples of the first kind of sexual selection (sexual struggle) were abundant in some mammals, like deers or elephant seals where males fight against each other for taking control of females. The same was true of the second kind (sexual choice), which was easily illustrated in birds, in which the often brightly coloured ornaments of males seemed to result from the continued action of the choice exerted by females, as for peacocks and paradise birds.

3.3 Matching Numbers, the Difficulty Begins

Darwin and Wallace's theories of natural selection were based on the principle that selection requires an excess of individuals (the famous "Malthusian" principle). Making a choice assumes the possibility of retaining some individuals and discarding others. In natural selection, this excess is provided by the reproductive potential of a species being larger than is necessary to maintain the population level, allowing the differential survival of individuals in the struggle for life (Darwin's "law of death"). For sexual selection Darwin admitted that selection did not act through death but through failure to reproduce. It is easy to conceive that in many species males can compete to inseminate females. An obvious objection was this: when the successful males have been allowed to mate, what bars the others from also doing so? This put in centre stage the proportion of the sexes ("sex ratio" in modern terms). If males and females are equally available at the age of reproduction, what will allow some males to mate more than the others? In species where male struggle occurs, successful males sometimes gather harems through which they monopolize females, as for stags or elephant seals, thus denying reproduction to other males. The answer is less evident in some other organisms, including monogamous birds. This argument will long be a major objection to Darwin's sexual selection, as shown by its use in the early twentieth century by T. H. Morgan in his refutation of Darwinism:

> Supposing that the females select the most beautiful males, then, since in the vast majority of higher animals the males and the females are in equal numbers, the others will also be able to unite with each other in pairs after this first selection has taken place. Nothing will therefore be gained in the next generation. (Morgan 1903, p. 214)

As in many other instances, Darwin had thought of the objection before anyone.

> I have not attempted to conceal that, excepting when the males are more numerous than females, or when polygamy prevails, it is doubtful how the more attractive males succeed in leaving a larger number of offspring to inherit their superiority in ornaments or other charms than the less attractive males. (Darwin 1871, part II p. 400)

Since sexual selection did not eliminate individuals through death, the simplest way to explain competition for access to mating was to imagine that one sex was limiting, thus the other was limited. The matching of male and female numbers in most species contradicted this explanation. Darwin found a solution, which appears in the two editions of *The Descent of Man*. If sexual choice is linked to an ability to leave a more vigorous progeny, then the first individuals to mate will still have an advantage over the others, even though the later would eventually be able to mate. This solution assumes a reward in both sexes.

> Such females, if they select the more attractive, and at the same time vigorous males, will rear a larger number of offspring than the retarded females, which must pair with the less vigorous and less attractive males. So it will be if the more vigorous males select the more attractive and at the same time healthy and vigorous females; and this will especially hold good if the male defends the female, and aids in providing food for the young. The advantage thus gained by the more vigorous pairs in rearing a larger number of offspring has apparently sufficed to render sexual selection efficient. (Darwin 1871, p. 271)

Darwin was somewhat satisfied with his reasoning, as shown by the fact that he mentions it as a mechanism acting both in humans and in the other animals, but he was not fully satisfied. In birds, this hypothesis was flying in the face of facts. Sexual dimorphism, which obviously proceeds from sexual selection, is at its pinnacle in birds, suggesting an asymmetry between the sexes, whereas this mechanism suggests the contrary. Darwin apparently doubted that he held the whole answer and maintained that a biased sex ratio would have been a better solution. But with its value being desperately even in most species, he admitted his disappointment:

> A large preponderance in number of the males over the females would be still more efficient; whether the preponderance was only occasional and local, or permanent: whether it occurs at birth, or subsequently from the greater destruction of the females; or whether it indirectly followed from the practice of polygamy. (Darwin 1871)

And the question recurred: why is extreme sexual dimorphism not associated with unbalanced sex ratios?

> A numerical preponderance of males would be eminently favourable to the action of sexual selection. Nevertheless especially birds, which are strictly monogamous, display strongly marked secondary sexual characters; whilst some few animals, which are polygamous, are not thus characterised. (Darwin 1871, p. 266)

Thus Darwin was led to address a new question: Why do males and females almost universally match in number? This led him to consider the action of natural selection on the sex ratio. Note that he considered sex ratio as a potential prerequisite of sexual selection, not as a consequence of it. Therefore, its value had to pertain to another cause than sexual selection itself, and he assumed it resulted from some as yet unidentified, yet universal, form of natural selection.

3.4 Current State of the Theory of Sex Ratio Evolution

The essentials of the theory of sex ratio evolution as it is currently conceived can be presented in simple words, and in three steps.

1. Step one (Figs. 3.1a–c). If the cost of producing a male or a female is the same, then a simple reasoning shows that the population sex ratio will adjust to a value of 1/1. Let's take a species in which all individuals are alike in size, in adaptation, and in developmental time, except that some are males and some are females. The number of matings effected by each sex is the same, since each mating involves a male and a female. Hence, if the numbers of males and females are equal (Fig. 3.1a), both sexes are equally "fertile", meaning that the probability of a given individual being involved in a mating is the same in both sexes. A sex which would be over-represented in the population (Fig. 3.1b) would be less "fertile", since the probability of a given individual being involved in a mating would be lower than for the other sex. In such a population, a parent producing more descendants of the rarer sex than the population mean would have more grandchildren than average. For instance, if there are 75 % males, it is advantageous to have more than 25 % daughters (that is between above 25 and 100 %). Note that this is true for a wide spectrum of proportions of the sexes in the progeny, including a one-to-one proportion of males and females. If the property of this parent to have a different offspring proportion is a heritable trait, this trait will rise in frequency as long as it benefits from the bias in the sex ratio. But this will not last, since it modifies the population. Its frequency will stabilize when

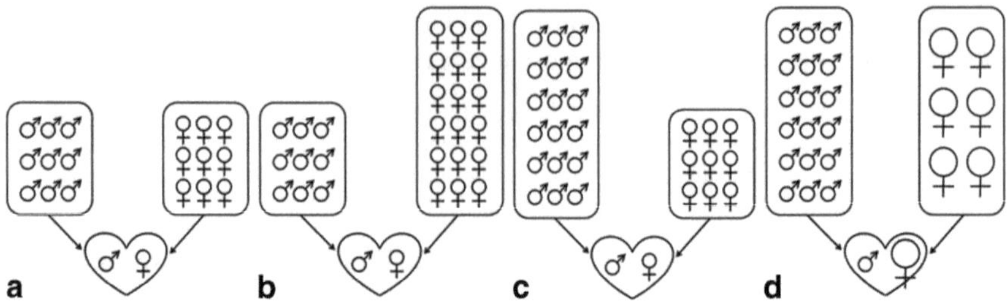

Fig. 3.1 A stepwise introduction to sex-ratio theory. Left to right: **a** Each mating involves one male and one female, therefore an equal number of matings are effected by members of either sex; in a population where males and females are in equal number, they have the same probability to mate; they are equally "fertile", all other factors being equal. **b** In a population where females are in excess over males, males have a higher probability to mate; they are more "fertile". In this population it is advantageous for a parent to produce either a progeny with a lesser deviation in sex-ratio, or with an equal number of each sex, or with an excess of the rarer sex. **c** Symmetrical case: if males are in excess over females, females are on average more fertile than males. It is advantageous for a parent to produce a lesser excess of sons over daughters, or an equal number of each, or a larger number of daughters. **d** If one sex is more costly to produce than the other, the optimal investment in each sex can be different. For instance, if three males are the same cost to produce as one female, it is advantageous to have more sons than average provided there is less than three males for one female in the population. If the population reaches ¾ males and ¼ females, then the expected fertility of three sons is equal to that of a daughter. The population is at equilibrium, and the sex of the progeny of a parent (whatever the proportion of its sons and daughters) is irrelevant to its fitness. If the proportion changes, then it becomes advantageous to produce more of the sex for which the average investment of the population is lower

the population sex ratio has the same value as the bias generated by the trait. For instance, if the sex ratio is of one male for three females, a variant producing one male for two females will have an advantage and rise in frequency in the population until the population sex ratio is of one male for two females. Then its advantage will stop. If the trait determines an excess of the rarer sex or a one-to-one sex ratio, its frequency will rise until the population sex ratio is one-to-one. Then, its frequency will remain at this value, and the population sex ratio will be stable. What is true for an excess of males holds true for an excess of females owing to our assumptions above that there are no other differences between the sexes than being "males" or "females" (Fig. 3.1c). This reciprocity means that a fluctuating sex ratio will tend to return to a one-to-one state whatever the direction of fluctuations (Shaw and Mohler 1953).

2. Step two (Fig. 3.1d). If the cost of producing a male or a female is not the same to some parent, then the population sex ratio will evolve towards an equalization of the investment of this parent in each sex. Let's suppose that all individuals are no longer alike in size, in adaptation, and in developmental time as was formerly assumed, and that new conditions of life make it possible to produce three males for the cost of one female. If the population sex ratio is still of one male per female, it becomes advantageous for a parent to use its resources for producing males rather than females. In turning females into males, for the cost of one

reproductive individual, it will have three! However, this condition will not last, since it will result in a new change in the population sex ratio, which will stop when the number of males is three times the number of females. Then a female is three times more fertile than a male. Thus, producing three males instead of one female is neither advantageous nor disadvantageous, and sex ratio evolution will stabilize at this point. Thus the actual control variable is not the number produced in each sex. The underlying factor determining the fitness of the parent is the balance of the parent's investment in its progeny of each sex, as compared to the average investment in each sex in the population.

3. Step three. In modern scientific literature, the evolution of sex is a domain of genetic research, with complex mathematical modelling (e.g. Karlin and Lessard 1986), encompassing a large number of strategies used by organisms for transmitting genes from one generation to the next. Species with separate sexes are only a fraction of actual cases, since reproduction in many plants and animals involves hermaphroditism, which can itself take several forms, and renders theory even more complex.

One fact was not apparent in the demonstration given above. In the guise of "access of individuals to reproduction", we were speaking of genes, and we implicitly assumed something of Mendelian inheritance: the presence of alleles transmitted unchanged across generations in the form of "diploid genes" (those genes which are borne by "autosomes", as opposed to "sex-chromosomes"). The genome includes different kinds of chromosomes, most of them being autosomes. In many species however, there are genetic elements which are transmitted only from mother to daughter. These include: mitochondrial DNA, so-called "B" chromosomes, and intracellular bacteria from the *Wolbachia* genus. Over evolutionary time, these elements most frequently tend to evolve in such a way as to bias Mendel's laws of inheritance towards the production of females. Why do they tend to behave this way? When a variant of these elements has the property to do so, it gets an advantage in its own transmission over alternative variants. Since we need considering only maternal lineages in this instance, we can easily understand that those elements producing more daughters than average will tend to overrun their competitors. X-chromosomes can also evolve this way. We all know that in some species, including humans, sex-determination has evolved into a system determining at random the sex of new eggs with a 50/50 probability of being male or female[1]. This part is played by "sex-chromosomes", like X- and Y-chromosomes in mammals. Strange as it may look, this system is under the control of autosomes. In some species, X-chromosomes inhibit the development of Y-bearing sperm cells, thus hijacking Mendel's laws to their own profit, since all surviving sperm cells will produce X-bearing sperm, resulting in all-daughter progeny. These cheating genes could lead to all-female populations, and to the eventual extinction of species. However, when males become rare, we are again in the process of sex ratio disequilibrium discussed

[1] This is a consequence of the fact that populations with an even sex ratio are in a stable state; see Maynard-Smith's (1982) concept of evolutionary stable strategies.

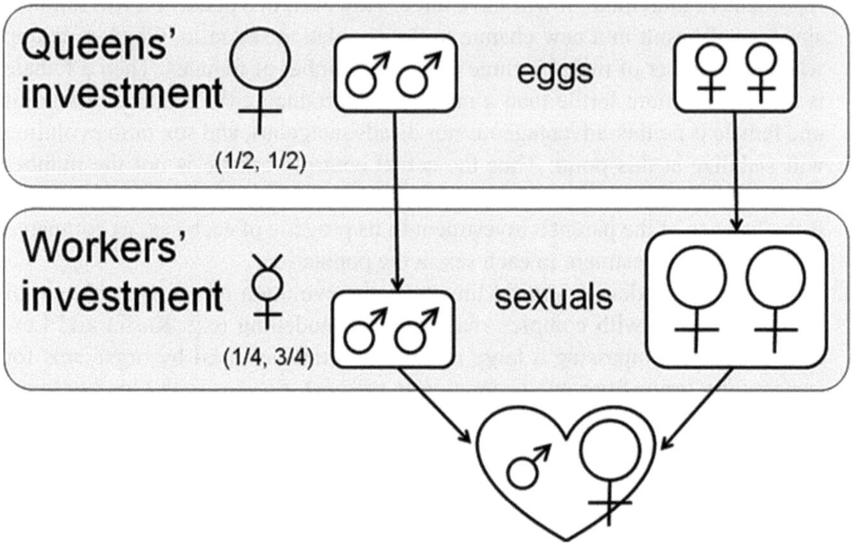

Fig. 3.2 Behavioral determination of the proportions of sexuals in a Mendelian world. In a hymenopteran nest, sex proportions at the adult stage are determined by social interactions, since female eggs can be reared either as workers or as gynes (reproductive females), thus genetics seems to be absent from the process; and yet Mendelian genetics still determine the effort put by the members of the colony into each sex. In a monogynous and monandrous species (when nests have a single queen, fertilized by a single male), queens are equally related to their sons and daughters, thus the numerical sex-ratio tends to be 1/1, whereas workers are three times more related to their sisters than to their brothers, thus the parental care they provide (and its conversion into biomass) tends to be female-biased. Thus even though a genetic switch is absent from the determination of sex, genetics ultimately determine the quantitative properties of the population of sexuals.

above. It is in the interest of compensatory genes, borne by autosomes, to re-establish an even sex ratio. This sometimes generates a continued sex ratio war, and its effects are recorded in Drosophila DNA even today (Derome et al. 2004, 2008).

The reader could also object that in some species, for instance in some turtles, sex is determined by the temperature at which an egg is incubated, and that even sex ratios seem to result from a benevolent environment rather than from genetics. But of course, behavioural adaptation is itself a result of evolution, and so is probably the parental care of mother-turtles. Nowhere is the interplay of genetics and behaviour so apparent as in the case of social insects (Fig. 3.2), in which mother-queens lay a large number of female eggs to produce workers. It is the behaviour of these workers which will ultimately decide how many of the female eggs will be reared as reproductive females rather than as workers; thus behaviour determines sex ratio. This will determine what is generally presented as a "conflict" between queen and worker behaviour. In ants, queens are genetically equally related to their sons and to their daughters, and thus tend to invest equally in each, whereas theory shows that in these species (at least in simple cases) workers are less related to their brothers than to their reproductive sisters, and thus tend to invest more in the latter. This apparent contradiction of interests is solved by the fact that queens invest through eggs, which are of equal size in the two sexes, whereas workers invest through their work, for which it is more costly to produce sisters than to produce sons. Thus an acceptable evolutionary balance is possible between queens and workers in the population.

To conclude on this technical part, we see that the current explanation of sex ratio evolution is in terms of genetic relatedness, even though a valid explanation may be put in terms of "access of individuals to mating" in the general case of an equal contribution of both sexes to their offspring ("diploidy").

3.5 Fisher's Theory

As will be seen below, Darwin's theory of sex ratio evolution went as far as step two. As early as 1884, his theory was put in mathematics up to step one by Carl Düsing (1884, See Edwards 2000). Düsing's account was well known in his time, and since Darwin had rejected his former views, *The Descent of Man* being generally known from the second edition, Carl Düsing was given full credit for his model. A reader could summarize it in 1903:

> Düsing (..) suggests that there are self-regulating influences of such a kind that, when one sex becomes less numerous, the conditions imposed in consequence on the other sex are such as to bring the number back to the normal condition. (Morgan 1903, p. 423)

Düsing's work on the statistics of sex ratio was widely known by those interested in quantitative biology (see the letter of Alphonse de Candolle to Francis Galton[2], 29 Sept. 1885) and his regulatory theory of sex ratio was known by those interested in research on heredity at the turn of the century (e.g. Thomson 1908). At a time when the relationship between the transmission and the development of characters was frequently obscured by the belief in the inheritance of acquired characters, some readers however misinterpreted his claim that sex ratio is regulated through factors like developmental time, the age of the parents etc., which were understood as primary causes rather than proximate factors, rendering his conclusions unreadable to most (see Delage 1895, p. 346), and Düsing's work was forgotten. Moreover research on this subject at the turn of the twentieth century went into a new direction, in parallel with investigations into the causes of heredity. It mostly involved

[2] "J'ai sur ma table un volume—malheureusement en allemand—qui contient le résumé de tout ce qu'on a réuni sur la proportion des sexes et le nombre des naissances en divers pays, chez l'homme et dans les animaux et les plantes. Le titre est: Die Regulierung des Geschlechtverhältnissen bei der Vermehrung der Menschen, Tiere and Pflanzen, von Carl Düsing, Dr Phil. Jena 1884. On rendrait service en traduisant cet ouvrage en anglais ou en français": I have on my table a book—in German unfortunately—containing a summary of everything having been collected on the proportions of the sexes and on the number of births in different countries, in man and in animals and plants. The title is: Die Regulierung des Geschlechtverhältnissen bei der Vermehrung der Menschen, Tiere and Pflanzen, von Carl Düsing, Dr Phil. Jena 1884. It would be helpful to translate this work into English or French. (Pearson 1924 p. 210)

cytological, embryological and experimental evidence. During this period, known as the "eclipse of Darwinism" (Huxley 1942; Bowler 1983), several schools—internalists, externalists and hereditarians according to Maienschein (1984)—drove an important debate on sex ratio.

In the twentieth century the theory of sex ratio evolution became widely known with the return to Darwin initiated by Fisher's *Genetical Theory of Natural Selection* (1930). Fisher went on to step two above. The argument was verbal, but was nevertheless difficult to read, as he put it in precise words that were the literal transcription of his abstract thinking as a mathematician[3]. The summary given by Fisher at the end of the chapter is however straightforward:

> Natural selection will tend to equalize the parental expenditure devoted to the production of the two sexes. (Fisher 1930 p. 162)

Fisher's book was seminal in reinterpreting Darwinian selection in the framework of Mendelism, and remained the main reference in evolutionary genetics for 30 years. Since Düsing's work was unknown, Shaw and Mohler (1953) derived the mathematical treatment that was missing in Fisher's book and unwittingly rediscovered Düsing's model in what became widely known as the Shaw-Mohler model of male and female brood size (see Maynard-Smith 1978; Crozier and Pamilo 1996).

The universal belief that Fisher's theory was first on the subject was probably reinforced by the importance played by Mendelian sex-determination in the victory of genetics over alternative theories of heredity. The X-Y sex-determination system in *Drosophila* (Morgan 1910), mirroring the W-Z system in *Abraxas*, provided the elegant proof that biological heredity was carried by chromosomes. This was also the first paper of a brilliant school (Morgan's Drosophila group) to investigate laws of recombination and to give the first insights into mechanisms of sex development (Bridges 1925). Sex ratio chromosomes in natural populations were also at the root of the first research project on the genetics of natural populations by Dobzhansky in 1935 (Provine 1981). So in a way, sex ratio "belonged" to Mendelism. To my knowledge, Darwin's brilliant insight in the first edition of *The Descent of Man* was first recognised by Sober (1984). The history of the misunderstanding was documented by Sober (1984, 2007) and Edwards (1998, 2000), who also rediscovered

---

[3] "Let us consider the reproductive value of these offspring at the moment when this parental expenditure on their behalf has just ceased. If we consider the aggregate of an entire generation of such offspring it is clear that the total reproductive value of the males in this group is exactly equal to the total value of all the females, because each sex must supply half the ancestry of all future generations of the species. From this it follows that the sex ratio will so adjust itself, under the influence of Natural Selection, that the total parental expenditure incurred in respect of children of each sex, shall be equal; for if this were not so and the total expenditure incurred in producing males, for instance, were less than the total expenditure incurred in producing females, then since the total reproductive value of the males is equal to that of the females, it would follow that those parents, the innate tendencies of which caused them to produce males in excess, would, for the same expenditure, produce a greater amount of reproductive value; and in consequence would be the progenitors of a larger fraction of future generations than would parents having a congenital bias towards the production of females. Selection would thus raise the sex ratio until the expenditure upon males became equal to that upon females." (Fisher 1930, p. 141).

Düsing's model, along with the long yet interrupted concern of a number of scientists around the proportion of sexes since the eighteenth century, which began with a debate about the role of providence in generation. Many people are still surprised to read the first edition of Darwin's book and to discover his priority[4]. Below I will show that Darwin actually went further than has been hitherto recognized.

3.6 Darwin's Theory of Sex Ratio Evolution

Here are the very simple words in which Darwin clearly expressed his deductions.

> Let us now take the case of a species producing (..) an excess of one sex -we will say of males-. Could the sexes be equalised through natural selection? We may feel sure, from all characters being variable, that certain pairs would produce a somewhat less excess of males over females than other pairs. The former, supposing the actual number of the offspring to remain constant, would necessarily produce more females, and would therefore be more productive. (..) Thus a tendency towards the equalisation of the sexes would be brought about. (Darwin 1871 p. 316)

In this fragment, Darwin reaches step one above. However it has generally been missed that just afterwards, he reaches step two, that is, the point Fisher had arrived at, and this by a very original explanation. Darwin remarks that in a species with an excess of males, an advantage can be obtained by parents in two different ways: (a) one is to produce the same number of offspring in different proportions (fewer sons and more daughters); (b) the other is to produce fewer sons without increasing the number of daughters, and use the energy so spared to produce more vigorous daughters. These will be more fertile, thus this excess will still be advantageous to the parent. In other words their progeny can be more fertile either because it is more numerous or because it is more vigorous.

> Those (parents) that produced few superfluous males would have one great advantage, namely that their ova or embryos would probably be larger and finer, or their young better nurtured in the womb and afterwards. (Darwin 1871, p. 317)

This conclusion has been overlooked by readers of Darwin, probably because he presents it as though the decisive reason to choose between the two models would be the advantage to the species. If evolution was leading parents to reproduce according to solution (a), "*our supposed species, he says, would by this process be rendered (..) more productive; and this would in many cases be far from an advantage*" (Darwin 1871, p. 316). But the alternative (b) mechanism so put forward decreases crowding and offspring mortality in the population as a whole. Presenting evolution as acting for the "good of the species" was usual in biology until the first half of the twentieth century, but fell in disrepute after the mid-1960s (Williams 1966)[5]. Another counterintuitive aspect in Darwin is that he seems to believe that if

---

[4] I was myself unaware of previous work on the subject when I discussed it (Veuille 2010).

[5] Sober (2007) notes: "As we will see, Darwin's explanation of even sex ratios invokes what is now usually classified as individual, not group, selection. But even so, it is well to remember that,

females are more numerous than males, they will remain unfertilized. This is "the assumption of monogamy that entails that some individuals must fail to reproduce if the sex ratio is uneven" (Sober 2007; Edwards 1998). Its corollary would be that the general fertility of the population would increase if the sex ratio is even. In modern language, the assumption that unmated females would fail to reproduce would be correct in strictly monogamous species where no "extra-pair copulations" occur[6], but a modern ecologist would expect this to be rare in nature. This does not ruin Darwin's sex ratio evolution model, but only his belief that an uneven sex ratio would decrease population fertility. Another idea of Darwin that may look strange to us is the kind of mutations he envisions: geneticists of the twentieth century would hardly believe that in a species with separate sexes the same phenotype could simultaneously lower the number of sons and increase the fertility of daughters.

Put in its historical context, and with its own approximations (which are in no way more extravagant than some models frequently used in population genetics textbooks nowadays) the principle eventually put forward by Darwin was mechanistic. Finally, the important conclusion remains that, as early as 1871, Darwin understood that the same reproductive effort of a parent could act either on the proportions of sons and daughters, or on the reproductive potential of each kind of offspring. In other words, he considered like Fisher that the controlling factor of sex ratio evolution was not simply the number, but the energy put in the progeny of either sex. This at least justifies that the title of Darwin's paragraph on this question is not simply:

> On the power of natural selection to regulate the proportional numbers of the sexes.

but

---

in other contexts, Darwin slides easily between talking about benefit to the species and benefit to individuals, eliding a distinction that post-1960s biology has found to be exceedingly important". Note incidentally that a reasoning in terms of selection acting on groups will appear in the same chapter, but in the second edition of *The Descent of Man*: "With animals living in herds or troops, in which the males come to the front and defend the herd, as with bisons of North America and certain baboons, it is conceivable that a male-producing tendency might be gained by natural selection; for the individuals of the better defended herds would have more numerous descendants" (Darwin 1874).

[6] It is sometimes wrongly believed that this assumption of monogamy led Darwin to overlook that males can mate with more than one female, and that this would increase male-male competition. But in any population with monogamous pairs, the number of extra-pair copulations effected by males is exactly the same as the number of extra-pair copulations effected by females, since it takes two mates for a mating, either legitimate or illegitimate. If the sex ratio is even, females are on average as untrue as males. But the variance can be different. Modern literature has stressed that the variance in mating success is larger in males than in females. I have put forward (Veuille 1982 Ph.D. thesis; Veuille 1986, Veuille and Mazeau 1988) that it is not necessary to assume that this larger variance results from a higher "eagerness" of males (Darwin 1871; Bateman 1948; Daly and Wilson 1978). It suffices to note that sexual competition stops with the fertilization of the last available oocyte; thus male mating success follows a Poisson distribution (in which the variance is equal to the mean), whereas female variance in mating success is virtually zero.

> On the power of natural selection to regulate the proportional numbers of the sexes, *and general fertility* (my emphasis).

Darwin will however continue along this line and move on to the new issue opened on "general fertility".

3.7 On General Fertility, or Darwin's "Density-Dependent Selection" Principle

The expression "density-dependent selection" belongs to modern literature. Darwin's principle has no name of its own, or maybe could we call it the principle of "*general fertility*", based on the title of this section of the book. It is a mere continuation of his former reasoning, and its deliberate inclusion in the book is indicated by the title of the section, as seen above. Of the five pages devoted to this section, three are on sex ratio ("*the proportional numbers of the sexes*"), and two on this new question derived from the former: can the fertility of a species increase indefinitely by the mere process of an increasing competition between individuals for reproduction? Maybe, some of Darwin's premises could seem obsolete. But since considering selection in a context other than the adaptation of the organism to its environment, Darwin was facing for the first time a consequence of the mechanistic process of intraspecific competition which probably looked counterintuitive to a naturalist: members of the same species are involved in a race for producing a maximum of offspring. If the increase in fertility is not associated with an increase in the adaptation of the species, the individuals so produced will tend to starve each other. Darwin was embarrassed by the fact that this was against the interests of the species.

> The only check to a continued augmentation of fertility in each organism seems to be either the expenditure of more power and the greater risks run by the parents that produce a more numerous progeny, or the contingency of very numerous eggs and young being produced of smaller size, or less vigorous, or subsequently not so well nurtured. (Darwin 1871, p. 319)

While Darwin as a naturalist could not admit that species may not be adapted to their environment, he reacted in accordance with his scientific principles by seeking a purely mechanistic explanation. In so doing, he was led to test the consistency of evolutionary theory: since evolution results from competition between individuals, how can it bring about mechanisms regulating population numbers? In other words, can it be advantageous for an individual to limit its own reproduction? At first glance this would seem to be a contradiction, since, while population decrease might be advantageous to all, it would not seem to be in the interest of an individual to limit its own progeny. Thus, seemingly, this behaviour cannot evolve. This question will be central to population biology a century later, in the 1960–1970s. It will launch the sociobiology debate when Wynne-Edwards (1962) will put forward group-selection as a mechanism of self-control of population numbers, which

Maynard-Smith (1982) will counter with kin-selection, an altruistic behaviour based on genetic selfishness. In a related issue, McArthur and Wilson (1967) will oppose "*r*-strategists" (species with an expanding mode of population growth) to "*K*-strategists" (those with stable, auto-regulated populations). After Levins (1968) has put forward models of adaptation involving trade-offs between several components, Roughgarden (1971) and Charlesworth (1971) will write mathematical models at different levels of generality through which density-dependent selection can adaptively promote the stabilization of population size. Darwin's explanation is purely verbal, but is very clear on the issue.

> The offspring indeed of the less fertile parents would partake of one great advantage; for under the supposed condition of severe competition, when all were pressed for food, it is extremely probable that those individuals which from some variation in their constitution produced fewer eggs or young, would produce them of greater size or vigour; and the adults reared from such eggs or young would manifestly have the best chance of surviving, and would inherit a tendency towards lessened fertility. (Darwin 1871, p. 319)

And he concludes the chapter:

> By these steps, and by no other ones a far as I can see, natural selection under the above conditions of severe competition for food, would lead to the formation of a new race less fertile, but better adapted for survival, than the parent-race. (Darwin 1871, p. 320)

The explanation is clear and valid. It does not rely on the good of the species. Maybe we should refrain from judging Darwin's expressions in the framework of our contemporary debates. Our disputes over group versus individual selection cannot be compared to the opposition he was facing in his time. Of the generations of naturalists trained in the belief that an almighty providence had adjusted species to their function in the bosom of universal harmony, he had been the first to envision the power of blind material causes, and to put forward the challenging idea that they were responsible for the equilibrium of nature. It was legitimate to reflect back and question in the words of his potential detractors whether all this would not result in an endless race to produce more and more offspring leading to a demographic collapse, and to give the Darwinian answer.

3.8 The Rejection of the Two Hypotheses by Darwin in 1874

There are relatively few changes in Darwin's thinking. He has been said to have progressively changed his views about inheritance between the first (1859) and the sixth Darwin (1872) and final edition of the *Origin of Species*, even though there are merely minor modifications. The "provisional hypothesis of pangenesis" which he put forward in 1868 is thought to have made him a proponent of "use and disuse" mechanisms of heredity. The desire of the evolutionary synthesis community to revive a mythical genuine Darwinism as expressed in the facsimile publication of the first edition of the *Origin of Species* (Mayr 1964), showed that his opinion had not changed so much[7]. There is no retractation on variation in Darwin.

The biggest of all changes in Darwin's thought is about the "proportions of the sexes". This is actually the deepest revision he ever made in his work[8]. It is abrupt. It happens in 3 years. He gives up a former hypothesis without any explanation. Moreover, this change is hidden. The second edition of the book opens with a list of the changes made to the first edition, and this change is concealed behind the understatement "*excess of males perhaps sometimes determined by selection*".

He actually deletes all of this section and replaces it with a discussion bearing exclusively on infanticide in humans, attempting to explain that infanticide results in an innate bias in sex ratio (see Veuille 2010). The title of this subsection is also less promising, since the once enthused "*On the power of natural selection to regulate the proportional numbers of the sexes, and general fertility*" is replaced by the more cautious "*The proportion of the sexes in relation to natural selection*", which sounds like a question mark and actually leads to a disappointing conclusion:

> I formerly thought that when a tendency to produce the two sexes in equal numbers was advantageous to the species, it would follow from natural selection, but I now see that the whole problem is so intricate that it is safer to leave its solution for the future. (Darwin 1874, print. 1882 vol II, p. 260)

The only evolutionary mechanism put forward is inspired by a fragment of Colonel Marshall's account published the year before on infanticide in the Todas of India. In short, an imaginary model assumes that if only one third of females survive infanticide before reproduction, this practice will create a bias in the sex ratio of the population, eventually favouring the birth of males. This is obviously the opposite of the principles he had put forward 3 years earlier: the sex ratio on which natural selection is expected to act is the sex ratio at the age of reproduction. A larger mortality of females in childhood would result in a bias in the operational sex ratio, eliciting a regulatory mechanism in favour of female births. We are thus caught in a

---

[7] In this facsimile edition, the editor introduced 30 new entries to "Lamarck J-B" in the index. Only two of them actually referred to pages where Darwin mentioned the name of Jean-Baptiste Lamarck. In the other 28 pages he candidly discussed the inheritance of acquired characters, weighting its positive and negative value.

[8] It is also the largest modification of *The Descent of Man*. Of the 554 pages devoted to sexual selection, 67 are on "principles of sexual selection". Thus the section of the book on mechanisms is relatively short, compared to empirical sections where he describes secondary sexual characters in the whole animal kingdom. In the "principles" chapter some changes are found between the two editions. In "*the male generally more modified than females", he* adds 59 lines to explain why males tend to vary more than females. In "*laws of inheritance", he* adds 23 lines to discuss the presence of female-linked traits in males (e.g. nipples in male mammals). The section on sex ratio stability is presented as a "supplement" of 21 pages, of which 16 are a discussion on various estimates of the sex ratio in different species, and 5 are on the action of natural selection on it. In the 16 pages of discussion he adds 55 lines in the subsection on "*sex ratio in man*" to discuss the larger mortality of illegitimate male infants. Finally, in the part where he presents his sex ratio evolution theory, he deletes everything (169 lines) and replaces it by a discussion bearing exclusively on infanticide in humans (253 lines).

contradiction. In this second edition, nothing is left of his earlier hypothesis, and it is not alluded to, except for the concluding words cited above. Likewise, Darwin's theory on general fertility, which was a consequence of his theory of sex ratio evolution is dropped, probably because it is no longer appropriate, being thought of as a consequence of his sex ratio theory.

3.9 Darwinian Inheritance

Why did Darwin suppress sex ratio evolution from his work? It would be of little help to try to understand what is meant when he says that the problem is too "intricate" to be dealt with. This is an understatement. Another question is why he published this hypothesis in the first edition, since it is probably one of the rare instances in his work where he published a reasoning, however brilliant, without a solid body of supporting facts. This also cannot be answered. We can just remark that in both editions the question arises in a "supplement", as though Darwin was just mentioning speculations which should not weaken the whole book.

But can we ourselves, from our perspective as contemporary scientists, say why he failed? There has been one misunderstanding from the very beginning. The mechanism of heredity he had in mind was probably not compatible with the hypothesis he was trying to clarify.

Thus a question to address is what Darwin meant by "variation". He made a difference between the origin of variations and the fact that some of them did not eventually survive life conditions whereas others did. This is the main difference between him and what will be called "Lamarckian" theories of evolution at the end of the nineteenth century. For Lamarckian schools, the origin of variation was also the mechanism of evolutionary change. Despite this, both schools—Darwin and most of the Lamarckians—shared general views on the mechanism of biological inheritance: the use or disuse of organs generated small quantitative changes during development, and these changes became hereditary. They were the raw material of evolution. These ideas were rather loosely written in Lamarck, who was a naturalist from another generation. They were more clearly presented later. Prosper Lucas's (1847, 1850) two-volume treatise on *Hérédité* gave a new meaning to the French word (*hérédité*: the social transmission of goods, estates and titles): that of biological transmission. In accordance with the ideas of French psychiatrists, his conception was that variations were pathological with respect to the normal state. Herbert Spencer (1867) was the first to present a version of Lamarckian inheritance in which variations were the normal state. "*Variation is co-extensive with heredity*" he said in a brilliant formula, while extending the new meaning of the French word to its English equivalent, "heredity"[9]. Darwin hardly ever had any strong conception of biological inheritance until he examined the question thoroughly and wrote a treatise on the subject. This was "*Variation in Plant and Animals*

[9] In his translation of Darwin's *Origin of Species*, Hoquet (2013, pub. Seuil, Paris) translates Darwin's "inheritance" by "*héritage*", thus avoiding using "*hérédité*".

under Domestication" (Darwin 1868). In it he prudently put forward the provisional hypothesis of "pangenesis", which is summarized in the theoretical chapter on sexual selection of *The Descent of Man*:

> This important distinction between transmission and development will be easiest kept in mind by the aid of the hypothesis of pangenesis, whether or not it be accepted as true. According to this hypothesis, every unit or cell of the body throws off gemmules or undeveloped atoms, which are transmitted to the offspring of both sexes, and are multiplied by self-division. They may remain undeveloped during the early years of life or during successive generations; their development into units or cells, like those from which they were derived, depending on their affinity for, and union with, other units or cells previously developed in the due order of growth. (Darwin 1871, p. 280)

In the sexual selection section of *The Descent of Man* (which represents about one half of the book), Darwin describes and discusses secondary characters throughout the whole animal kingdom, and often refers to a series of empirical laws, which are broadly inferred from the phenomena of heredity, concerning characters limited to sex or to a special part of the life cycle. That he considered these laws as provisional is shown by his frequent reference to the "unknown laws of variation".

From the point of view of Lamarckians, their conception was more logical than Darwin's. If variations were acquired as a mechanism of active adaptation of individuals during the developmental process, they were adapted from the outset, and no further selection was necessary. Darwin believed that developmental variation and natural selection often went the same way, but he nevertheless retained selection as the endpoint of the evolutionary process. Nowhere is it more visible than in the case of sexual choice, since females select variability among males in a way that can sometimes decrease the survival of the later:

> It is evident that the brilliant colours, top-knots, fine plumes, &c., of many male birds cannot have been acquired as a protection; indeed they sometimes lead to danger. That they are not due to the direct and definite action of the conditions of life, we may feel assured, because the females have been exposed to the same conditions, and yet often differ from the males to an extreme degree. (Darwin 1871, p. 234)

Like breeders, females selected variation as a raw material that was enhanced by variations in the conditions of life, but which they used to their own profit.

Darwin frequently made an implicit assumption which was rarely expressed since it was then widely admitted: it is the idea of "blending inheritance", whereby the offspring were thought to be intermediate between their parents, as though they were a mix of them. This mistake was swept away only in 1900 with the advent of Mendelism and its principle of segregation of characters, whereby meiotic reduction precedes the union of haploid gametes, these being mere random drawings of the parental alleles. With Mendelism, heredity is particulate (genes are distinct entities) and transmission is semi-conservative between each parent and their common offspring. It may be blending inheritance, rather than the assumption of monogamy, which Darwin had in mind when referring so often to the characteristics of "certain pairs". In which case his trouble must have been at its deepest, since one can hardly figure out a mechanism whereby sex-determination would ensue from a mix of the parents, not to speak of use and disuse.

3.10 Could Darwin Have Not Withdrawn Sex Ratio Evolution From His Work?

Everybody agrees that sex is not heritable. Males do not beget males and females do not beget females. In birds, mammals, and many insects, sex chromosomes are a token allowing sex to be drawn a random at each birth. Darwin apparently assumed that some pairs had a "tendency" to produce some proportion of the sexes, in which case strong recurrent selection would have been required to counterbalance blending effects. There are reasons to believe that he was deeply embarrassed. For instance, sex-biased infanticide (the focus of the new chapter in the second edition of *The Descent of Man*) cannot lead to a heritable tendency of survivors to generate more of the unaffected sex, since our heredity is not influenced by the death of our fellow conspecifics. On the contrary, if juveniles are killed in one sex, this will become the rarer sex at maturity, and, according to Darwin's first edition of *The Descent of Man*, producing this sex will be advantageous to the parents. But Darwin endorsed Colonel Marshall's suggestion that female infanticide in the Todas led to a surplus of male births in the population. It would be interesting to understand why he made this mistake, which he had not made 3 years earlier.

The "unknown laws of variation" seemed to offer no way for an evolution of sex ratio through natural selection in the schemes he could envision. He must have felt as if in a labyrinth. Darwin also knew that polygamy could not by itself bias the sex ratio, and very cleverly cites the example of horses:

> Hardly any animal has been rendered so highly polygamous as our English race-horse, and we shall immediately see that their male and female offspring are almost equal in number. (Darwin 1871, p. 303)

It is not surprising that he withdrew his account on the proportions of the sexes and general fertility. The ways of nature seemed unknowable.

As noted by Sober (2007), "*Darwin does not state his reasons for retracting*". Maybe we will never know. Neither will we probably ever know why he once decided to publish a hypothesis which, however perceptive it may look today, and however "Darwinian" it may seem for our Darwinian culture, must have been very difficult to conceive in his time. It is only a marvel for us that he published it, letting us admire how deep his proper reflection could go towards solving some of the most complex issues of evolutionary biology, at least as temporary hypotheses. Moreover, he did not simply retract, though we may be too inclined to see it in such terms because we are more attentive to how far he was from us than to the difficulties of research at a time when experimental biology was in its infancy. He deleted his former chapter, but also added a longer one on infanticide in humans. He says why he chose this subject: because there are more data in humans, and because "*there is reason to suspect that in some cases man has by selection indirectly influenced his own sex-producing power*". This was probably an illusion, but one to tell us something of his method. Being disappointed by his hypothesis, he was trying to gather new data, in search of firm ground.

In introducing the book where he rediscovered the subject, Fisher stated that Mendelian heredity was particulate and that this made the difference between the new era and that of Darwin: blending inheritance had been refuted. It appears that Darwin had gone as far as could be gone, given the poor understanding of hereditary mechanisms in his generation. The matter remained speculative, and publishing on it looked premature. A man who had been patient enough to delay publication on natural selection for so many years until it could be understood; a man who had so many books to finish, could withdraw reflections which were not ripe enough to be put in the shape of a comprehensive theory. The wanderings of his thought in the second edition of *The Descent of Man* suggest he himself had difficulties to follow the thread of his analyses. We don't know what he had on mind when he withdrew his chapter, but he was surely wise to do so. We are only happy that he had not come to this conclusion immediately.

Acknowledgments I am most grateful to Thierry Hoquet and Annabel Whibley for their help in the preparation of this manuscript. My research is supported by UMR 7205 (Muséum National d'Histoire Naturelle, Centre National de la Recherche Scientifique, Ecole Pratique des Hautes Etudes).

References

Bateman AJ (1948) Intra-sexual selection in *Drosophila.* Heredity 2:349–368

Bowler PJ (1983) The eclipse of Darwinism, Anti-Darwinian evolution theories in the decades around 1900. The Johns Hopkins University Press, Baltimore

Bridges C (1925) Sex in relation to chromosomes and genes. Am Nat 59:127–137

Charlesworth B (1971) Selection n density regulated populations. Ecology 52:469–474

Crozier RH, Pamilo P (1996) Evolution of social insect colonies. Sex allocation and kin-selection. Oxford University Press, Oxford

Daly M, Wilson MI (1978) Sex, evolution and behaviour: adaptations for reproduction. Duxbury Press, Duxbury

Darwin CR (1859) On the origin of species by means of natural selection. John Murray, London

Darwin CR (1868) Variation in plants and animals under domestication. John Murray, London

Darwin CR (1871) The descent of man, and selection in relation to sex, 1st ed. John Murray, London

Darwin C (1872) The origin of species by means of natural selection, or the preservation of favoured races in the struggle for life. 6th ed. Murray. 458 p

Darwin CR (1874) The descent of man, and selection in relation to sex, 2nd ed. John Murray, London

Delage Y (1895) La structure du protoplasma et les théories sur l'hérédité et les grands problèmes de la biologie générale. Reinwald & Cie, Paris

Delage Y (1903) L'hérédité et les grands problèmes de la biologie générale. Deuxième édition revue et augmentée. Reinwald Schleicher frères & Cie, Paris

Derome N, Métayer K, Montchamp-Moreau C, Veuille M (2004) Signature of selective sweep associated with the evolution of *sex ratio* drive in Drosophila simulans. Genetics 166:1357–1366

Derome N, Baudry E, Ogereau D, Veuille M, Montchamp-Moreau C (2008) Selective sweeps reveal a two-locus model for *sex ratio* meiotic drive in Drosophila simulans. Mol Biol Evol 25:409–416

Düsing C (1884) Die Regulierung des Geschlechtsverhaltnisses bei der Vermehrung der Menschen, Tiere und Pflanzen. Jenaische Zeitschr für Naturw 17:593–940

Edwards AWF (1998) Natural selection and the sex ratio: Fisher's Sources. Am Nat 151(6):564–569
Edwards AWF (2000) Carl Düsing (1884) on the regulation of the sex ratio. Theor Popul Biol 58(3):255–257
Fisher RA (1930) The genetical theory of natural selection. Clarendon, Oxford
Huxley J (1942) Evolution the modern synthesis. Allen & Unwin, London
Karlin S, Lessard S (1986) Theoretical studies on sex ratio evolution. Princeton University Press, Princeton
Levins R (1968) Evolution in changing environments, Princeton University Press. 120 p
Lucas P (1847) Traité philosophique et physiologique de l'hérédité naturelle dans les états de santé et de maladie du système nerveux avec l'application méthodique des lois de la procréation au traitement général des infections dont elle est le principe, vol 1. Baillière, Paris
Lucas P (1850) Traité philosophique et physiologique de l'hérédité naturelle dans les états de santé et de maladie du système nerveux avec l'application méthodique des lois de la procréation au traitement général des infections dont elle est le principe, vol 2. Baillière, Paris
Maienschein J (1984) What determines sex? A study of converging approaches 1880–1916. ISIS 75:457–480
Maynard-Smith J (1978) The evolution of sex. Cambridge University Press, London
Maynard-Smith J (1982) Evolution and the theory of games. Cambridge University Press, London
Mayr E (1964) Charles Darwin on the origin of species: a facsimile of the first edition. Harvard University Press, Cambridge
McArthur RH, Wilson EO (1967) The theory of island biogeography. Princeton University Press, Princeton, 203 p
Morgan TH (1903) Evolution and adaptation. McMillan Co, New-York
Morgan TH (1910) Sex-limited inheritance in *Drosophila*. Science 12:120–122
Pearson K (1924) The life, letters and labours of Francis Galton, vol II: researches of middle life. Cambridge University Press, Cambridge.
Provine WB (1981) Origin of the genetics of natural populations series. In: Lewontin RC, Moore JA, Provine WB, Wallace B (eds) Dobzhansky's genetics of natural populations I-XLIII. Columbia University Press, New York, pp 1–86
Roughgarden J (1971) Density-dependent natural selection. Ecology 52:453–468
Shaw RF, Mohler JD (1953) The selective significance of the sex ratio. Am Nat 87:337–342
Sober E (1984) The nature of selection. MIT Press, Cambridge
Sober E (2007) Sex ratio theory, ancient and modern: An eighteenth-century debate about intelligent design and the development of models in evolutionary biology. In Jessica Riskin (ed.) Genesis Redux: Essays in the history and philosophy of artificial life. University of Chicago Press. pp 131–62
Spencer H (1867) Principles of biology vol I. Williams and Norgate, Edinburgh
Thomson J (1908) Heredity. John Murray, London
Veuille M (1986) Natural variation in sexual behaviour in *Drosophila*. In: Médioni J, Vaysse G (eds) Genetic approaches to behaviour. Privat, Toulouse, pp 121–128
Veuille M (2010) Darwin and sexual selection: one hundred years of misunderstanding. In special issue "Another Darwin". CR Biol 333:145–156
Veuille M, Mazeau S (1988) Genetic variability of sexual behaviour in natural populations of *Drosophila melanogaster*. Behav Genet 18:389–403
Williams GC (1966) Adaptation and natural selection. Princeton University Press. 307 p
Wynne-Edwards VC (1962) Animal dispersion in relation to social behaviour. Olivier and Boyd, Edinburgh

Chapter 4
Sexual Selection in the French School of Population Genetics: Claudine Petit (1920–2007)

Jean Gayon

Abstract This paper analyses Claudine Petit's contributions to the experimental study of sexual selection. It examines the two sources of Petit's work on the "advantage of the rare" in *Drosophila melanogaster*: the Tessier school of experimental genetics, with its methodology founded upon population cages, and the Dobzhansky/Mayr American school, which considered it mainly in relation with the problem of reproductive isolation. The evolution of Petit's research is examined in detail, as well as the important role that her experimental work had in the issue of frequency-dependent selection. Petit's final overtly non-adaptive interpretation of sexual selection is examined, in contrast with Dobzhansky's conception.

Keywords Sexual selection · Experimental population genetics · Population cages · Advantage of the rare · Frequency-dependent selection · History of science · Bösiger · Dobzhansky · Petit · Tessier

This paper analyses Claudine Petit's contributions to the experimental study of sexual selection. In 1982–1983, I spent a year in her laboratory of evolutionary genetics as a student. I was then mostly ignorant that sexual selection had been so important in her career. However, I was impressed by the quality of the teaching that her lab offered in evolutionary biology, something not unique, but quite exceptional in France at that time. Claudine Petit was one of the pupils of Georges Tessier (1900–1972), who founded a productive school of population genetics in the 1940s. Other prominent population geneticists who completed their PhDs under Tessier were Maxime Lamotte (1920–2007), who obtained international recognition for his work on random genetic drift in wild populations of the snail *Cepaea nemoralis*, Charles Bocquet (1918–1977), who made extensive studies on the geographical polymorphism of *Sphaeroma serratum*, and Ernst Boesiger (1914–1975), who studied heterozygosity in natural populations of *Drosophila* melanogaster and its relationship with relation with sexual selection. Although not the most productive scientists among those mentioned here, Boesiger is probably the best known in the

J. Gayon (✉)
Institut d'histoire et de philosophie des sciences et des techniques,
Université Paris 1 Panthéon-Sorbonne, 13 rue du Four, 75006 Paris, France
e-mail: jean.gayon@gmail.com

T. Hoquet (ed.), *Current Perspectives on Sexual Selection,* History, Philosophy and Theory of the Life Sciences 9, DOI 10.1007/978-94-017-9585-2_4

anglophone literature because he was a friend of Theodosius Dobzhansky and published several books with him (Dobzhansky and Boesiger 1968, 1983).

I examine here the scientific context within which Claudine Petit initiated her work on sexual selection (The Tessier school of experimental population genetics, but also the influence of the Dobzhansky and Mayr's approach to sexual isolation and sexual selection), and the development of her own work.

Since Petit's contribution consisted in studying sexual selection in the fruit fly with the tools of experimental population genetics, I will first locate this work within the context of Philippe L'Héritier and Georges Tessier's work in the 1930s, the first in population genetics in France (for more information, see Gayon and Veuille 2001). Philippe L'Héritier (1906–1982) and Georges Tessier were two mathematically trained biologists. In 1932, when he was 26, L'Héritier went for several months to USA with a Rockefeller grant to study genetics. There he discovered the existence of population genetics when attending the 6th International Congress of Genetics at Ithaca, where he was particularly impressed by the talks of Ronald Fisher and Sewall Wright. Fascinated by their mathematical genetical approach to evolution, he returned to France with the project of studying the demographic evolution of different strains of *Drosophila* in a controlled experimental environment. When he came back to France at fall 1932, he designed the first "demometer" or "population cage". Because he was a skilled amateur carpenter, he had no difficulty in making himself several samples of this unprecedented laboratory apparatus.

Population cages were wooden boxes (50 cm × 30 cm × 15 cm), with glass on the top. At the bottom, 21 removable vials contained food for the flies. Every day, a vial of fresh food (yeast) was introduced to replace the oldest one. The 21 vials corresponded to the development of two generations of Drosophila at 26°C. Females laid their eggs in the last available vial of food, which became crowded within a few hours. Flies were periodically counted after anesthetization by CO_2, and reintroduced into the box afterwards. At the beginning, L'Héritier and Teissier used the population cage to study the demographic growth of a genetically homogeneous population, and observe whether an equilibrium was reached or not. Teissier, who was primarily interested in phenomena of growth of all sorts (organs, organisms, populations of cells) immediately saw that the population cage was more appropriate than Raymond Pearl's procedure of transferring indefinitely a growing collection of flies from one bottle to another or several other bottles (Pearl 1925, 1927). L'Héritier and Teissier's first result was that a population of flies could be indefinitely maintained at a level of approximately 400 adults. Since a new vial of food allowed females to deposit approximately 2000 eggs every day, there was a tremendous larval competition. Following this experiment, L'Héritier and Teissier carried out an impressive amount of work in just 5 years (1933–1938). In 1936 they ran experiments on no fewer than 36 populations simultaneously. Their general objective was to test the predictions of the models of theoretical population genetics (mainly Fisher and Wright), and of some models of theoretical ecology (especially Gause's principle of competitive exclusion). In 1938, Sewall Wright visited L'Héritier and Teissier's laboratory at the École Normale Supérieure (Paris). There, at Teissier's request, he wrote probably his best account of the shifting balance theory

(Wright 1939; Provine 1986), and he brought back to US the notion of population genetics. This led to a joint paper with Dobzhansky, the very first paper ever made with the help of a population cage outside France (Wright and Dobzhansky 1946).

To illustrate the spirit of Teissier and L'Héritier's experimental approach to population genetics, I will here describe two major results. In 1934, they published the result of their very first experiment, which placed in competition two strains of *Drosophila* (*wild* and *Bar*), initially present in equal numbers. The curve published in this paper (L'Héritier and Tessier 1934) describes the evolution of the frequency of the *Bar* gene over five months. They observed a progressive elimination of the mutant gene, but the *rate* of elimination diminished over time, suggesting a possible equilibrium (Fig. 4.1). The authors hypothesized that competition was lower when the mutant was present at a low frequency. Since the experiment accidentally came to a premature end, the two biologists repeated the same experiment with two population cages, over 600 days instead of 150. They did indeed observe that the mutant did not become extinct, stabilizing at approximately 2%. "It can be concluded, without any assumptions about the nature of the factors determining the superiority of the normal gene over *Bar* in the struggle for life, that this superiority tends to disappear when *Bar* has become rare in the population. If this superiority cancels out before the total disappearance of *Bar*, populations should show an indefinite state of stability, and would now show only random variations in their genetic make up." (L'Héritier and Teissier 1937a). Thanks to my friend and colleague Michel

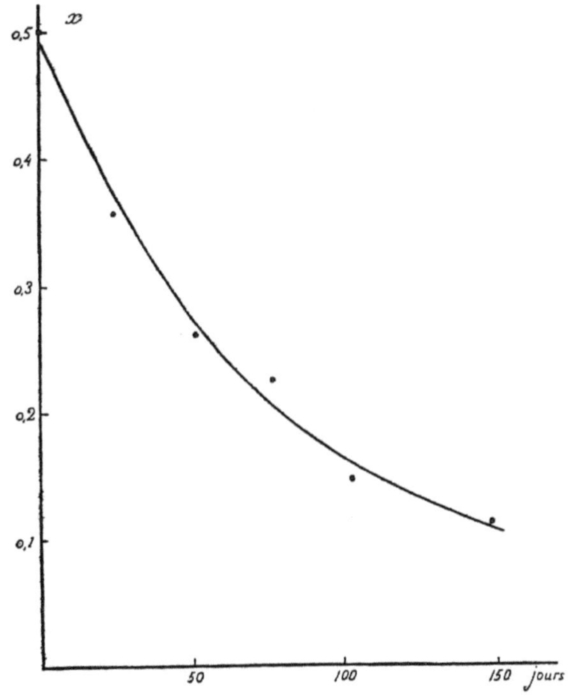

Fig. 4.1 Curve of elimination of *Bar* in a mixed population of *wild* and *Bar*. Abscissa: time (days). Ordinate: frequency of gene *Bar*. (L'Héritier and Tessier 1934)

Veuille, who found in his laboratory the raw data of this experiment, I have been able to see the tables of data, and the remarkable curve drawn from them: during the last months of the experiment, the curve becomes indeed strictly flat. However, the frequency of *Bar* being very low, there was a significant risk of total elimination of the mutant. L'Héritier and Tessier's experimental protocol could not in fact discriminate between selective and stochastic elimination. However, this remarkable experiment remained a model for Teissier's pupils in the 1940s and 1950s. Claudine Petit, furthermore, found in it the inspiration for her hypothesis of the "advantage of the rare" (or frequency-dependent selection) in her first work on sexual selection in flies.

In the same time as the *Bar* experiment was conducted, L'Héritier and Tessier began similar work with a mixed population of *wild* and *ebony* (L'Héritier and Tessier 1937b). Over 2 years, the frequency of the mutant decreased from 0.90 to 0.14, and seemed to evolve towards an equilibrium, although the results were stranger than in the case of *Bar* (see Fig. 4.2). In the conclusion of the paper, the authors proposed an explanation of the observed phenomenon through Fisher's model of heterosis (i.e. selective advantage of the heterozygote).

At the time, these results were spectacular. They had a tremendous impact on the French scene, because, in a deeply Lamarckian country, they showed that Darwin's theory of natural selection was not just rhetoric and speculation, but could be experimentally investigated. They also made L'Héritier and Tessier widely recognized among the members of the rather limited club of population geneticists in

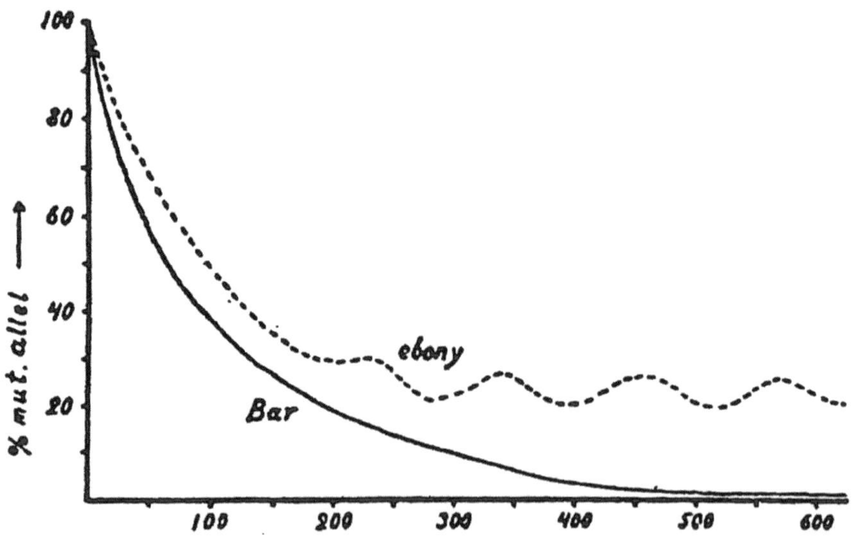

Fig. 4.2 Selective elimination of mutant genes in mixed populations of *Bar/wild* and *ebony/wild*. Representation of L'Héritier's and Tessier's synthesis (1937c) by Timofeeff-Ressovsky (1940). Abscissa: number of days. Ordinates: % of mutant allele. Curiously, Timofeeff-Ressovsky gives an erroneous reference (good title, error on journal title and pages), mixing up the synthetic paper (1937c) and the two papers of the same year on *Bar* and *ebony*. (1937a and b)

the world at the time. In 1939, Nikolay Timofeeff-Ressovsky, who was then the leading biologist in the domain of experimental population genetics (significantly earlier than Dobzhansky), mentioned these experiments of L'Héritier's and Teissier's in the famous Edinburgh Conference published by Julian Huxley in 1940 as *The New Systematics* (Timofeeff-Ressovsky 1940, p. 81–82). Quoting L'Héritier and Teissier (1937c), he gave a striking diagram (Fig. 4.2) representing simultaneously the results of the two experiments of competition—*Bar* against *wild*, and *ebony* against wild (Timofeeff-Ressovsky 1940, p. 80). L'Héritier attended the conference, but was not asked to publish in Huxley's volume. Nevertheless, the very fact that he was invited, as well as Timofeeff-Ressovsky's quotation, show that the two young French biologists' work was known among population geneticists. Other facts attest to this. I have already mentioned Sewall Wright's stay in Paris in Teissier's laboratory. Julian Huxley's close relationship with Teissier, because of their joint paper on the terminology of allometry, was another connection (Huxley and Teissier 1936a, b, c).

The scientific methodology that I have described is exactly that which inspired the group of young biologists who began working in population genetics under Teissier after WWII. By then, L'Héritier and Teissier had agreed to work along different lines. L'Héritier, although he had a leading role in the mid-1930s, decided to work on a strange phenomenon of non-Mendelian inheritance discovered in the course of his experiments with his population cage (a phenomenon that was eventually attributed to the *sigma* virus). Teissier took over the work in population genetics, and attracted a number of young biologists to this new field; I mentioned the best-known earlier. The spirit of this school was to compare as rigorously as possible the prediction of theoretical population genetic models with experimental data obtained either in the wild or with population cages. Another feature of Teissier's school, exceptional in the 1930s and 1940s, was its focus upon the level of genes, not gene complexes (like Dobzhansky), or phenotypes.

Claudine Petit was one of Teissier's students who decided to work in this spirit. She had not the sophisticated and brilliant curriculum vitae of most of her colleague. Born in 1920, she had begun as a schoolteacher in 1940. Then, after the German invasion, she engaged in the Resistance. She was caught and imprisoned, as her father before her, but she succeeded in escaping. She was condemned to 20 years of hard labor by a French court in 1943 in Chartres, and condemned to death *in absentia* by a German court in 1944 in Amiens. Since she was not caught again, she escaped both these sentences, and she remained in hiding till the end of the war (Petit 2000). After the war, she spent a few months at the Central School of Journalism (supported by the French Communist Party). Encouraged by her party to find a job, she was recruited as a young researcher at the CNRS (*Centre national de la recherche scientifique*), and entered George Teissier's laboratory in October 1945. As she herself often declared to me, her past in the French Resistance, her strong Communist commitment, and a rather difficult personal itinerary probably convinced Teissier to accommodate her with generosity. In 1946, she was appointed Assistant at the Faculty of Science of Paris, with teaching duties, but pursuing her research under Teissier. Here she made her entire career, ending as director of the laboratory of evolutionary genetics of the renamed University of Paris 7, after the dissolution of the University of Paris in 1969.

The young Claudine wanted to work in the field of population genetics. The reasons for this are unclear. It is likely that, with her poor academic curriculum, her modest origin, her fame as a Resistant, and active membership in the Communist Party, she felt attracted by Georges Tessier, a prominent figure in the Resistance and in the Party, a famous scientist, who was also Director General of the CNRS when she entered his laboratory. As he always did with his students, Teissier allocated a subject to Claudine, and, since she said she wanted to work in population genetics, he put her at work on sexual selection. She accepted and began working on the mutant *Bar*, looking for evidence of sexual selection—a possible component of the strong selective handicap of this fly. Till her death in 2007, Claudine Petit remained extremely grateful to Teissier. Among Teissier's students, she was probably the one who remained most faithful to Teissier's methodology and model organism (*D. melanogaster*).

Claudine Petit's entire list of publications is somewhat modest with respect to today's standards. However, the obsessive pressure for publication was not what it has now become. Including everything (that is, 36 articles in scientific journals, as well as chapters in collective books, communications in the proceedings of scientific meetings, and encyclopaedic articles), this list contains 56 titles[1]. Among these, seven are obituaries or historical papers. The 49 remaining titles are scientific contributions, which can be classified as follows:

- 12 whose title includes "sexual isolation".
- 14 whose title contains "sexual selection" or "sexual competition".
- 9 that bear upon subjects that rely heavily on Claudine Petit's work on sexual selection (advantage of the rare, mating systems)
- 12 papers dealing with problems indirectly related to sexual isolation and sexual selection (*e.g. Drosophila*'s sexual behavior, genetic structure of population)

In addition, there wert also two textbooks, both on genetics and evolution. The last one, written with Emile Zuckerkandl, was quite successful in France in the 1970s and 1980s (Petit and Prévost 1967; Petit and Zuckerkandl 1976).

We are therefore looking at a scientific career of 40 years almost entirely devoted to sexual isolation and sexual selection. This not trivial, especially in the case of sexual selection, if one thinks of the rarity of the works devoted to sexual selection, especially among geneticists, in the mid twentieth century. Petit's work was a pioneering contribution and was quoted by a number of international specialists, notwithstanding Dobzhansky's skepticism, which lasted until one of his own pupils, Lee Ehrman, did similar work in the late 1960s (Eherman 1966). Whatever that may be. This work offers an interesting opportunity for examining how sexual selection became a problem for experimental population genetics.

As noted above, when Claudine Petit began her research as a doctoral student, she was highly dependent on the methods and models elaborated by L'Héritier and Tessier in experimental population genetics. But, as she observed at the beginning of her PhD dissertation, published as a special issue of the *Bulletin biologique de la*

[1] Source: personal communication, shortly before Claudine Petit's death.

France et de la Belgique, L'Héritier and Teissier postulated random mating. What if this assumption were lifted; what consequences would there be if homogamy and sexual choice were to be taken into account? (Petit 1951a, p. 393). This way of locating herself relative to her supervisor's work was typical of Teissier: when they came to him and expressed to work in population genetics, he suggested to them to lift one of the conditions that he had himself neglected in his experimental work. To Maxime Lamotte, he recommended finding a suitable subject species to evaluate the role of random genetic drift; to Charles Bocquet, he suggested working on isolation through distance (whence Bocquet's clever choice of *Sphaeroma serratum*, a small crustacean that ranges all along the coasts of western Europe—only one geographic dimension to consider!); Ernst Boesiger was advised to work on heterosis; in the same mode, Petit was put to work on models implying non-random mating.

Once this general orientation was fixed, Petit looked at the existing literature. She never quoted Ronald Fisher's theoretical consideration on sexual selection, but, from the beginning, she was aware of Dobzhansky's and Mayr's experiments on sexual isolation together and separately (Dobzhansky and Mayr 1944, Dobzhansky and Streisiger 1944, Dobzhansky 1944, Mayr and Dobzhansky 1945, Mayr 1946a, b). This literature was obviously essential to her, and constituted the second inspiration of her work, together with L'Héritier and Teissiers methodology of population cages. This is why she wrote that sexual selection began being considered by population genetics as early as 1938: this was the year of the first paper ever published on the subject (Dobzhansky and Koller 1938). The close association of "sexual selection" and "sexual isolation" is characteristic of the American Dobzhansky/Mayr school. Like these authors (esp. Dobzhansky), Petit associated the two terms in a number of papers before 1970. The American school (Dobzhansky, Mayr, Patterson) had a simple methodology. They used to confront a male with two females from two different species, and they observed that the male most often chose a mate of his own type. They used the expression "sexual selection" for this phenomenon.

Petit was however not interested, or not primarily interested, in reproductive isolation, and she did not work on individuals from different species in the key papers that led to her PhD dissertation in 1958. An excerpt of a paper published in 1953, which includes the word "isolation" in its title, shows well that the purpose was not to study reproductive isolation: "Sexual isolation between races, lineages or mutants of a given species, has been recently studied by the school of Dobzhansky and Mayr. But the effects of this phenomenon upon the evolution of populations have rarely been studied; moreover, prior theoretical work on populations supposed that mating was random, any male having equal chances to fertilize any female" (Petit 1951b, p. 2482; my translation).

This early paper shows well both the saturation of Petit's vocabulary by the Dobzhansky school. But it also indicates, from the very beginning, her independence from both the American way of using "sexual selection" with the purpose of treating the problem of the origin of reproductive isolation, and the current work done by her supervisor. In reality, if one looks at what she effectively did in her early papers (1949, 1951a, b, 1952, 1953, 1954, 1955, 1958), it can be seen that:

(1) Petit explicitly worked on what she called “the sexual selection coefficient” (ratio of the number of males of a given genotype that copulates with females of a given genotype). (2) She did not work on geographical races (much less species) of *Drosophila*, but on laboratory mutants. (3) She was not primarily interested in reproductive isolation, but in the dynamics of populations where mating was not random. (4) She did not concern herself with male choice, but treated only female choice. To support this, she relied upon her own experiments. Her bibliographies, however, indicate she was aware of authors such has Rendel (1945) and Bateman (1948).

It was only after 10 years of work that Claudine Petit began clearly to dissociate the expressions “sexual isolation” and “sexual selection”. Still, in her PhD dissertation (Petit 1958), perhaps the best scientific contribution she ever wrote, she says that her study did not aim to discover cases of recent isolation of geographic races, but to discover in the lab “the genetic mechanisms that control isolating variations [*les variations d'isolement*]” (Petit 1958, pp. 251–252). Retrospectively, this sounds strange. In fact, the entire doctoral dissertation is devoted to the problem of the factors that determine the variation of the “sexual coefficient”. This tendency to confound “sexual isolation” and “sexual selection” was common in all population geneticists who spoke of sexual selection at the time, not only Dobzhansky and his collaborators, but also Mather and Harrison (1949) and Bruce Wallace (1950). After 1960, this kind of approximation disappeared in Petit's publications, to be replaced by subtle reflections about the nature of the relation between sexual selection and sexual isolation, and homogamy, subjects to which I will return in my conclusion. I will now analyze Petit's methodology and main results on sexual selection over her entire career.

4.1 Methodology

Petit's method of studying sexual selection was quite simple and elegant, and unique when she invented it. Basically, she combined two techniques. One consisted in assembling “synthetic population” in population cages. These synthetic populations were composed of two strains of *Drosophila melanogaster*, most generally a wild type and a mutant. The flies were put together for a rather short time (from a few hours to 5 days), so as to identify a possible mating bias as a component of selection. The other technique was based upon the extraction of females, which were each then introduced into a different test tube, where they released their fertilized eggs. Thanks to appropriate genetic markers, Petit could infer what type of males the female had copulated with.

The main objective of the experiments was to establish the curve of variation of the “sexual selection” (K), defined as the ratio between the probability [P] of fertilization by male [A], and the probability (Q) of fertilization by male [B]. If the frequencies of males [A] and [B] in the population are p and q, then it can be shown that $K = P/Q \times q/p$ (for details, see Petit 1951, p. 8–15). This coefficient measures the intensity of sexual selection. If $K = 1$, it means that all males have equal access to females, and there is no sexual selection.

4.2 *Bar/wild* Experiments

Petit's first experiments were made on a mixed population of *Bar* and *wild*, the material on which L'Héritier and Tessier had begun their own experiments on natural selection in 1934. Petit published her results in 1951. She plotted *K* agains the relative numbers of the two types of males in the population. Repeating the experiment form any initial proportion of males, she obtained the graph given in Fig. 4.3. The curve obtained shows that (1) *wild* males are always advantaged (*K* always >1). (2) *K* varies in function of the number of mutant males in the population. (3) When the number of males is >50%, *K* is constant.

In light of this result, Petit reproduced L'Héritier and Teissier's experiments of 1934 and 1937 (described earlier in this paper). L'Héritier had observed that an equilibrium seemed to be reached when the mutant *Bar* became rare (~1%; see Fig. 4.2). With no justification, they had attributed this result to a variation in the value of the selection coefficient. And, since they had identified only one selective factor, larval competition, they had hypothesized that larval competition disappeared when *Bar* became rare. In 1951, Claudine Petit reproduced L'Héritier and Teissier's experiment, and obtained a similar curve of elimination of *Bar*. But she explained it by a combination of two processes: variation of the value of the coefficient of larval competition (σ), and variation of the value of the coefficient of sexual selection (*K*). Since she had been able to find the values of both, she was able to compare the theoretical evolution of the population under several conditions, and the observed evolution. Figure 4.4 reproduces a graph that summarizes the results.

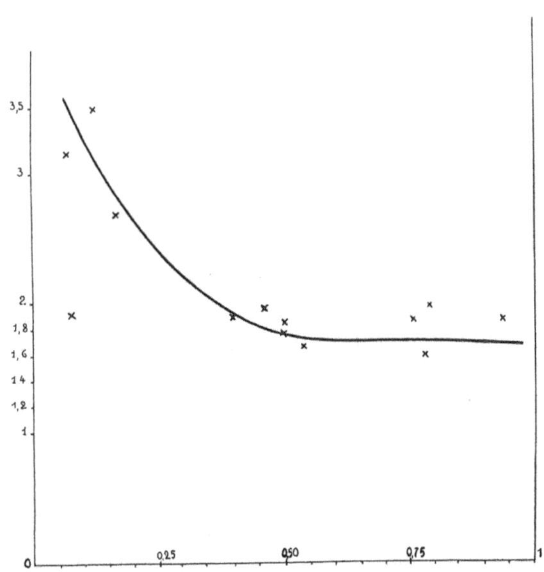

Fig. 4.3 Evolution of the sexual coefficient *K* in a mixed population of *Bar* and *wild*, in the course of the evolution of the population (Petit 1951, p. 403). Abscissa: frequency of *wild* males. Ordinates: *K* (Probability of fertilization by a *wild* male/ Probability of fertilization by *Bar*). The graph shows that *K*, although always >1, *decreases* when the frequency of *wild* males *increases*

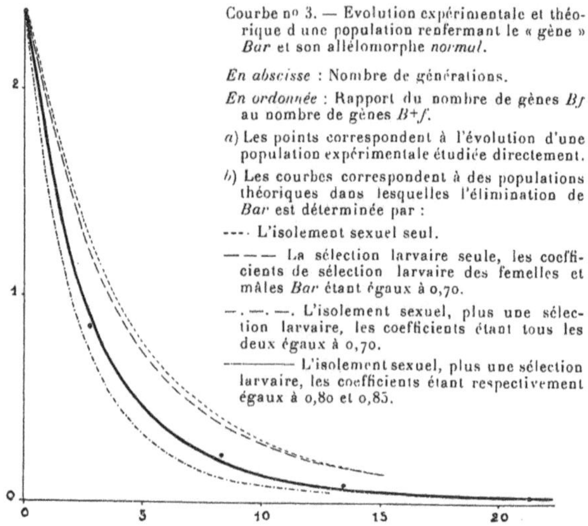

Fig. 4.4 Comparison of experimental and theoretical evolution of a mixed population containing the *Bar* and the *wild* gene (Petit 1951, p. 413). Abscissa: number of generations. Ordinates: ratio of the frequencies of *Bar* gene over *wild* gene. *Bold dots* represent the experimental population effectively observed. The *dotted lines* represent three theoretical populations. (The line with smaller dashes (---) "Sexual isolation" (understand: sexual selection) alone, the line with bigger dashes (– – –) Larval selection alone, the dotted dash line (–·–·–) Sexual isolation and larval selection in combination)

To sum up, these complex experiments showed that sexual selection changed in function of the composition of the population. Their contribution to L'Héritier and Teissier's famous experiment on *Bar* was at least as important as the idea of larval competition, quite an interesting result. *Bar* is indeed handicapped in more than one way.

4.3 *White/wild* Experiments

After the *Bar/wild,* Claudine Petit began working on a synthetic population of *white* and *wild*, a more favourable system that gave spectacular results (Petit 1954, 1958). The evolution of a *white/wild* mixed population is quite similar to that of an *ebony/ wild* population reported by L'Héritier and Teissier in 1937, that is establishment of an equilibrium at a level where the mutant stabilizes at a rather high frequency (see Fig. 4.2). This fact was known in Tessier's surroundings, and was puzzling, because *white* is an extremely handicapped mutant. It is blind, and it has severe limitations in terms of his capacity for moving, using its wings, etc. What Petit was quick to observe with her methodology was that *white* has a strong mating advantage when it becomes rare in the population (<40 %). Figure 4.5 reproduces the diagram which revealed the phenomenon. Figure 4.6 reports the results of the same experiments, as well as other similar ones carried out after 1954; this diagram was reproduced many times in subsequent publications, and gave Claudine Petit an international

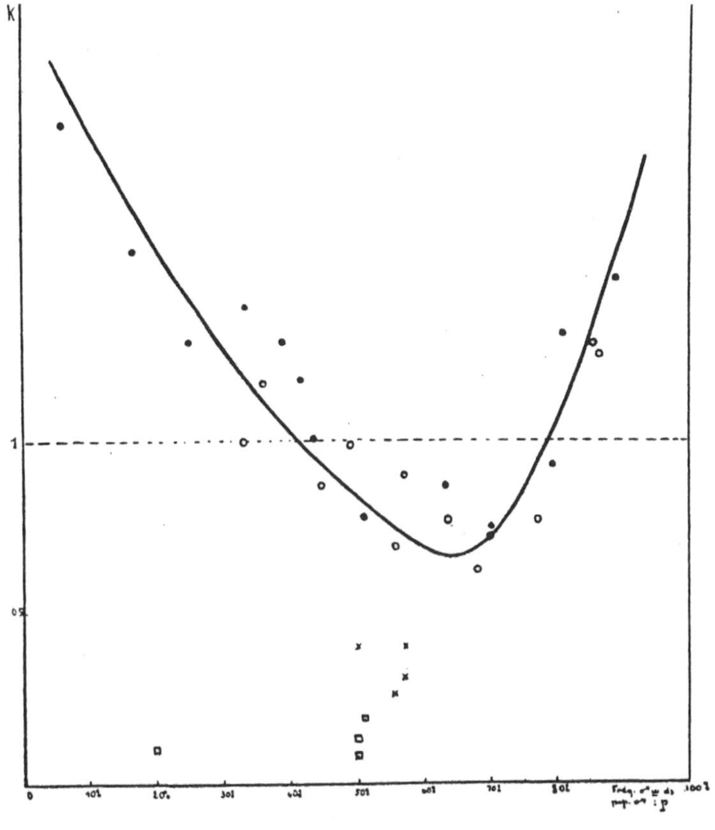

Fig. 4.5 Variation of the coefficient of "sexual isolation" in a mixed population *white/wild*. Abscisa: frequency of *w* males in the populations of males. Ordinates: coefficient of sexual isolation, *K*. The various points indicated (*circles, squares,* etc.) refer to four experiments made from 1953 and 1954

reputation in the 1960s. This figure was first given in 1954, then reproduced many times in subsequent papers. Note a small but conceptually significant difference between the two diagrams. The *K* coefficient, originally designated an "isolation coefficient", was renamed the "sexual coefficient" in 1958. Both diagrams show that the mutant is advantaged when it is either "rare" (<40%) or abundant (>80%). In between, the *wild* type has an advantage. Therefore an intermediate equilibrium is possible with respect to sexual selection. However, it never occurs that *white* eliminates *wild* in a population cage (L'Héritier and Teissier 1936; Petit 1954, 1958). This results from many handicaps affecting *white*, beside its peculiar behaviour in a mating situation. Therefore, in contrast with the Petit's previous experiments on a *Bar/wild* mixed population, sexual and natural selection exerted opposing forces in the *white/wild* experiments.

In her 1958, Claudine Petit went further. She methodically examined a number of factors that could affect the shape of the curve of variation obtained for the sexual coefficient: effect of the density of population, effect of temperature, effect

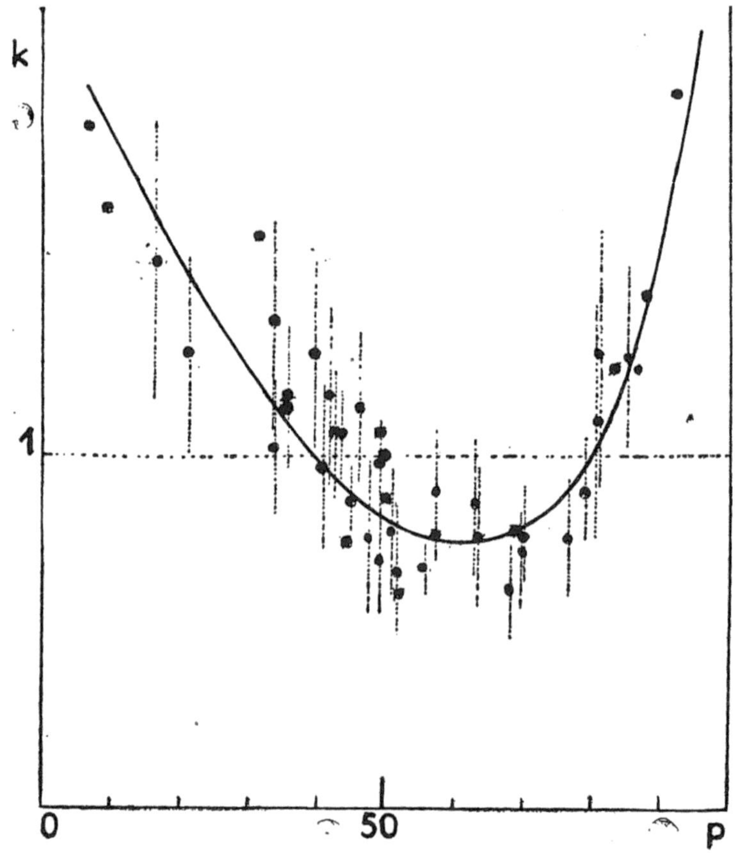

Fig. 4.6 Same experiment as in Fig. 4.5, graph given in Petit (1958). The original caption says: "Variation of the coefficient of sexual selection in function of the proportion of *white* males". *Dark circles* summarize the results obtained in 1953-1954. Crosses summarize further experiments. Dashed lines represent the standard deviation

of genetic context (revealed by the replacement of one or several chromosomes). The general conclusion of the inquiry was that sexual competition is "an extremely complex phenomenon", and a "plastic phenomenon" that depends highly upon the physical environment, the demography of the population, and the genetic context. The sexual coefficient was not, therefore, an intrinsic property of a given strain.

4.4 Further Developments

After her remarkable results about sexual competition in *D. melanogaster*, Petit remained attached to the same experimental system (*white*/wild), but her work evolved along two main directions.

First, she first developed a systematic argument in favour of the "advantage of the rare", and, more generally, of frequency-dependent selection, as a major factor explaining the genetic polymorphism observed in nature (*e.g.* Ehrman and Petit 1969; Petit and Anxolabehere 1968; Petit 1970, 1974; Petit and Nouaud 1975). Lewontin's *The Genetic basis of evolutionary change* (Lewontin 1974) echoes this argument. In this book, Lewontin claimed that, although the advantage of the rare is an important phenomenon, it couldn't suffice to explain the magnitude of genetic polymorphism.

Secondly, Claudine Petit pursued the study of sexual behaviour that she had begun in her PhD. She entered into a controversy over the real mechanism that confers an advantage to either the wild type or the mutant in function of the composition of the population, and thus acquired a reputation as a specialist of the genetics of behaviour. Two psycho-physiological phenomena are involved in the mating behaviour of Drosophila. One is sensorial, the other was locomotory. From a sensorial point of view, neither olfaction or vision is relevant. What counts are the mechanoreceptors of the females (Johnston's organ). This organ, located on the legs of the flies, is sensitive to air vibration. Females are sensitive to the beating of wings of males. If this beating does not occur, the female is not receptive. Therefore, female choice could be based upon differences among females with respect to mechanoreception. The locomotory aspect of mating behavior is dance. The males vibrate their wings and dance in the front of the female. However the *white* mutant is unable to dance around. He is just able to beat his wings and weakly dance behind the female (See Fig. 4.7). What kind of advantage, then, could, the mutant have when rare in the population?

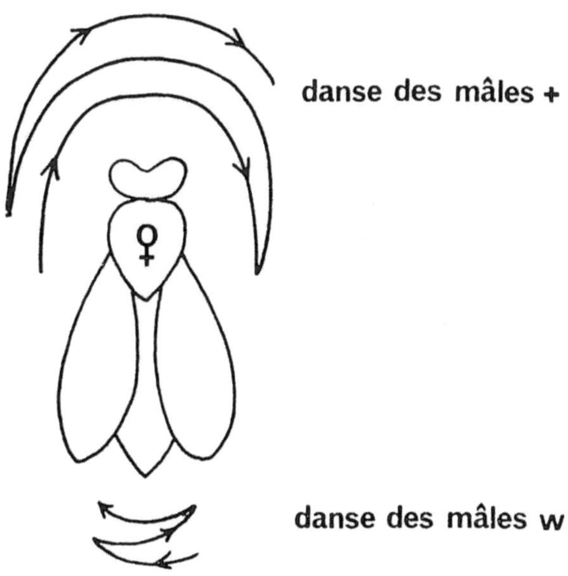

Fig. 4.7 Schematic representation of the dance of wild and *white* male flies (Petit 1974, p. 435). Wild males dance in front or on the sides of the female in order to attract her attention. Mutants have a poor dance limited to small movements behind the females.

Petit ultimately asserted (Petit 1974) that the advantage is not properly a behavioral mechanism. It is not either an adaptation designed by a selective process. Rather, it is an "ecological" constraint, which she described in terms of competition for space in the context of courtship. Here is the proposed explanation: If the male mutant is rare, he will be advantaged because the male will stand behind the female and jump over her while two or three males will dance around her. If the mutant male is abundant, several mutants will take positions behind the female and impede one another's access to her: then a wild fly will more likely jump over the female. It is as simple as this. For Petit, this process does not require to hypothesize a genetically determined behavior aiming at something. It is just a question of accessibility to a resource (space in this case). If a given resource is abundant relative to the rare type, then the rare type will be favored. Therefore, as Petit proposed, the advantage of *white* is just a case of the more general case of partitioning of the ecological space of resources. There is no innate mechanism, there is just a spatial constraint determined by the composition of the population (itself a particular dimension of the environment). Oddly, this is exactly the general explanation of the advantage of the rare that Lewontin proposed in his 1974 book, in the section devoted to frequency-dependent selection. Since Lewontin's book and Petit's chapter were published the same year, it is hard to know whether one of the authors influenced the other. Whatever the case may be, Lewontin wrote that Petit's work on sexual selection was the best-known case of the advantage of the rare.

To conclude, I will return to several conceptual ambiguities that I have mentioned in passing in this paper: sexual selection *vs.* sexual isolation, sexual selection *vs.* natural selection, and sexual selection *vs.* homogamy All have something to do with Dobzhansky, and also with Dobzhansky's best ally in France on the subject of sexual selection, Ernst Bösiger. They may also be related to Darwin's original notion of sexual selection.

At the very time when Claudine Petit was developing her own studies on sexual selection in flies, another pupil of Georges Tessier, Ernst Bösiger, was also working on sexual selection. In 1968, Dobzhansky and Bösiger co-authored a book entitled *Essais sur l'évolution*, which contains a full chapter on sexual selection (Chap. 10, pp. 90–118). Retrospectively, the connection between the Teissier school and the Dobzhansky school (initiated in 1938 on the occasion of Sewall Wright's visit to Teissier's lab in Paris) illuminates the terminological association between "sexual selection" and "sexual isolation" in Petit's early papers that I have underlined a number of times in my paper. Looking at Bösiger and Dobzhansky's 1968 chapter on sexual selection reveals what was at stake in her work, something that she discovered only progressively, when she moved away from Dobzhansky's conception of sexual selection. It will be useful, then, to recall what the conception of sexual selection shared by Dobzhansky and Bösiger was.

Bösiger wrote his PhD dissertation on heterosis in natural populations. Intrigued by the high number of recessive genes in natural populations of *Drosophila melanogaster*, he claimed, and actually showed, that one of the mechanisms that maintain a high degree of polymorphism was sexual selection, insofar as male heterozygotes are more fertile and are able to inseminate many more females than monozygotes. This fitted well with Dobzhansky's view that genetic polymorphism is not so much a "load" due to imperfect selection, but rather something that is positively maintained by one or various sorts of balanced selection. I have no idea how Bösiger and Dobzhansky came to enter into their close professional and friendly relationship, but the intellectual convergence between the two of them certainly contributed to their collaboration.

Let us now look to what Bösiger and Dobzhansky actually say. Their chapter begins with a quotation from Darwin's *Descent of Man, and Selection in Relation to Sex*:

> Sexual selection depends on the success of certain individuals over others of the same sex in relation to the propagation of the species; whilst natural selection depends on the success of both sexes, at all ages, in relation to the general conditions of life. The sexual struggle is of two kinds; in the one it is between the individuals of the same sex, generally the male sex, in order to drive away or kill their rivals, the females remaining passive; whilst in the other, the struggle is likewise between the individuals of the same sex, in order to excite or charm those of the opposite sex… (Darwin 1871, p. 339)

After this quotation, Dobzhansky and Bösiger distance themselves, and say that it is necessary today to adjust Darwin's concept of natural selection to "present knowledge". According to them, three new mechanisms, unknown to Darwin, should be included in the domain of sexual selection. A first mechanism is interspecific isolation. The authors claim that sexual selection has played a role in the construction of adaptive mechanisms prohibiting mating between geographic races that have diverged. A second "mechanism" is the maintenance of genetic polymorphism by sexual selection. Dobzhansky and Bösiger mention a number of authors, among them Petit. They conclude that sexual selection may not act simultaneously with natural selection to maintain polymorphism. They also mention Bösiger's hypothesis that heterosis favors male vigor. The third mechanism is speciation through intraspecific sexual selection. This "mechanism" refers to Dobzhansky's hypothesis of reinforcement of reproductive isolation through ethological processes.

These three claims illuminate both the reasons why the young Petit confounded "sexual isolation" and "sexual selection", and also what was original in Petit's approach to sexual selection. She confounded the two notions because Dobzhansky's school had been her first and most important influence. Dobzhansky could not conceive that sexual selection had no adaptive effect, and, since he believed that the main adaptive effect was to reproductively isolate populations that had previously diverged, the two concepts and the two terms tended to fuse one into the other, or at least to be interchangeable.

However, the comparison between Petit and Bösiger/Dobzhansky also reveals what was distinctive in Petit's approach. In reality, the process that she progressively identified and explained had nothing adaptive about it. As she convincingly showed, sexual selection in favor of mutants such as *Bar* or *white* in artificial populations is not based either upon adaptive devices preventing interfertility (Dobzhansky), or male vigor (Bösiger), but upon a pure phenomenon of competition, whose outcome depended upon contextual circumstances of different sorts, especially the composition of the population. Therefore, in a sense, Claudine Petit's work on selection was more "Darwinian" than Dobzhansky's. It emphasized the aspect of pure competition in a "natural process of selection" (sexual selection) different from conventional "natural selection", that is, independently of any adaptive effect.

References

Bateman AJ (1948) Intra-sexual selection in *Drosophila*. Heredity 2:349–368

Darwin C (1871) The descent of man and selection in relation to sex, vol. 2. John Murray, London

Dobzhansky T (1944) Experiments on sexual isolation in *Drosophila*. III. Geographic strains of. *Drosophila stutervanti* Proc Natl Acad Sci U S A 30:335–339

Dobzhansky T, Koller PC (1938) An experiment on sexual isolation in *Drosophila*. Biol Zbl 58:589–607

Dobzhansky T, Mayr E (1944) Experiments on sexual isolation between geographic strains of *Drosophila willistoni*. Proc Natl Acad Sci U S A 30:238–244

Dobzhansky T, Streisiger G (1944) Experiments on sexual isolation in Drosophila. II. Geographic strains in *Drosophila prosaltans*. Proc Natl Acad Sci U S A 30:340–345

Dobzhansky T, Boesiger E (1968) Essais sur l'évolution. Masson, Paris

Dobzhansky T, Boesiger E, (1983) Human culture: a moment in evolution (edited and completed by Bruce Wallace). New York University Press

Eherman L (1966) Mating success and genotypes frequency in *Drosophila*. Anim Behav 14:332–339

Ehrman L, Petit C (1969) Sexual selection in *Drosophila*. Evol Biol 3:177–223

Gayon J, Veuille M (2001) The genetics of experimental populations: L'Héritier and Teissier's populations cages. In: Singh R, Krimbas C, Paul D, Beatty J (eds) Thinking about evolution: historical, philosophical, and political perspectives. Cambridge University Press, Cambridge, pp. 77–102

Huxley JS, Teissier G (1936a) Terminology of relative growth. Nature 137:780–781

Huxley JS, et Teissier G (1936b) Terminologie et notation dans la description de la croissance relative. Comptes Rendus des Séances hebdomadaires de la Société de Biologie 121:934–937

Huxley JS, Teissier G (1936c) Zur Terminologie des relativen Grössenwachstums. Biol Centr 56(:):381

Lewontin R (1974) The genetic basis of evolutionary change. Columbia University Press, New York

L'Héritier P, et Teissier G (1934) Une expérience de sélection naturelle, Courbe d'élimination du gène *Bar* dans une population de *Drosophila melanogaster*. Comptes rendus des séances hebdomadaires de la Société de biologie 117:1049–1051

L'Héritier P, et Teissier G (1936) Contribution à l'étude de la concurrence larvaire chez les Drosophiles. CR Soc Biol 122:264–267

L'Héritier P, et Teissier G (1937a) Elimination des formes mutantes dans les populations de Drosophiles. Cas des *Drosophiles Bar*. Comptes rendus des séances hebdomadaires de la Société de biologie 124:880–882

L'Héritier P, et Teissier G (1937b) Elimination des formes mutantes dans les populations de Drosophiles Cas des *Drosophiles ebony*. Comptes rendus des séances hebdomadaires de la Société de biologie 124:882–884

L'Héritier P, et Teissier G (1937c) L'élimination des formes mutantes dans les populations de Drosophiles. Soixante-dixième Congrès des Sociétés Savantes 297–302

Mather K, Harrison BJ (1949) The manifold effects of selection. Heredity 3:1–52 and 131–162

Mayr E (1946a) Isolation between *Drosophila pseudoobscura* and. *Drosophila persimilis* their hybrids Proc Natl Acad Sci U S A 32:128–137

Mayr E (1946b) Experiments on sexual isolation in Drosophila. VII. The nature of isolating mecahnisms beteen *D. pseudoobscura* and *D. Persimilis*. Proc Natl Acad Sci U S A 32:128–137

Mayr E, Dobzhansky T (1945) Experiments on sexual isolation in *Drosophila*. IV. Modification of the degree of isolation between *D. pseudoobscura* and *D. persimilis* and of sexual preferences in. *D prosaltans*. Proc Natl Acad Sci U S A 31:75–82

Pearl R, (1925) The biology of population growth. Knopf, New York

Pearl R (1927) The growth of populations. Q Rev Biol 2:532–548

Petit C (1949) L'isolement sexuel chez *Drosophila melanogaster*: étude du mutant *Bar*. Bulletin de la Société française de génétique 3:7–9

Petit C (1951a) Le rôle de l'isolement sexuel dans l'évolution des populations de *Drosophila melanogaster*. Bulletin biologique de la France et de la Belgique 85(4):392–418

Petit C (1951b) Le rôle de l'isolement sexuel dans l'évolution des populations de *Drosophila melanogaster*. Comptes rendus de l'Académie des sciences 232:1482–2484

Petit C (1952) La sélection sexuelle: étude théorique et expérimentale de la compétition entre *Bar* et *sauvage*. Bulletin de la Société française de génétique 4:8–9

Petit C (1953) Les variations du coefficient d'isolement entre des souches isogéniques de *Drosophila melanogaster*. Comptes rendus de l'Académie des sciences 237:203–204

Petit C (1954) L'isolement sexuel chez *Drosophila melanogaster*: étude du mutant *white* et de son allzlomorphe *sauvage*. Bulletin biologique de la France et de la Belgique 88:435–443

Petit C (1955) Le déterminisme génétique de l'isolement sexuel. Comptes rendus de l'Académie des sciences 241:521–522

Petit C (1958) Le déterminisme génétique et physiologique de la competition sexuelle chez *Drosophila melanogaster*. Bull biologique de la France et de la Belgique 92:248–329. (Thèse de Doctorat d'État)

Petit C (1970) Quelques facteurs responsables de l'avantage du type rare chez *Drosophila melanogaster*. Comptes rendus de l'Académie des sciences 270:1719–1922

Petit C (1974) L'avantage du rare, facteur du maintien du polymorphisme. In: Lamotte M. (éd) Le polymorphisme dans le règne animal. Mémoires de la Société Zoologique de France, Paris, pp 417–442

Petit C (2000, unpublished) Souvenirs: la famille Petit dans la Résistance. Typographed manuscript, 79 pages. The foreword mentions 1997 as date of beginning of the manuscript. At the end of the text, the author mentions the date of completion, February the 20th, 2000. Personal communication

Petit C, Prévost G (1967) Génétique et evolution. Hermann, Paris

Petit C, Anxolabehere D (1968) Frequency-dependent selection and larvo-nymphal competition. Proc 12th International Congr of Genet 1:228

Petit C, Nouaud (1975) Ecological competition and rare type mating advantage in. *Drosophila melanogaster* Evol 29:763–776

Petit C, Zuckerkandl E (1976) Evolution; Génétique des populations. Evolution moléculaire. Hermann, Paris

Provine WB (1986) Sewall wright and evolutionary biology. The University of Chicago Press, Chicago

Rendel JM (1945) Normal and selective matings in *Drosophila subobscura*. J Genet 46:187–302

Timofeeff-Ressovsky N (1940) Mutations and geographical variation. In: J. Huxley (ed) The new systematics, pp. 73–135

Wallace B (1950) An experiment on sexual isolation Drosophila Information Service 24: 94–96

Wright S (1939) Statistical genetics in relation to evolution. Hermann, Paris

Wright S, Dobzhansky T (1946) Experimental reproduction of some of the changes caused by natural selection in certain populations of *Drosophila Pseudoobscura*. Genetics 31:125–156

Part II
Current Challenges

Chapter 5
Sexual Selection: Is Anything Left?

Joan Roughgarden

Abstract The biological definition of what counts as sexual selection has gradually changed since Darwin introduced the concept. This paper reviews a sequence of definitions that reflect increasing knowledge of the diverse ways animals participate in reproductive social behavior. Later definitions accommodate more data than Darwin's primary formulation. The latest approach, suggested in a catalyst meeting held at the National Evolution Synthesis Center in July 2013, distinguishes fertility selection as a density-dependent process from sexual-selection as a frequency-dependent process. The former consists of reproductive social behavior focussed on increasing the size of the reproductive pie, the later on securing a larger share of a given reproductive pie. The approach of social selection advocated here focusses on how to increase the size of the pie rather than on how to monopolize a larger share of a fixed pie. Social selection reverses the logic of sexual selection by starting with offspring production and working back to mating, and by starting with behavioral dynamics and working up to gene pool dynamics. In social selection courtship can potentially be deduced as a negotiation leading to an optimal allocation of tasks during offspring rearing. Mating pairs may form "teams" based on the reciprocal sharing of pleasure. The parent-offspring relation can be managed by the parent considered as the owner of a "family firm" whose product is offspring. The cooperation in reproductive social behavior may often evolve as a mutual direct benefit through individual selection rather than as some form of altruism requiring kin or multi-level selection.

Keywords Sexual selection · Social selection · Social infrastructure selection · Anisogamy · Teams · Pleasure · Family firm · Cooperation · Mutual direct benefit · Nash Bargaining Solution

J. Roughgarden (✉)
Hawaii Institute of Marine Biology, University of Hawaii, 5241 Wili, Road, Kapaa, HI 96746, USA
e-mail: joan.roughgarden@hawaii.edu

T. Hoquet (ed.), *Current Perspectives on Sexual Selection,* History, Philosophy and Theory of the Life Sciences 9, DOI 10.1007/978-94-017-9585-2_5

5.1 Introduction

As readers are well aware, Darwin's (1871) theory of sexual selection has been debated since its origin. Milam (2010) has reviewed the disagreements about sexual selection that began with Wallace (1871) and continued through the early to mid 1900's. Early critics contested the ability of animals to make "choices" about mates and whether animals have "minds". The mid 1900's witnessed an eclipse of attention to sexual selection, although female choice *per se* interested the Modern Synthesis architects as a mechanism contributing to reproductive isolation between species (Mayr 1963). During the 1970's, interest in sexual selection to explain the evolution of ornaments, armaments and mating behavior resurfaced (Parker et al. 1972; Trivers 1972; Dawkins 1976). Today though, sexual selection is yet again open to debate (Roughgarden 2009; Clutton-Brock 2009; Roughgarden and Akçay 2010; Shuker 2010; Rosvall 2011). The biological disagreements are now about what animals are choosing for, and who is doing the choosing rather than whether animals are capable of making choices.

Philosophical disagreement continues on whether and how sexual selection might be conceptually distinct from natural selection (eg. Millstein 2011). In contrast, this article focusses on how within biology the idea of sexual selection has changed in recent years. This article also briefly presents the alternative approach of social selection. In this article both sexual selection and social selection are taken as alternative mechanisms within an overall genetical selection process that includes natural selection, the later being understood as lumping viability with fertility selection.

This conference has asked whether anything is now left of sexual selection. My answer is no, nothing is left of the original sexual selection although something might be found in the guise of restated definitions of sexual selection. More importantly, an expanded field of inquiry not limited to Darwin's conceptualizations that I have termed "sexual selection studies" is prospering.

5.2 Sexual Selection and its Changing Definitions

Sexual selection may be interpreted as a specific hypothesis about how ornaments and armaments evolved together with what objectives mating and courtship are intended to accomplish. Sexual selection may also be interpreted as a "master narrative" for reproductive social behavior as presented by Parker and Pizarri (Chap. 7 this volume).[1] This article critiques sexual selection as a specific hypothesis that

[1] Parker and Pizarri present an "evolutionary cascade" beginning with the origin of sexual reproduction, continuing through the evolution of anisogamy, and culminating in the emergence of sexual selection, sexual conflict and characteristic sex strategies. They contend this sequence has "remarkable logical beauty" and that sexual selection is a "logical imperative" because "sexual selection emerges deductively as the logical consequence of this evolutionary succession." I disagree. The Parker and Pizarri master narrative begins with anisogamy as an outcome of a

has been continually morphing. It does not focus on sexual selection as a master narrative nor does it critique the large field of what I have termed "sexual selection studies." Settling on what a specific hypothesis states is a necessary condition for assessing its truth, for otherwise the hypothesis can always be (and has been) re-stated to accommodate any facts that are discovered. The way in which the concept of sexual selection has needed to be continually modified reveals a trend away from role-based definitions of sexual selection.

5.2.1 Sexual Selection 1.0: Darwin 1871

Darwin (1871) asserted that the following universal sex roles: "Males of almost all animals have stronger passions than females," [p. I.272] and "The female, on the other hand, with the rarest of exceptions is less eager than the male. As the illustrious Hunter long ago observed, she generally 'requires to be courted;' she is coy" [p. I.273].

Darwin claimed that female choice during courtship causes the evolution of ornaments, now called inter-sexual selection: "Many female progenitors of the peacock must, during a long line of descent, have appreciated this superiority; for they have unconsciously, by the continued preference for the most beautiful males, rendered the peacock the most splendid of living birds" [p. II.141].

Darwin further claimed that competition among males for access to females, now called intra-sexual selection, causes the evolution of armaments: "In the same manner as man can improve the breed of his game-cocks by the selection of those birds which are victorious in the cockpit, so it appears that the strongest and most vigorous males, or those provided with the best weapons, have prevailed under nature, and have led to the improvement of the natural breed or species" [p. I.258].

Darwin summary of reproductive social life in animals is one of continuing conflict: "On the whole there can be no doubt that with almost all animals, in which the sexes are separate, there is a constantly recurrent struggle between the males for the possession of the females" [p. I.260].

"primordial sexual conflict" between competing gametes. This origin to their narrative is problematic. An alternative possible origin of anisogamy relies on cooperation among gametes to maximize fertilization rates. Union in an anisogamous population among many micro- with a few macro-gametes achieves more surviving zygotes than union in an isogamous population with an intermediate number of same-sized gametes (cf. Roughgarden and Iyer 2011, also termed the "Kalmus" effect). This cooperative advantage, rather than one gamete type parasitizing the investment of the other gamete type, may underlie the origin of male and female gametes. This means that the entire evolutionary cascade can veer off in a sexual-cooperation direction or a sexual-competition direction depending on unknowable initial conditions. Hence, the outcome of such a cascade is an empirical contingency, not a logical necessity. Furthermore, the Parker and Pizarri master narrative is not the only proposed master narrative. In *The Genial Gene* I sketched an alternative master narrative based on cooperative interactions at each step in the cascade that appears to accord better with the facts (cf. Roughgarden 2009, esp. pp. 235–248). These alternative master narratives should be subjected to test.

One of Darwin's other passages however, seems to define sexual selection more generally than a specific set of claims about the roles of males and females and their concomitant social life. When distinguishing sexual selection from natural selection, he wrote "We are, however, here concerned only with that kind of selection, which I have called sexual selection. This depends on the advantage which certain individuals have over other individuals of the same sex and species, in exclusive relation to reproduction" [p. I.256].

Is Darwin's definition of sexual selection a specific hypothesis about the evolutionary causes of ornaments and armaments, or simply a hypothesis that a general distinction can be drawn between an advantage in relative reproduction *vs.* an advantage in adaptation to the environment? The answer is not clear although his application of the idea of sexual selection was solely in the context of the sex-role narratives of his day.

Therefore, this article singles out as Sexual Selection 1.0, Darwin's original hypothesis that:

> SexS 1.0: Generally throughout nature, competition among males for mates, and choice of mates by females, causes the evolution of armaments and ornaments in males.

5.2.2 Sexual Selection 1.1: Adding Genes

Because SexS 1.0 was framed before genes were discovered, it has been restated in light of genetics. As Coyne (2004) articulates, "We now understand … Males, who can produce many offspring with only minimal investment, spread their genes most effectively by mating promiscuously … Female reproductive output is far more constrained by the metabolic costs of producing eggs or offspring, and thus a female's interests are served more by mate quality than by mate quantity." Here, "quality" means heritable genetic quality. Thus, Sexual Selection 1.1 is the hypothesis that:

> SexS 1.1: Because of their low investment in sperm, males spread their genes most effectively by mating promiscuously whereas females, because of their high investment in eggs, spread their genes most effectively by selecting to mate with the genetically best male, resulting in the evolution of ornaments and armaments in males.

Tracing sex-role difference to a difference in parental investment departs from Darwin's view.

In 1871 he wrote, "We are naturally led to enquire why the male in so many and such widely distinct classes has been rendered more eager than the female, so that he searches for her and plays the more active part in courtship. It would be no advantage and some loss of power if both sexes were mutually to search for each other; but why should the male almost always the the seeker?" [p. I.273] Further, "the male element is invariably brought to the female; and we can see the reason; for the ova, even if detached before being fertilized and not requiring subsequent nourishment of protection, would be, from their larger relative size, less easily transported than the male element" [p. I.274]. Thus, "It would, therefore, be a great advantage to such animals, as their organisation became perfected, if the males when ready to emit the fertilising element, were to acquire the habit of approaching the female as closely as possible" [p. I.274] Continuing, "in order that they should become more efficient seekers, they would have to be endowed with strong passions" [p. I.274].

In 1874, Darwin expands, "The female has to expend much organic matter in the formation of her ova, whereas the male expends much force in fierce contests with his rivals, in wandering about in search of the female, in exerting his voice, pouring out odoriferous secretions, etc.: and this expenditure is generally concentrated within a short period" [p. 232]. Therefore, "On the whole the expenditure of matter and force by the two sexes is probably nearly equal, though effected in very different ways and at different rates" [p. 233].

Although a single sperm is smaller than a single egg, according to Darwin the parental investment of both male and female winds up being about the same. Thus, Darwin emphasized the sexes differed in the kind, not the quantity, of their reproductive investment.

To Darwin, the explanation of the male sex role lies in what is needed to transport sperm and not in how cheap sperm are relative to an egg. Thus, SexS 1.1 is not merely an updating of the original SexS 1.0 to include genes, but a substantive revision in regards to the cause of why males and females differ in their mating behavior.

5.2.3 Sexual Selection 1.2: Accounting for Sex-Role Reversal

Both SexS 1.0 and 1.1 are contradicted by the phenomenon of sex-role reversal.

As Darwin (1874) wrote, "With birds there has sometimes been a complete transposition of the ordinary characters proper to each sex; the females having become the more eager in courtship, the males remaining comparatively passive, but apparently selecting the more attractive females, as we may infer from the results. Certain hen birds have thus been rendered more highly colored or otherwise ornamented, as well as more powerful and pugnacious than the cocks" [p. 233]. To explain the evolution of ornaments in females rather than male, Darwin reverses the narrative but gives no guide or explanation as to when such a transposition of roles should occur.

Today, sex role reversal is attributed to a reversal in the operational sex ratio (OSR) (Emlen and Oring 1977) which is the ratio of the number of males ready to mate to the number of females ready to mate. Typically the OSR is biased towards the male (i.e., more males than females available for mating). The supposition is that the sex showing the lower parental investment is more available while the sex showing the higher investment is busy with the offspring care. The male's sperm represents an initially lower parental investment than the female's egg, thereby initially biasing the OSR toward males. However, changing the availability and need for resources for each sex may result in the male supplying more of the parental investment overall than the female. If so, the operational sex ratio shifts to favor females rather than males, resulting in sex role reversal (Gwynne and Simmons 1990;

Simmons 1992; Fosgren et al. 2004). To take this explanation for sex-role reversal into account, the definition of sexual selection now becomes modified, as Sexual Selection 1.2, to be the hypothesis that:

> SexS 1.2: The sex with the higher proportion of individuals available to mate, usually but not necessarily the male, spreads its genes most effectively by mating promiscuously whereas the other sex, usually but not necessarily the female, spreads its genes most effectively by selecting to mate with the genetically best partner, resulting in the evolution of ornaments and armaments in the promiscuous sex.

The existence of sex-role reversal breaks any necessary connection between gamete size and sex role, and sex role is determined by the OSR. Some local ecological argument must explain why the OSR is biased toward the male or female, with the default being a male-bias because of sperm size relative to egg size.

5.2.4 Sexual Selection 2.0: Shuker's Consensus Definition

The success of *any* role-based definition of sexual selection is challenged by further considerations, however. In many non sex-role reversed species, females have ornaments and armaments including secondary sexual characters such as brightly colored plumage or pelage, elaborate ornaments, and weaponry (Andersson 1994; Amundsen et al. 1997; Amundsen 2000; Kraaijeveld et al. 2007). Moreover, females do not lack for competition with one another, typically for access to breeding territories or resources needed for rearing offspring (Clutton-Brock 2009) and modes of competition among females have recently been summarized by Rosvall (2011) and Stockley and Bro-Jørgensen (2011). Furthermore, in many non sex-role reversed species male mate choice is as pronounced as female mate choice. Johnstone et al. (1996) point out that the logic behind the connection between choosiness and the OSR involves counteracting tendencies such that "nature abounds with bi-parental-care species in which both sexes are ostentatiously plumed or brightly colored". Moreover, Bonduriansky (2001) writing about male choice in insects, also concludes that non-sex-role reversed male choosiness (partial sex role reversal) is not as much related to the OSR as to existence of a large female mate quality variance present in "systems where female fitness increases with each copulation". These findings challenge the adequacy of SexS 1.2 that relies on the OSR as the criterion to determine sex roles.

Many other phenomena appear to belie any role-based version of sexual selection's definition (Roughgarden 2004, 2009). Monomorphism occurs in species where both male and female are identical in external appearance, leading to the question of whether dimorphism or monomorphism is the norm. Mating initiation and frequency is much more extensive than needed for offspring production, suggesting in cases not explained as sperm competition, that mating serves social purposes. Template (or gender) multiplicity refers to species in which there are more than one phenotype per sex, so a "norm" for the sex cannot be identified. Typical examples include species featuring a territorial male morph and a non-territorial male morph who somehow obtains copulations from the territorial male by mating with females in his territory or under his control. Transgender presentations pertain to species in which some individuals of one sex resemble the majority of individuals in the other sex, as in sunangel hummingbirds, suggesting that the phenotypic presentation of a male or female, including their ornaments, may be more related to signaling some social role other than a role directly in mating. Many species also feature extensive same-sex sexuality which again points to a social function of mating not related directly to fertilization. However, these diverse phenomena reviewed in Roughgarden (2004, 2009) are frequently downplayed as exceptions to sexual-selection's sex-role templates.

Studies have also appeared since 2004 on species that were not thought to be exceptional to standard sex-role templates but that nonetheless refuted sexual selection. Examples include the collared flycatcher in which a 24 year study of 8500 marked individuals found no genetic correlation between mate choice and ornament size (Qvarnström et al. 2006), the blue tit in which offspring fitness depended on environmental rather than genetic effects (Hadfield et al. 2006), the peacock in which peahens did not express any preference for peacocks with more elaborate trains similar to other studies of galliforms showing that females disregard male plumage (Takahashi et al. 2008) with further studies with peacocks continuing to report equivocal results (Hale et al. 2009; Dakin and Montgomerie 2011), a meta-analysis of extra-pair parentage in birds revealing that 60% show an absence of genetic benefits to extra-pair matings (Akcay and Roughgarden 2007). And the classic studies of Bateman (1948) purporting to demonstrate in *Drosophila* that males attain a higher fitness through quantity of mating whereas female attain a higher fitness through quality of mating have been thoroughly discredited by four independent critiques (Tang-Martinez and Ryder 2005; Dewsbury 2005; Snyder and Gowaty 2007; Gowaty et al. 2012).

Because of these difficulties with role-based definitions of sexual selection, Roughgarden (2005, 2007, 2009) argued that sexual selection as a general hypothesis about how mating behavior causes the evolution of ornaments and armaments had been falsified by the data, that it was "wrong", did not occur at all, and was a "myth". In response Shuker (2010) argued that behavioral ecology had long since moved beyond role-based definitions of sexual selection and that Roughgarden's critique was misdirected and misrepresented the current state of animal behavior research. However, Parker and Pizarri (Chap. 7, this volume) illustrate the hold that the traditional sex roles have on sexual selection theory.[2] Nonetheless, Shuker articulated a definition of sexual selection here labeled Sexual Selection 2.0:

[2] Parker and Pizarri endorse the Darwin-Bateman Paradigm (DBP) for typical male and female sex roles, contending not only that these sex roles are true but also are logically necessary, and that the Darwin-Bateman rationale for these sex roles provides "a powerful heuristic tool and the most parsimonious explanation for a bewildering diversity of biological patterns". Again I disagree. The Darwin/Bateman Paradigm is not heuristic. It has not helped to understand female choice or the great many species in which the classic DBP sex roles do not occur. Adherence to the DBP requires an elaborate system of fabricated mimicries and assumptions about deceit to account for animals who do not act, or do not send the right signals, in accordance with DBP expectations. As a

SexS 2.0 Sexual selection is the selection of traits arising from competition for mates.

Shuker went on to say that this definition was a "standard definition of sexual selection" and that in his perception, it "represents the consensus among evolutionary biologists of what sexual selection is." It remains to be seen whether this definition is a consensus. It is endorsed by Rosvall (2011) but not by Ah-King (2011) or Gowaty (2011). For my part, at the time it was proposed I endorsed this definition, primarily because, as detailed below, Shuker offered testable criteria to ascertain whether sexual selection is taking place.

Female choice *per se* is not sexual selection according to SexS 2.0, nor does female choice for direct benefits count as sexual selection because, as Shuker (2010) writes, sexual selection is "a component of an overall natural selection process".

SexS 2.0 is a considerable departure from previous definitions (hence the 2.0 designation). As Shuker writes, "sexual selection is not dependent on what have been termed 'sex roles' (who chooses, who competes and so on)." Theories about sex roles are now located in a subject termed mating systems theory. Shuker writes "a test of our understanding of mating systems, [is] not a test of the fundamental process of sexual selection. I reiterate this point because clearly our grasp of mating systems is not as complete as we perhaps thought". And, "I am aware that this separation may seem a little convenient, a nice way of hiding sexual selection away from some inconvenient truths". SexS 2.0 is insulated from discoveries about mating roles for either sex.

Now, some phenomena might be some consequence of competition for mates and thereby qualify as sexual selection under SexS 2.0 but that did not qualify as sexual selection under SexS 1.x. In particular, sperm competition leading to the correlation of testes size with number of females each male mates might be an example of SexS 2.0 and yet would be beyond the purview of SexS 1.x because testes size *per se* is typically neither an ornament nor armament (Harcourt et al. 1981; Birkhead 2000; Pizzari and Foster 2008; also Parker and Pizzari in this volume). However, because males who mate with many females have a higher demand their sperm supply, it is not clear if the testis size in such males is larger than that needed to supply the demand. Any excess in testis size beyond that needed to meet the demand is what would represent the outcome of sperm competition, and measuring such an excess would be difficult.

If SexS 2.0 is affirmed as the consensus definition, a task will be to acquaint the rest of the academy as well as the general public about this progressive step of disowning universal sex roles. SexS 1.0 is still accepted, I think, by social scientists including anthropologists and psychologists, as well as by philosophers including ethicists and theologians, as a statement of nature authorized by contemporary evolutionary science. SexS 1.0 is the basis of an essentialist biological theory of human genders that turns up in the popular media, nature shows, and dating services and is often criticized

dominant and hegemonic paradigm, the DBP has had a chilling effect on innovation. Furthermore, parsimony is not a principle of logic. Whether a proposition is true or not is independent of whether it is simple or complicated. A parsimonious hypothesis may be so simple as to be simplistic, as indeed the DBP may be judged.

within the humanities as a sexist Victorian relic. This very widespread misimpression about what evolutionary science says about mating should be rectified.

5.2.5 Problems with Sexual Selection 2.0

Improving the definition of sexual selection does not make it true. The truth of sexual selection, even by SexS 2.0, is an empirical matter, and the possibility always exists that some selection force other than competition for mates may account for the evolution of ornaments, armaments and other traits in particular cases. Moreover, Shuker (2010) offers two criteria sufficient for empirically falsifying SexS 2.0: "For sexual selection not to occur in a population, there either has to be no scope for competition (partners as resources are not limiting, and all partners are of equal quality), or the outcome of any competition for mates is totally random with respect to the traits expressed by individuals, such that successful partnerships represent a random sample of pairs of individual phenotypes (and thus genotypes)", here referred to as the "equal quality" and "random pairing" criteria respectively. Both of Shuker's criteria may often be satisfied.

For inter-sexual mate competition, choice by one sex, say females quickly weeds out any bad genes in males that are signaled by ornament attractiveness. Thereafter, the "equal quality" criterion is satisfied because the weeding of bad genes has caused males to become genetically equal. This equality of male genetic quality removes any benefit to females of continued choosiness. Unless bad genes in males can be continually replenished, female choice for male genetic quality should quickly disappear as a expensive waste of time and energy. This situation is called the "paradox of the lek". As of 2007 nine distinct hypotheses to circumvent the paradox of the lek had been proposed (Miller and Moore 2007). In 2008, two more attempts at circumvention appeared as well (Harris et al. 2008; Kotiaho et al 2008). The growing literature of the last dozen years aimed at "resolving" the paradox of the lek suggests the paradox may be difficult or impossible to resolve.

The "random pairing" criterion is also satisfied because according to population-genetic theory, the fitness difference between a "good" and "bad" male resulting from the accumulation of weakly-deleterious mutations is too small to be detectable (Roughgarden 2009, pp. 53–54). Hence pairing will be at random with respect to the weakly-deleterious mutation load. Of course, strongly deleterious mutations are immediately detectable and do not require revelation through the display of ornaments or use of armaments.

The possibility that SexS 2.0 may not explain ornaments, armaments and courtship behavior generally invites consideration of additional hypotheses for these phenomena. One approach advocated by Prum (2010) as a "null model" argues that the ornaments are not adaptive. A preference for an arbitrary trait, say red color, can feed on itself so to speak, leading to the evolution of exaggerated characters combined with a strong preference for such characters—a "Fisherian runaway process" (Fisher 1915, 1930). As Prum write, "An arbitrary trait is neither honest nor dishonest because it does not communicate any information that can be untruthful or falsified. Arbitrary traits do not indicate anything other than availability to mate." Prum argues that the colorful ornaments in manakins are consistent with a Fisherian runaway process. Although the Fisherian runaway process has received theoretical validation (Lande 1981; Kirkpatrick 1982) it has also been extensively criticized, and Prum acknowledges "There is not a single, generally accepted, textbook example of an arbitrary secondary sexual trait in any organism". If a runaway process accounts for an exaggerated male trait, then the female preference for that trait should be correspondingly strong. Yet where data exist, such as for the collared flycatcher mentioned earlier (Qvarnström et al. 2006), female preference for the ornament (badge size in males) is not heritable and the genetic correlation between female choice and male badge size is zero, suggesting that the ornament in this species has nothing to do a runaway process.

5.2.6 Sexual Selection 3.0: NESCent Definition #3

To clarify the definitional status of sexual selection, in July 2013 I organized a catalyst meeting at the National Evolutionary Synthesis Center (NESCent) in Durham, North Carolina. I coined the phrase "sexual selection studies" to serve as an umbrella referring to studies of courtship and mating, parent-offspring relations, family organization, and the interrelations among these. The meeting was attended by 34 participants reflecting a diversity of ages, nationalities, and disciplines, some of whom are contributors to the present volume. Two thirds of the participants brought special experience from their research and teaching in some area of sexual selection studies and one third brought perspectives from other areas of evolutionary biology and from the social sciences and humanities. The participants were not able to arrive at a consensus definition of sexual selection, did not accept Shuker's definition (SexS 2.0), and disagreed on many issues pertaining to sexual selection as recorded in the meeting's final report (Roughgarden et al. 2013).

Here I wish to bring forward and endorse one of the definitions for sexual selection proposed by some of the meeting's participants:

> SexS 3.0 Sexual selection is the component of selection that results from differential fertilization success among genotypes within a sex and that does not change total fertility.

This definition does not specify sex roles, does not specify the identity of the processes that might produce sexual selection, and refers to fertilizations and not matings. This later feature implicitly recognizes that many matings do not result in fertilizations and that the function of many instances of mating may be social rather than to effect a fertilization.

Of the definitions considered by the NESCent group, this definition is the most explicitly genetical. It regards sexual selection as a component of a genetical selection process. In this definition the differences in fertilization success are assignable to different genotypes, implying that the differences are heritable. In this definition, the presence or absence of sexual selection is assayed in terms of genetical change and not in terms solely of behavior.

Most importantly, this definition introduces the distinction between a frequency-dependent process wherein genotypes differ in their share of a fixed reproductive pie *vs.* a density-dependent process wherein genotypes differ in the size of their reproductive pie. According to SexS 3.0, sexual selection pertains solely to competition for a larger share of a fixed pie, and not to changing the size of the pie.

This definition is consistent with a possible reading of Darwin's (1871) passages intended to clarify the distinction between sexual selection and natural selection: "The males of many oceanic crustaceans have their legs and antennae modified in an extraordinary manner for the prehension of the female; hence we may suspect that owing to these animals being washed about by the waves of the open sea, they absolutely require these organs in order to propagate their kind, and if so, their development has been the result of ordinary or natural selection" [p. I.256]. But, "if the chief service rendered to the male by his prehensile organs is to prevent the escape of the female before the arrival of other males, or when assaulted by them, these organs will have been perfected through sexual selection, that is by the advantage acquired by certain males over their rivals. But in most cases it is scarcely possible to distinguish between the effects of natural and sexual selection" [p. I.257].

In Darwin's first scenario, grasping organs that prevent being washed out to sea while mating in ocean surf correspond to the outcome of fertility selection (increasing the size of the pie). In Darwin's second scenario, grasping organs that allow monopolizing a female to exclude other males correspond to the outcome of sexual selection. Darwin lumps fertility selection together with viability selection to form natural selection. So, in definition SexS 3.0, sexual selection is distinct from fertility selection, and thereby from natural selection as well, but both sexual selection and natural selection are still components of a common overall genetical selection process. If this definition becomes widely adopted, new statistical methodology will be needed to partition the overall genetical selection process into its sexual-selection, fertility-selection, and viability-selection components.

I single this particular definition out of those considered at the NESCent meeting because the approach I have been advocating focusses on increasing the size of the pie—it is about fertility selection rather than sexual selection. As such, it offers an alternative to sexual selection to account for phenomena involved in courtship and parent-offspring relations.

5.3 The Social Selection Alternative

I have sketched a new and different approach to determine the adaptive function of ornaments and armaments that I have termed "social selection", a contraction from the longer phrase "social infrastructure selection" and not to be confused with "social selection" *sensu* West-Eberhard (1983) who uses the phrase as a synonym for "social competition". Social infrastructure selection hypothesizes, to borrow Darwin's words, that the "chief service rendered" by reproductive social behavior is to fashion a social infrastructure that produces an increased number of offspring.

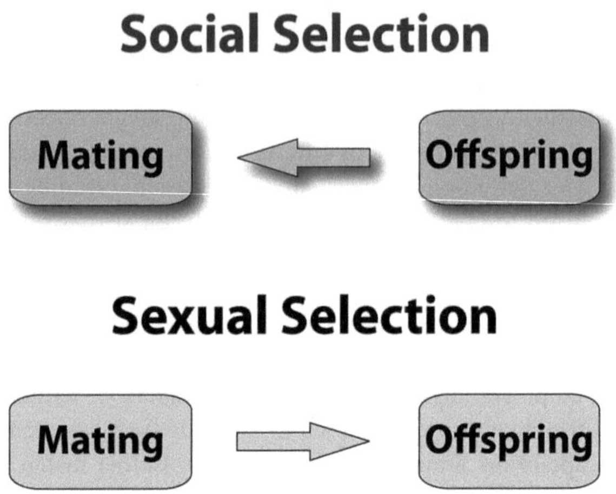

Fig. 5.1 Direction of inference for mating systems theory

Social selection theory is developed differently from sexual selection in two respects, as diagrammed in Figs. 1 and 2 taken from Roughgarden (2012a). First, the modeling of mating systems starts from the stage of offspring production and works back from there to earlier life-history stages, including courtship. Second, the modeling of behavior starts with cooperative as well as competitive game theory and works up from there to gene pool dynamics. The logical progression in social selection is the opposite to that in sexual selection and is motivated by the modeling approaches in ecology for life-history theory and for mechanism-based and individual-based models for population dynamics.[3]

Concerning Fig. 5.1, social selection views mating systems as a stage in an optimal life history. Biologically, this approach would first consider the local ecological situation into which the offspring are released from their parents and determine what actions by them will yield to each the largest number of its offspring at that time. Then second, the ornaments, signaling and negotiation needed during courtship to bring about those actions would be calculated. Thereafter, once these solutions were in hand, one could then play the predictions forward and observe the optimal courtship followed by the optimal parental care.[4]

[3] Gowaty, (Chap. 6 this volume) contends that social selection is not an alternative to sexual selection. She sees sexual selection as a subset of social selection not an alternative. Gowaty is confusing my idea of social *infrastructure* selection with social selection *sensu* West-Eberhard (1983). Of course, sexual selection is a subset of West-Eberhard's social selection because her idea of social selection was explicitly posed as a generalization of sexual selection to begin with. Sexual selection is obviously not a subset of social *infrastructure* selection as diagrammed in Figs. 1 and 2.

[4] As Gowaty (Chap. 6, this volume) notes, I have suggested in particular that " 'fancy traits, such as some elaborate secondary sexual characteristics, like … a peacock's tail, a rooster's comb, wattle and cockle-doddle-do facilitate male-male interactions' rather than male-female interactions posited by Darwin". Specifically, the ornaments may function as admission tickets to power holding cliques. She goes on to state that "I can imagine testing a classical idea about male-male competitive dynamics over number of mates in the absence of cliquey coalitions and male-male competitive dynamics in the presence of cliquey coalitions." I would welcome such a test.

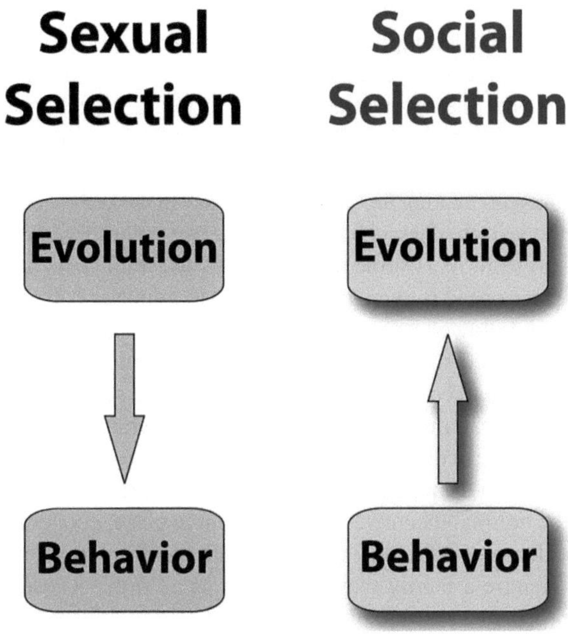

Fig. 5.2 Top-down vs. bottom-up modeling of social evolution

Turning to Fig. 5.2, social selection envisions a bottom-up logic for the evolution of behavior. Population ecology has successfully derived equations for population dynamics from properties of individuals, an approach termed variously, "individual-based models", "individual oriented models", "agent-based models", or "mechanism-based models". Population-dynamic models that are developed bottom up from individuals to populations offer a more relevant and testable theory than the venerable equations of population ecology such as the logistic or Lotka-Volterra equations that feature a top-down logic. One can think of Fig. 2 as illustrating separate dynamics for the phenotypic and genotypic levels of organization in evolution. Rather than thinking of the phenotype as a static product of the genotype, in social selection the phenotype is assumed to have its own dynamic, not only the solitary dynamic realized during development and morphogenesis, but also a social dynamic carried out within the lifespans of the interacting individuals. This interactive phenotypic dynamic shapes the properties and capabilities of the animals just as surely as morphogenesis shapes anatomy. Yet the rules for the dynamical process by which the phenotypes interact ultimately derives from the genotype with its own separate dynamics, and indeed these levels are coupled with feedback to and from each other. However, the dynamics of phenotype interaction need not, and presumably generally does not, mirror the dynamics of gene pool. In particular, the competitive dividing-the-pie dynamics of evolution in a gene pool need not be mirrored in the dynamics among phenotypes that may involve expanding the pie as much or more than dividing the pie.

Two schemes of social organization seem particularly relevant to courtship, mating, parental care and family life: a "team" and a "firm".

A "team" consists of participants who take coordinated actions to achieve a team objective. During teamwork, animals are hypothesized to experience pleasure from physical and/or vocal intimacy. The pursuit of this pleasure is hypothesized to be the proximal motivation for animals to participate in teamwork (Roughgarden et al. 2006; Roughgarden 2009, 2012b). It is further hypothesized that the act of cooperation itself is pleasurable. As a human analogy, consider the difference in pleasure between making an "Alley-Oop pass" in basketball compared with making two foul shots, both of which yield the same two points. Of course every member of a basketball team feels some pleasure when two foul shots are successful, but the pleasure experienced is even greater if the two points are obtained with a beautiful acrobatic pass followed by a teammate's dunk at the basket.

A firm is a collection of animals working together because one member is in a position to control incentives that align the self-interests of others to cooperate. For example, parents and offspring form a hierarchy in which parents control the food. Parents can dispense incentives that lead the offspring to cooperate with one another and with the parents themselves (Roughgarden and Song 2014; Akçay 2012). When the optimal incentives have been put in place, cooperation occurs because of a coincidence of individual interests. Each member's objectives are purely self-regarding. Because a family may be viewed as a "firm" whose product is offspring it is appropriate to turn to management science and the economic theory of the firm to see how a family might be organized to maximize its offspring production.

The family-firm approach confirms predictions that offspring should signal honestly to their parent (Grafen 1990; Godfray 1991; Godfray and Johnstone 2000), but offers a different derivation, and shows that if the parent implements "optimal incentive policies", the parent and chicks work honestly together as a team to maximize the fitness produced by the nest. Unlike genetic approaches, the family-firm approach in the behavioral tier can be extended to predict the time of weaning or fledging—this amounts to the time for "spinning off" divisions from the parent corporation. This approach agrees with Alexander's (1974) perspective that emphasizes parental control of the parent-offspring interaction. In contrast to Alexander however, who visualized continuing parent offspring conflict dominated by the parent, according to this theory the parent resolves conflict by how it allocates resources to the offspring. This theory fundamentally disagrees with the Trivers/Parker/Mcnair perspective (Trivers 1972; Parker and Mcnair 1978, 1979; Mcnair and Parker 1978, 1979) which features unresolved and continuing parent-offspring conflict as well as dishonesty in parent-offspring signaling whereby offspring continually psychologically manipulate their parents, especially at the time of weaning or fledging.

A biological firm is not necessarily harmonious as the preceding discussion might suggest. For example, in some ecological circumstances the offspring might know their value to the parent and be able communicate it to the parent. (The parent might value the size of the offspring and the offspring might know its own size and thus be able to communicate it.) If so, a parent's best interest would be served by setting incentives to ensure honest communication and to minimize the inefficiency of lying and conflict. In other circumstances offspring might not know their value to the parent and thus not be able to communicate it to their parent even if they were attempting an honest report. (The parent might value the prospective resource capturing ability of an offspring, which would be unknown to the offspring itself until it learned how to forage effectively.) In this case a parent's best interest would be served by allowing conflict between itself and the offspring and competition between the sibs with one another to reveal their capabilities. The inefficiency of conflict would be the cost to the parent of obtaining information about the offsprings' value to it that would not otherwise be knowable. This contrast between whether the information about their value to the parent is communicable or not is perhaps a better way to view the distinction between "offspring control" and "parent control" (Parker and Macnair 1978, 1979; Macnair and Parker 1978, 1979) because in either case, the parent does control the resources made available to the offspring. However, if the offspring know their what the parent values about them then the parent's interest is served by setting incentives for honest communication whereas if the offspring are ignorant of their value to the parent, the parent's interest is served by witnessing the outcome of offspring squabbling.

Pleasure-based teamwork and incentive-structured firms offer mechanisms by which cooperation can evolve as a direct benefit. That is, cooperation realized through teamwork or working in a firm is not altruism, and its evolution is consistent with, but does not require, kin/multilevel selection or other evolutionary processes that cause the evolution of traits that benefit the receiver but disadvantage the donor. Mutual direct benefits, including pleasure-based teamwork and working in incentive-structured firms, may be more important overall in explaining cooperative behavior than altruism-based explanations.

Overall, the underlying rationale of parental teams and parent-offspring firms is to increase the number of offspring produced. In this sense, social selection is focussed on increasing the size of the pie. Social selection interprets the behavior pursuant to these relationships in terms of fertility selection rather than sexual selection, which is focussed on strategies to monopolize ownership within a reproductive pie of fixed size. Of course, the pattern of reproductive social behavior in a particular species might include components both social selection (fertility selection) as well as sexual selection (frequency-dependent selection), but the supposition to the research program of social selection is that the importance of sexual selection has been greatly exaggerated, in large part because of the lack of an alternative hypothesis to consider.

5.4 Conclusions

This conference has asked whether anything is now left of sexual selection. My answer is no, nothing is left of the original sexual selection. The sequence of biological definitions of sexual selection makes clear that Darwin's original narratives of male and female sex roles and his interpretation of the motivations behind their actions during courtship have been largely abandoned by researchers from the 1970's on.

The deeper question before us is whether sexual selection even offers a useful approach to the study of mating and other aspects of reproductive social behavior. Is the best starting point the mating episode in an organism's life history, or the offspring-rearing episode? If the former, then allegiance to a sexual selection approach would appear worthwhile. If the later, then sexual selection should be abandoned as a scientific relic that has outlived its usefulness. Similarly, should social behavior strategies still be framed as games between genotypes whose outcome is visualized in terms of an Evolutionarily Stable Strategy (ESS, Maynard Smith 1982), or should social behavior strategies be seen from the perspective of cooperative as well as non-cooperative game theory and incorporate conceptualizations from the theory of the firm in economics? If former, then the present top-down approach to social-behavior theory still has legs. If the later, then a new generation of bottom-up theories for the development and evolution of social behavior can be anticipated.

The abandonment of sexual selection in its original definition signifies a healthy and vibrant discipline. What I have termed "sexual selection studies" offers an umbrella under which the hypotheses that replace the original sexual selection may reside.

Acknowlegements I thank Thierry Hoquet for organizing the conference and for editing this volume. I also thank him for his translation of *The Genial Gene* into French. I thank Erol Akçay, Priya Iyer and Jeremy Van Cleve for their insight and counsel over the years.

References

Ah-King M (2011) Female sexual selection in light of the Darwin-Bateman paradigm. Behav Ecol. doi:10.1093/beheco/arr109

Akçay E (2012) Incentives in the family II: Behavioral dynamics and the evolution of non-costly signaling. J theor Biol 294:9–18

Akçay E, Roughgarden J (2007) Extra-pair paternity in birds: review of the genetic benefits. Evol Ecol Res 9:855–868

Alexander RD (1974) The evolution of social behaviour. Ann Rev Ecol Syst 5:325–383.

Amundsen T, Forsgren E, Hansen L (1997) On the function of female ornaments: male bluethroats prefer colourful females. Proc R Soc Lond B 264:1579–1586

Amundsen T (2000) Why are female birds ornamented? Trends. Ecol Evol 4:149–155

Andersson M (1994) Sexual Selection. Princeton University, Princeton

Bateman AJ (1948) Intra-sexual selection in *Drosophila*. Heredity 2(3):349–368

Birkhead TR (2000) Promiscuity: an evolutionary history of sperm competition. Harvard University, Cambridge

Bonduriansky R (2001) The evolution of male mate choice in insects: a synthesis of ideas and evidence. Biol Rev 76:305–339

Coyne J (2004) "Charm schools" (Review of *Evolution's Rainbow* by J. Roughgarden). *Times Literary Supplement*. 30 July 2004.

Clutton-Brock T (2009) Sexual selection in females. Anim Behav 77(1):3–11

Dakin R, Montgomerie R (2011) Peahens prefer peacocks displaying more eyespots, but rarely. Anim Behav 82:21–28

Darwin C (1871) The descent of man and selection in relation to sex. Volumes I. and II. John Murray, London

Darwin C (1874) The descent of man. Cromwell, New York

Dawkins R (1976) The selfish gene. Oxford Univ., Oxford
Dewsbury D (2005) The Darwin-Bateman paradigm in historical context. Integr Comp Biol 45:831–837
Emlen S, Oring LW (1977) Ecology, sexual selection, and the evolution of mating systems. Science 197:215–223
Fisher RA (1915) The evolution of sexual preference. Eugenics Rev 7:184–191
Fisher RA (1930) The genetical theory of natural selection. Clarendon, Oxford
Fosgren E, Amundsen T, Borg AA, Bjelvenmark J (2004). Unusually dynamic sex roles in a fish. Nature 429:551–554
Godfray HCJ (1991) Signaling of need by offspring to their parents. Nature 352:328–330.
Godfray H, Johnstone R (2000) Begging and bleating: the evolution of parent-offspring signalling. Phil Trans R Soc Lond B 355:1581–1591
Gould S, Lewontin R (1979) The spandrels of San Marco and the Panglossian paradigm: a critique of the adaptationist programme. Proc R Soc Lond B Biol Sci 205(1161):581–598
Gowaty P (2011) What is sexual selection and the short herstory of female trait variation. Behav Ecol. doi:10.1093/beheco/arr113
Gowaty P, Kim Y, Anderson W (2012) No evidence of sexual selection in a repetition of Bateman's classic study of *Drosophila melanogaster*. Proc Nat Acad Sci U S A 109:11740–11745. www.pnas.org/cgi/doi/10.1073/pnas.1207851109
Grafen A (1990) Biological signals as handicaps. J Theor Biol 144:517–546
Gwynne DT, Simmons LW (1990) Experimental reversal of courtship roles in an insect. Nature 346:172–174
Hadfield JD, Burgess MD, Lord A, Phillimore AB, Clegg SM, Owens IPF (2006) Direct versus indirect sexual selection: genetic basis of colour, size and recruitment in a wild bird. Proc R Soc B 273:1347–1353
Hale M, Verduijn M, Møller A, Wolff K, Petrie M (2009) Is the peacock's train an honest signal of genetic quality at the major histocompatibility complex? J Evol Biol 22:1284–1294
Harcourt AH, Harvey PH, Larson SG, Short RV (1981) Testis weight, body weight and breeding system in primates. Nature 293:55–57
Harris WE, McKane AJ, Wolf JB (2008) The maintenance of heritable variation through social competition. Evolution Int J org Evolution 62:337–347
Johnstone R, Reynolds J, Deutsch J (1996) Mutual mate choice and sex differences in choosiness. Evolution 50:1382–1391
Kirkpatrick M (1982) Sexual selection and the evolution of female choice. Evolution Int J org Evolution 82:1–12
Kotiaho JS, LeBas NR, Puurtinen M, Tomkins JL (2008) On the resolution of the lek paradox. Trends Ecol Evol 23:1–3
Kraaijeveld K, Kraaijeveld-Smit F, Komdeur J (2007) The evolution of mutual ornamentation. Anim Behav 74:657–677
Lande R (1981) Models of speciation by sexual selection on polygenic traits. Proc Natl Acad Sci USA 78:3721–3725
Macnair MR, Parker GA (1978) Models of parent-offspring conflict. II. Promiscuity. Anim Behav 26:111–122
Macnair MR, Parker GA (1979) Models of parent-offspring conflict. III. Intra-brood conflict. Anim Behav 27:1202–1209
Maynard Smith J (1982) Evolution and the theory of games. Cambridge University Press, Cambridge
Mayr E (1963) Animal species and evolution. Harvard University Press, Cambridge
Millstein R (2011) Sex and sensibility: The role of social selection (book symposium on Joan Roughgarden's *The genial gene*). Metascience 20:258–264
Milam E (2010) Looking for a few good males: female choice in evolutionary biology. Johns Hopkins University Press, Baltimore
Miller CW, Moore AJ (2007) A potential resolution to the lek paradox through indirect genetic effects. Proc R Soc B 274:1279–1286

Parker G, Baker R, Smith V (1972) The origin and evolution of gamete dimorphism and the male-female phenomenon. J Theor Biol 36:529–553
Parker G, Macnair M (1978) Models of parent-offspring conflict. I. Monogamy. Anim Behav 26:97–110
Parker G, Macnair M (1979) Models of parent-offspring conflict. IV. Suppression: evolutionary retaliation by the parent. Anim Behav 27:1210–1235
Pizzari T, Foster KR (2008) Sperm sociality: cooperation, altruism, and spite. PLoS Biol 6:e130
Prum RO (2010) The Lande-Kirkpatrick mechanism is the null model of evolution by intersexual selection: implications for meaning, honesty, and design in intersexual signals. Evolution 64:3085–3100
Qvarnström A, Brommer JE, Gustafsson L (2006) Testing the genetics underlying the co-evolution of mate choice and ornament in the wild. Nature 441:84–86.
Rosvall K (2011) Intrasexual competition in females: evidence for sexual selection? Behavioral Ecology. doi:10.1093/beheco/arr106
Roughgarden J (2004) Evolution's rainbow: diversity, gender and sexuality in nature and people. University of California Press, Berkeley CA
Roughgarden J (2005) The myth of sexual selection. Calif Wild 53:18–23
Roughgarden J (2007) Challenging Darwin's theory of sexual selection. Daedalus 136:1–14
Roughgarden J (2009) The Genial Gene: Deconstructing Darwinian Selfishness. University of California, Berkeley
Roughgarden J (2012a) The social selection alternative to sexual selection. Phil Trans Roy Soc B doi:10.1098/rstb.2011.0282
Roughgarden J (2012b) Teamwork, pleasure, and bargaining in animal social behavior. J Evol Biol doi:10.1111/j.1420-9101.2012.02505.x
Roughgarden J et al (2013) Sexual Selection Studies: Progress, Challenges, and Future Directions. Final Report from a NESCent Catalyst Meeting. Durham NC, July 15–17, 2013. Revision 1.1, December 11, 2013
Roughgarden J, Oishi M, Akçay E (2006) Reproductive social behavior: cooperative games to replace sexual selection. Science 311:965–969
Roughgarden J, Akçay E (2010) Do we need a sexual selection 2.0? Anim Behav 79:E1–E4
Roughgarden J, Iyer P (2011) Contact, not conflict, causes the evolution of anisogamy. In: Togashi T, Cox P (eds) The evolution of anisogamy: a fundamental phenomenon underlying sexual selection. Cambridge UK, Cambridge University Press, pp 96–110
Roughgarden J, Song Z (2014) Incentives in the family I: The family firm, an evolutionary/economic theory for parent-offspring relations. In: Narvaez D, Valentino K, Fuentes A, McKenna J, Gray P Ancestral landscapes in human evolution: culture, childrearing and social wellbeing. Oxford University Press, New York, pp 221–240
Shuker D (2010) Sexual selection: endless forms or tangled bank? Anim Behav 79:E11–E17
Simmons LW (1992) Quantification of role reversal in relative parental investment in a bush cricket. Nature 358:61–63
Snyder B, Gowaty, PA (2007) A reappraisal of Bateman's classic study of intrasexual selection. Evolution Int J org Evolution 61:2457–2468
Stockley P, Bro-Jørgensen J (2011) Female competition and its evolutionary consequences in mammals. Biol Rev 86:341–366
Takahashi M, Arita H, Hiraiwa-Hasegawa M, Hasegawa T (2008) Peahens do not prefer peacocks with more elaborate trains. Anim Behav 75:1209–1219
Tang-Martinez Z, Ryder TB (2005) The problem with paradigms: Bateman's worldview as a case study. Integr Comp Biol 45:821–830
Tickell WLN (2000) Albatrosses. Yale University Press, New Haven Conn
Trivers RL (1972) Parental investment and sexual selection. In: Campbell B (ed) Sexual selection and the descent of man. Aldine Publishing, Chicago, pp 136–179
Wallace AR (1871) Darwin's 'The descent of man and selection in relation to sex'. The Acad 1871:177
West-Eberhard MJ (1983) Sexual selection, social competition and speciation. Quart Rev Biol 58:155–183

Chapter 6
Standing On Darwin's Shoulders: The Nature of Selection Hypotheses

Patricia Adair Gowaty

Abstract "Standing on Darwin's Shoulders" is about the information we need to reject or confirm selection hypotheses. Sexual selection remains important, and contrary to very old and recent claims that Darwin's ideas cannot explain the origins of fancy traits, Darwin was not wrong, even if other hypotheses might also explain fancy traits. The problem seems to be that many readers seem to lack appreciation for the power of a good, that is, a well-constructed, testable hypothesis. To be a good hypothesis does not mean that it needs to be the truth, but only testable (or potentially testable). My main point is that what's needed to resolve the recent state of conflict over selection of whatever name is to say as precisely as possible the details of the requisite assumptions so we can get on with testing the fundamentals of how selection works. The fundamentals of selection hypotheses are assumptions about (1) heritable variation among the units of selection, (2) the environmental and/or social circumstances that (3) affect differential probabilities of survival and reproductive success of the units of selection. Selection is simple. For example, if traits of individuals are variable, if traits are heritable, if environments (abiotic, biotic, social, etc.) vary so that some individuals have a higher probability of survival or reproduction in the environment they experience because of their trait variation, natural selection has occurred. Likewise, if individuals within a sex in the same species vary, if traits are heritable, if social environments vary so some individuals have more or better mates than others and thus greater probabilities of reproductive success, sexual selection has occurred. Thus, sexual selection hypotheses are not "systems of belief" or an associated chain of cascading ideas as Roughgarden recently claimed, but rather deductive hypotheses, admirably vulnerable

---

"the manner in which the individuals of either sex or of both sexes are affected through sexual selection cannot fail to be complex in the highest degree" (Darwin 1871, t. I, p. 296).

P. A. Gowaty (✉)
Department of Ecology and Evolutionary Biology and Institute of Environment and Sustainability, UCLA, Los Angeles, CA, USA
e-mail: gowaty@eeb.ucla.edu

Smithsonian Tropical Research Institute, Washington, DC, USA

Polistes Foundation, Belmont, MA, USA

T. Hoquet (ed.), *Current Perspectives on Sexual Selection,* History, Philosophy and Theory of the Life Sciences 9, DOI 10.1007/978-94-017-9585-2_6

to empirical tests of within sex differential fitness associated with "reproductive competition"—a hypothesis against which we can test nature. If we see more or more clearly than Darwin, it is because we stand on his broad and high shoulders!

Keywords Natural selection · Sexual selection · Hypothetico-deductive hypotheses · Darwin

6.1 Introduction

If creative scholars are lucky enough to garner any attention at all, their ideas are often controversy prone. Perhaps few ideas have been so consistently controversial as Darwin's about sexual selection. Even the co-discover of natural selection (Darwin and Wallace 1858), Alfred Russel Wallace differed with Darwin (Wallace 1889) on the on the evolutionary orgins of bright coloration (see Hoquet and Levandowsky, this volume). The conference 'What's left of sexual selection?' that inspired this book was organized to evaluate and rebut recent slings and arrows directed at Darwin that had been bubbling up since the publication of *Evolution's Rainbow* (Roughgarden 2004) and the author's claim that Darwin was wrong. One could be cynical about the origin of all the recent fuss (Roughgarden et al. 2006b) including sometimes even the perspectives of modern darwinians (Buss 2006; Dall et al. 2006; Day et al. 2006; Ghiselin 2006; Hurd 2006; Lessells et al. 2006; Miller 2006; Pizzari et al. 2006; Roughgarden et al. 2006a; Shuker and Tregenza 2006; Stewart 2006). Critics and defenders could just be defending turf, miffed that others are insufficiently familiar with their modern scholarship. Less cynically, there could actually be something wrong with the claim Roughgarden made, when she cried, "Darwin was wrong". I favor the last explanation because, as I discuss below at length, offering an alternative hypothesis does not invalidate an existing hypothesis.

To know what it means when someone says "Darwin was wrong about sexual selection", we need to know what sexual selection is.

Does not everyone know by now what sexual selection is? Possibly not, as there was a 2010 symposium at the International Society for Behavioral Ecology meeting in Perth called 'What is sexual selection?' Given that set of experts who were expected to know what sexual selection is, it is not so surprising that Roughgarden and her critics were also worked up about what is sexual selection and what counts as an alternative to sexual selection. It would appear from all the noise that there is something difficult to grasp. Yet, recasting recent controversies over hypotheses of "sexual selection" and "not sexual selection" in terms of their assumptions (premises, first-principles) exposes the considerable strengths of most selection hypotheses (Gowaty 2011). Keeping assumptions of ideas in the foreground points out ways to explicitly test the veracity of specific first-principle, deductive hypotheses. This essay is about the information we need to reject or confirm any selection hypothesis. What I attempt to clarify are the semantic minefields of some of the recent high profile discussions that motivated this volume, but which have implications for how we think about and test ideas about selection.

Among the things I emphasize are that (i) Modern selection hypotheses are more often than not well-formulated, testable first-principle deductive statements in the best sense of the Darwinian tradition. (ii) Darwin gave us the formula for constructing novel selection hypotheses, perhaps even ones he did not think of and ones about problems he explicitly put off to the future. (iii) Darwin also gave us more than one hypothesis ("definition') of sexual selection, (iv) suggesting that he, at least, knew that his first-principle deductive approach for framing selection hypotheses was robust to variation in the required assumptions, which is (v) one of the reasons that modern-day evolutionists, standing on Darwin's shoulders, see more clearly all the time. (vi) The Cheshire Cat (Carroll 1865) has been at work, so that our naming conventions for selection hypotheses are not standardized, providing opportunities for semantic confusion and a lot of mischief. (vii) What might resolve the recent state of conflict over selection of whatever name is to say, formally state, as precisely as possible the details of the requisite assumptions so we can get on with testing the fundamentals of how selection works.

Doing that simple thing—attending to assumptions–facilitates the empiricist's path. When one examines the premises (assumptions, first principles) of deductive hypotheses of any sort, it is obvious that the assumptions provide the pathways to potential tests and possible rejection of ideas. A side-by-side list of the premises of different selection hypotheses (Table 6.1) reveals some arguments as being about naming conventions, rather than about the utility or power of a given explanation for this or that. It seems we have been arguing about what they, you, or I *call* a particular hypothesis. Arguments over naming conventions tell us nothing about the strengths and weaknesses of the assumptions of any hypothesis. The muddle of

Table 6.1 Darwin's formal statements of selection

Assume variation among the units of the level of selection	*Assume* environmental circumstances of advantage or disadvantage to units of the level of selection because of between unit variation	*Assume* advantages and disadvantages affect component(s) of fitness	*Predict* changes in particular traits or in functional categories of traits	"Name"
Within- population variation among *individuals*	Access to resources	Survival	Traits facilitating resource acquisition	Broad-sense natural selection
Within-population variation in *males only*	"Struggle" between the males for possession of the females"	RS	Traits facilitating "possession of females"	Narrow-sense sexual selection (Darwin 1859, p. 103)
Within population variation confined to one sex	Within-sex "struggles" over reproductive advantage/disadvantage	RS	Traits facilitating behavioral or physiological within-sex "reproductive contests"	Broad-sense sexual selection (Darwin 1871, t. I, p. 256)

naming conventions has become largely a distraction keeping us from the task at hand: How does selection operate?

To make my arguments, I review (1) the inferential chain in the development of Darwin's ideas about natural selection, (2) Darwin's narrow and broad sense definitions of natural and sexual selection, and the meanings of (3) units of selection, (4) the social circumstances of selection, (5) and the components of fitness affected by selection. I illustrate my points with (6) some modern views of selection hypotheses, (7) including Fisher's astute but largely forgotten idea about benefits for choosers, and (8) try to show how all these forgotten or overlooked contributions resulted in recent controversy.

6.2 The Origins of Darwin's Selection Hypotheses

Consider Mayr's (1977) description of the chronology of intellectual experiences that fueled Darwin's imagination and his discovery of the theory of evolution via natural selection. First, Darwin moved from a typological, Platonic ideal of species to a belief in the "importance of individual differences and the reality of the variation within a population". Thus came the first principle of selection hypotheses: *Assume variation among the units in the level of selection of interest,* where levels of selection may be groups, species, populations, demes, sexes, individuals, genes, etc.). Darwin also wrote much about "the struggle for existence", which he inferred from the principles of Malthus and the mathematics of high intrinsic reproductive rates that never went to their logical conclusion, e.g., the planet is not covered in elephants. Darwin used these facts of nature to buttress his assumption that the circumstances of nature/environment challenged individuals, so that many individuals died or failed to reproduce. He thereby inferred there was a struggle for existence and that the environmental and social circumstances associated with "struggle" were seemingly endless. One might characterize *On the Origin of Species by Means of Natural Selection, or the Preservation of Favoured Races in the Struggle for Life* (Darwin 1859) as a description of the many ways that animals "struggle for" resources, space, protection from the elements, to avoid predation or disease, or sometimes against changed conditions. Darwin often used the term "wedge" to refer to environmental or social circumstances that correlated with differential survival or reproduction probabilities among individuals. Thus, we have the second and third necessary assumptions that characterize all selection hypotheses: *Assume an environmental circumstance that correlates with a component of fitness to affect differential survival or reproduction probabilities among the units within the level of selection.* Of course, for much of Darwin's discussion he was interested in variation among individuals within a population: the level of selection being the individuals (see Ruse, this volume).

Thus as moderns have come to see Darwin's ideas about natural and sexual selection, (Darwin 1859, 1871) the hypotheses are logical, rule-like statements with assumptions about the units of selection, the environmental (or social) circumstances that affect the components of fitness so that differential survival or reproduction occur. When we do as Darwin did and assume that traits are variable and heritable, any selection hypothesis becomes one of evolution via selection.

6.3 Darwin's Broad and Narrow Definitions of Natural and Sexual Selection

Darwin said "natural selection: a power which acts solely by the preservation of profitable variations in the struggle for life". This is an extremely broad declaration fitting the realized ambition of Darwin's logic about natural selection.

Darwin's (1859, 1871) assertions about sexual selection are particularly interesting given modern debates about what is and what counts as sexual selection. As I noted previously (Gowaty 2011, p. 1146): In 1859, Darwin's definition of sexual selection was an extremely narrowed definition of natural, not artificial, selection. The units of selection, the environments (physical, biotic, or social) that affect components of fitness were each narrowed with respect to the usual statements of natural selection. Remember that in 1859, he was making distinctions between artificial selection and natural selection, and his topic was not sexual selection per se. He differentiated sexual selection as a type of selection having to do with selection within a sex rather than within all individuals of a population. "Sexual Selection… depends, not on a struggle for existence, but on a struggle between the males for possession of the females; the result is not death to the unsuccessful competitor, but few or no offspring" (Darwin 1859, p. 88).

In 1871, Darwin was older and wiser, and his definition then of sexual selection was undeniably more mature and much broader than his 1859 definition. The more mature definition was in fact the first broad-sense definition of sexual selection: "Sexual selection…depends on the advantage which certain individuals have over other individuals of the same sex and species, in exclusive relation to reproduction" (1871, t. I, p. 256). Here he expanded the scope of sexual selection when he included environmental circumstances yielding individual advantage/disadvantage relative to reproduction, not just competition over numbers of mates. Notably, he also expanded the application of the term "within-sex selection" to include not just reproductive advantages/disadvantages among males, but among females. And, just to keep the ledger straight: Broad-sense sexual selection is a narrow-sense natural selection hypothesis (i.e., not artificial selection), as Darwin (1871) made clear. Table 6.1 contains Darwin's definitions about what counts as selection from the broadest natural selection hypothesis to more restricted sexual selection hypotheses. These are a continuum as Darwin called "sexual selection" a type of "selection". As seems obvious from looking at Darwin's different selection definitions, one can start broadly and later narrow, limit, or make more concise any assumption, or start narrowly and later broaden any assumption. What's handy is that the narrower and more precise a selection hypothesis is, the easier it is to test it (and potentially reject a particular flavor of selection as the explanation for this or that).

Darwin named his hypotheses of sexual selection not in terms of the environmental circumstances that affected differential fitness, but in terms of the units of selection environmental circumstances affected. Likewise, Hamilton (1963, 1964) named his idea of kin-selection also on the unit of selection. Others seem to think that "sexual selection" was more about the sexy—doing it—part, having to do with mating success. West-Eberhard (1979) named her social selection hypotheses of the evolution of elaborate traits and signals in terms of the social situations yielding advantage or disadvantage. Both naming conventions, the one based on the units of selection and the one based the environmental circumstances of selection, are fine with me: I see no necessity to standardize naming conventions. What does exercise me is the need to be as explicit as one can be about each assumption of any selection hypothesis. If we are explicit about our assumptions, it won't matter what we name the idea. Being explicit guides us to attend to all of the essentials in deductive selection hypotheses (units of selection, environmental circumstances, and components of fitness through which differential reproductive success and survival accrue).

6.4 The Units of Selection

Darwin's fascination with breeders' attention to within-species variation in traits is notorious. In his quest to learn about standing variation of farmed and domesticated animals, he even had printed for broad distribution his many questions about breeder behavior and breeders' targets of artificial selection. The answers later became part of *On the Origin of Species by Means of Natural Selection* (Darwin 1859). Neophyte students of evolution sometimes admit to boredom while reading Darwin on, say, variation in pigeons, but they are also the students who do not quickly understand that individual variations—genetic, cultural, epigenetic—are the stuff of evolutionary change. If no variation exists, no evolutionary change is possible. In fact, Darwin's chapters on variation are essential to the development of the idea of natural selection. Under artificial selection animal and plant breeders choose (often consciously) among their stocks particular individuals as breeders because of some trait variant such as a plumage length or color. In stunning contrast, under natural selection there is no actor, there are no "targeted" traits, there is no "target" of selection. In contrast to artificial selection, natural selection just happens. If there is no variation among the individuals in the unit of selection, even if some individuals have greater survival or reproductive success, there is no natural selection. Natural selection implies a linkage between the environmental circumstances that advantage or disadvantage individuals in the unit of selection according to available variation so that some individuals have greater survival or reproductive success than others. However, the traits that remain after an episode of selection are often very hard to predict. Selection is differential survival and reproductive success of variable individuals in terms of their environmental circumstances. To repeat, neither natural selection nor sexual selection is goal-directed. Selection just happens. It is possible to predict the function of traits that may be left after selection, but it is much harder to predict the resulting phenotypes.

In his search to understand the extent and limits of variation, Darwin attended most closely to variation within populations of interbreeding animals. He began his thinking about selection at the level of individuals within populations. One can see this in his definitions of natural selection that he contrasted to artificial selection, which was "selection by man". (One might call what Darwin called *artificial selection* "farming", because we now know that other creatures besides humans purposely manipulate and control for their own benefit the breeding and survival of other species. So, I am inclined today to describe the hypothesis of artificial selection as a type of natural selection acting among individuals of a population through another species' manipulation and control of breeding and survival of the variants among the "farmed" species, so that selection acts through differential reproductive success and survival of individuals.)

It is obvious in *On the Origin* (1859) or *The Descent of Man* (1871) that Darwin's selection hypotheses mostly were about individuals within populations. Nonetheless, Darwin did cast hypotheses about group selection, and even kin-selection, and when he discussed such ideas he always seemed careful to say what was the unit of selection. Lewontin's (1970) influential paper on units of selection is a useful guide to the concept. Group selection and kin selection are hypotheses whose names indicate the level of selection whose members vary and though which between group agonism or within kin group collaboration yield advantage and disadvantage. What is important is not the name of a hypothesis, but its assumptions.

6.5 The Social Circumstances of Selection

Most hypotheses that investigators call "sexual selection" assume that social circumstances affecting number of mates *among males* (female preferences and male-male competition) and thus variance in reproductive success *among males* is the whole story. Even long before the well-published papers of Clutton-Brock (2007), others (Blackwell 1875; Sherfey 1966; Hrdy 1981; Gowaty 1981; Johnson and Hubbell 1984; Gowaty et al. 1989) had argued that male preferences and female-female competition also occur and are circumstances of selection with results on differential female reproductive success, i.e., selection among females. In addition, others cast equally useful selection hypotheses that posited that differential reproduction occurs among individuals within a sex linked to other types of social interactions (Fig. 6.1) including those that are "nice" or "nasty" (Gowaty 1996). Social circumstances like variation in the quality of potential mates, not just the number of potential mates, could have effects on differential reproductive success of individuals (Altmann 1997). Environmental circumstances of selection may be socially or ecologically mediated within-sex resistance to others' control of reproductive decisions (Gowaty 1992, 1996, 1997b, 2003b). So that resistance to others' control of reproductive decisions is a circumstance of within-sex selection that can have effects through behavioral or physiological interactions of individuals of the same or opposite sexes. *Female Control* (Eberhard 1996) remains a useful source

Key Challenges
to Narrow Sense Sexual Selection

	Nice/Nasty		
Social Circumstances		**Within-sex**	**Between-sex**
	Nice = affiliative	Coalitions Friendships	Mate Preferences Cooperation Helping
	Nasty = antagonistic	Combat Dominance & Counter-tactics	Rejection, Conflict Control & Manipulation Resistance

Fig. 6.1 Social circumstances of selection affecting within sex variation in fitness are common. Many more than listed here are possible. (Modified from Gowaty 1997)

for specific clues about the circumstances of selection of female resistance. In fact, in the dynamic bouts of between-sex conflict, what appears as most dramatic—the behavioral and physiological control and resistance dynamics between females and males—is the tip of the iceberg, because there is not just one, but two episodes of sexual selection that result. Sexual conflict produces not just the usual male-male type of "sexual selection", but also female-female selection. So, the circumstances of sexual conflict are male-female behavioral and physiological tugs-of-war, but who wins the fitness battle among females is one selective event and who wins among males is another (Gowaty 1997b). In sexual conflict the components of fitness that affect trait variation among females may be quite different from the components of fitness affecting trait variation among males. I called these male-female tugs of war "sexual dialectics" (Gowaty 1997b).

One of the points of the paragraph above is that sexual conflict—usually some social interaction—has effects on differential survival and reproductive success—on individuals within a sex. How might we bin this type of selection? The unit of selection is within sex: that is within females and separately within males, so with one naming convention it is a type of "sexual selection". Because the selection results because of social interactions between males and females, it fits the rubric of "social selection". But, wait, it's even more interesting in that the components of fitness affected by the social interactions include both survival and reproductive success: Others committed to other naming conventions might want to claim it is "natural selection" because of consequences for survival or "sexual selection" because of consequences for reproductive success. Commitments to naming conventions are opportunities for mischief and confusion. Perhaps we might call the sexual conflict idea "Cheshire" for short.

6.6 The Components of Fitness in Selection

Fisher (1930) put the Darwinian discussion of variation in survival and reproductive success firmly in demographic context, which is required given that fitness (survival and reproductive success) is a concept of relative variation among the units of selection. He urged investigators to evaluate fitness costs in terms of survival variation among individuals, and he emphasized that relative fitness depended on reproductive value, which is itself a function of whether populations are growing, retracting, or in stasis, not simply an age-associated characteristic. Empiricists are practical, of course, and many have emphasized that individuals may "trade off" earlier fitness components (number of sperm, ovules, mates, or eggs laid and offspring born) against later fitness components (number of larvae, pupa, or better yet, number of offspring reaching reproductive age). Many empiricists measure the relative output of eggs as a measure of fitness assuming perhaps incorrectly that the number of eggs is always positively associated with number of adult offspring. There are many reasons to doubt that always-positive association (Gowaty 2008), and thus, it seems most reasonable to use estimates on the number of adult offspring or their health as the best measure of an individual's reproductive success.

6.7 Some Modern Views of Darwin's Ideas

By definition, specific assumptions of selection hypotheses are everything. Selection is like addition or subtraction (Dennett 1995). If traits of individuals (a level of selection) are variable, if traits are heritable, if environments vary (the circumstances that sort among the variants) so that some individuals survive or reproduce (fitness components) better than others because of their trait variants, that is selection. If traits of individuals—within populations or within-demes or within-sexes or within-families—are variable, if the traits are heritable, if social interactions (e.g., aggression, affiliations, collaborations, competitions) vary so that some of the units reproduce more, produce healthier offspring, or survive better because of their trait variation relative to their social interactions, that is often called *social selection.* Likewise, if there is variation among individuals within a specified sex, and if there are non-random environmental/social circumstances that affect the number and/or quality of their mates so that their reproductive success varies, that is *sexual selection*, which is a sub-type of "social selection". Therefore, most hypotheses called "sexual selection" are simple hypotheses of among male differential fitness due to female preferences (e.g., sensory biases) and male-male combat, interference, etc. None of the above is complex. It is relatively easy to cast first-principle, deductive hypotheses of differential number or quality of mates produced via cooperative or affiliative interactions of either same or opposite sex interactions (Fig. 6.1) (Gowaty 1992, 1996, 1997a, b, 2003a, b; Gowaty and Buschhaus 1998). As investigators detail greater specificity of each assumption, the testability of the hypotheses will increase. If we simply listed our assumptions about units of selection, about the environmental and social circumstances of selection, and the components of fitness that environmental circumstances affect, we would thereby capture the sufficient and necessary details of any selection hypothesis. Greater specificity increases clarity of alternatives, and suggests paths for differentiating among hypotheses. Selection hypotheses are, after all, really and truly like addition and subtraction.

6.8 Back to the Future and More Controversy

Fisher's (1930) view of sexual selection is more encompassing and interesting than that for which he gets so much modern credit. What strikes me when I read Fisher is not his famous runaway hypothesis explaining remarkable, bizarre traits usually in males, but the fact that he stressed that preferences evolve too.

> ...The tastes of organisms, like their organs and faculties, must be regarded as the products of evolutionary change, governed by the relative advantage which such tastes may confer, it appears...that occasions may be not infrequent when a sexual preference of a particular kind may confer a selective advantage, and therefore become established in the species (Fisher 1930, p. 136).

He said that preferences could evolve within a sex when variation among individuals in preferences exist and when environmental circumstances affect differential fitness of individuals because of the variation in their preferences. He also noted *that preferences might have no effect on traits of preferred individuals,* if for example, natural selection maintained the discriminated differences in the opposite sex. He also reasoned that sexual preferences might be more widespread than fancy secondary sexual characters.

Here's a selection hypothesis from Fisher's general reasoning for the evolution of discriminatory ability, which he emphasized *may or may not be tightly linked* to the evolution of a particular trait: Assume within population between-individual variation in discriminatory senses exists. Assume ecological circumstances affect the advantages or disadvantages of discriminatory senses. Assume that because of variation in discriminatory senses and ecological circumstances that differential fitness among individuals in the unit of selection occurs. This is a hypothesis my colleagues and I tested in mice, ducks, and flies using an experimental protocol that had us draw those to be discriminated at random with respect to observable trait variation. When we found individuals with repeatable preferences—after testing each discriminator's preference for one discriminatee or the other (in arenas that allowed discriminator and discriminatees to hear, see, smell, but not touch one another)—we randomly assigned discriminators to enforced pairs with the discriminatee they did or did not prefer. Then, we measured components of reproductive success—the number of eggs laid per day or offspring born, the number of adult offspring, and the percent egg-to-adult or pup-to-adult survival. In mice and flies, the species in which we tested preferences of female and male discriminators, percent egg to adult survival was lower when discriminators were in enforced pairs with individuals they did not prefer. Thus discriminators gained fitness benefits when paired with those they preferred. In these tests, both sexes discriminated and both gained fitness benefits from their preferences while both sexes suffered fitness deficits when in enforced pairs with a partner they did not prefer. Females in both species traded off number of pups born or eggs laid with the numbers of adult offspring. Related experiments in flies revealed that discriminatees did not vary in "quality" (i.e., lifespan variation), suggesting that the advantage of preference on offspring viability was not a function of absolute quality of the discriminatees, but an interaction effect of the discriminator and discriminatee. We speculate that the sensing of the immune coding alleles of potential mates mediates contributions of each partner to offspring health via genes, maternal and paternal epigenetic effects. The results of our experiments in a suite of very-distantly related species (Gowaty 2008; Gowaty et al. 2007) are consistent with Fisher's expectation that preferences for opposite-sex potential mates necessarily evolve even if preferences have weak or no effect on reproductive success of those with discriminated traits (fancy or otherwise). In other words, choosers may have preferred the naturally selected traits of potential partners, and thereby gained fitness rewards that would affect the mate preferences in both sexes. Naturally selected mate preferences for naturally selected traits could produce within-sex variation in numbers of mates (which some consider "the sign" of sexual selection), demonstrating again why our naming conventions are sometimes confusing, misleading, and downright mischievous.

The expectations of Fisher's first hypothesis above are very different from the modern, canalized expectations of Fisher's better-known second hypothesis of strongly linked female preferences and dramatic, bizarre, attractive traits in males. Fisher was careful to say that whenever males with certain traits leave more offspring than males with other traits, there are two types of advantages: (1) those accruing through natural selection and (2) those associated with intensity and directionality of female preferences. The second advantage could "run-away" in such a way that exaggerated trait variation becomes a mismatch with other environmental circumstances so that bearer's suffer reduced survival. Fisher's inference of the costliness of exaggerated traits was that selection would be periodic, with stasis in most such traits most of the time.

> In most existing species the runaway process must have been checked, and we should expect that the more extraordinary developments of sexual plumage were not due like most characters to a long and even course of evolutionary progress, but to sudden spurts of change. (p 137).

To restate Fisher's runaway idea: The unit of selection is among males within a species; the environmental circumstances of selection are females' tastes for males, which advantages males with the most elaborate and fancy traits, so that males with fancy traits have more mates, and more offspring. The fitness payout for females presumably is that any sons she has are "sexy" and any daughters inherit her sensitivities. Fisher also explained that investigators should simultaneously evaluate the enhanced reproductive success or survival of those with preferences when one is considering the reproductive success and survival of the preferred individuals.

I took that to mean that the within-sex variation in fitness payouts of preferences needed evaluation just as the within-sex variation in the fitness of having a fancy trait needed evaluation. Fisher's exhortation emphasized there are two linked selective processes in his runaway idea. Importantly, the environmental circumstances of selection on those with preferences and those with lavish traits are not the same, nor are the components of fitness, and the single fitness linkage of sexy sons may be invisible during periods of stasis.

6.9 Much Ado About Nothing: A Modern Controversy

Roughgarden's hypothesis for the evolution of fancy traits is that the traits are badges of inclusion in elite clubs and in which club members kill other same-sex individuals who lack the club insignia. Roughgarden's social selection hypothesis assumes males vary in traits, that the social circumstance of lacking an insignia induces adorned males (those with the insignia) to aggress against unadorned males reducing their survival probability. Roughgarden said Darwin was wrong about the evolution of fancy traits, because her idea seems more convincing to her, a stronger circumstance of selection than the seemingly weak selection exerted through mate choice. However, Roughgarden's narrow sense "social selection" is no challenge to Darwin's principles upon which all selection hypotheses rest, nor do they violate the principles of what many investigators consider sexual selection. A good way for males to eliminate rivals from the future classes of mated mates is to kill them and if rivals are killed early enough they may enter the class of zero mates for that cohort. Because this narrow sense social selection hypothesis is within-sex and depends upon the social circumstances of a male-male competitive interaction, it appears to fit well enough almost everybody's intuitive sense of narrow sense sexual selection. Moreover, when Roughgarden presents what she says is her hypothesis that fancy traits, such as some elaborate secondary sexual characteristics, like "…a peacock's tail, a rooster's comb, wattle and cockle-doddle-do facilitate male-male interactions" rather than male-female interactions posited by Darwin (1871), she is non-controversial. There are many examples of exuberant traits that function as same-sex signals (Alonso-Alvarez et al. 2004; Berglund et al. 1996; Brumm and Todt 2004; Furlow et al. 1998; Gotmark 1993; Hagelin 2002; Mateos and Carranza 1997; Matos and McGregor 2002; Parker and Ligon 2002). Thus, the basic idea was around for a long time before *Evolution's Rainbow*.

That the origins of exuberant traits lie in their value as signs of social inclusion fits into a long-term discussion of social interactions besides female choice and male-male combat (e.g., see Smuts and Smuts 1993). Sexual coercion is another social circumstance of sexual selection that can account for within-sex differential reproductive success. To repeat: Darwin (1871) defined sexual selection as "a kind of selection having to do with reproductive competition between individuals of the same sex…" He said that sexual selection "depends on the advantage which certain individuals have over other individuals of the same sex and species, in exclusive relation to reproduction" (Darwin 1871, t. I, p. 256). He used the word "competition" to include combat between males and an affiliative, friendly behavior of females towards some males and called it "mate choice". The "competition" Darwin spoke of sometimes depended on cooperative and friendly behavior including being preferred. Social inclusionary traits are ones that same-sex preferences might mediate. Because the presence and absence of social inclusionary traits yield advantages for some and disadvantages for others, within-sex differential reproductive success occurs (the "competition" that Darwin emphasized) and so does differential survival. Thus, social inclusionary traits can evolve via sexual selection. Again, many define sexual selection as occurring when individuals expressing sex-limited heritable traits (morphology, physiology, or behavior) leave more descendants because they had more or better mates than individuals of the same sex who lack the trait or have a lesser expression of the trait. That "secondary sexual traits are admission tickets to power-holding cliques, social inclusionary traits, condition indicators of capacity to supply direct benefits" (Roughgarden 2007) fits into commonly discussed social circumstances of sexual selection. Whenever secondary sexual traits are a social circumstance advantaging membership in the "power holding clique", and whenever not being in the clique means that an individual will surely not breed, clique membership mediates one level of *within-sex differential reproduction* and it is sexual selection by the definitional sense that most understand. What is controversial is the claim that social inclusionary trait evolution is an alternative to sexual selection. It is not clear how this would be so, and because this is not clear, many (Stewart 2006; Lessells et al. 2006; Hurd 2006; Ghiselin 2006; Day et al. 2006; Dall et al. 2006) dismiss Roughgarden's ideas as not different from conventional ideas in sexual selection. Also important is that even before Roughgarden claimed it as her own, West-Eberhard's view of social selection subsumed Roughgarden's. Claiming that Roughgarden's ideas forsake Darwin is much ado about nothing.

Given that one's naming conventions for selection hypotheses is on the social circumstances that sort among variants in the units of selection, it makes sense that West-Eberhard's (1979, 1983, 1984) and others say that social selection subsumes sexual selection, and accounts for the evolution of dramatic signals that serve and do not serve within-sex reproductive competition, and affecting both reproductive success and survival. All of the most famously and recently contested social circumstances of within-sex competition over reproductive success are social, but not all circumstances of social selection are about within-sex competitive reproduction, so I have no difficulty with "social selection subsuming sexual selection" claims. Social selection is a type of natural selection at the level of individuals, unless of course, one broadens West-Eberhard's view of social interactions to those between species. Obviously, Darwin's narrow-sense sexual selection from female choice and male-male contests are social circumstances of selection, so the concept of social selection subsumes sexual selection, but perhaps not when one talks about physiological mechanisms of, say, between-male competition.

6.10 Tests of Alternative Selection Hypotheses

As most elementary discussions of scientific logic say: it is important to test all the parts in the chain of inference, not just some of the parts. To me it means testing each of the assumptions, preferably against reasonable alternatives. Therefore, I can imagine testing a classical idea about male-male competitive dynamics over number of mates in the absence of cliquey coalitions and male-male competitive dynamics in the presence of cliquey coalitions. How often investigators actually will do such testing of assumptions of alternative hypotheses is a problem for the future. My impression is that, when the assumptions are clear, empiricists are eager to test them using strong inference methods (Platt 1964).

6.11 Conclusion

Recent arguments over sexual selection seem semantic rather than substantive, perhaps based only on muddled or missing statements of assumptions. We have only to stay clear of muddied shoals to continue to build on Darwin's legacy to understand how social circumstances affect differential fitness.

Acknowledgments I thank Thierry Hoquet for his helpful comments and his patience. I thank Sarah Hrdy, Steve Hubbell, and Therese Markow for comments on an earlier draft. I acknowledge support during the preparation of the manuscript from NSF award IOS-1121797 and from previous NSF awards for the support of my empirical research discussed herein.

References

Alonso-Alvarez C, Doutrelant C, Sorci G (2004) Ultraviolet reflectance affects male-male interactions in the blue tit (*Parus caeruleus ultramarinus*). Behav Ecol 15:805–809

Altmann J (1997) Mate choice and intrasexual reproductive competition: contributions that go beyond acquiring more mates. In: Gowaty P A. (ed) Feminism and evolutionary biology: boundaries intersections and frontiers. Chapman and Hall, New York

Berglund A, Bisazza A, Pilastro A (1996) Armaments and ornaments: an evolutionary explanation of traits of dual utility. Biol J Linnean Soc 58:385–399

Blackwell ALB (1875) The sexes throughout nature. New York, G. P. Putnam's Sons

Brumm H, Todt D (2004) Male-male vocal interactions and the adjustment of song amplitude in a territorial bird. Anim Behav 67:281–286

Buss DM (2006) Debating sexual selection and mating strategies. Science 312(5774):690–691

Carroll L (2003) Alice's adventures in wonderland. Penguin, UK

Clutton-Brock T (2007) Sexual selection in males and females. Science 21:882–1885

Dall SRX, McNamara JM, Wedell N, Hosken DJ (2006) Debating sexual selection and mating strategies. Science 312:689–689

Darwin C (1859) On the Origin of species by means of natural selection. John Murray, London

Darwin C (1871) The descent of man and selection in relation to sex. John Murray, London

Darwin C, Wallace A (1858) On the tendency of species to form varieties; and on the perpetuation of varieties and species by natural means of selection. J Proc Linnean Soc London Zool 3(9):45–62
Day T, Houle D, Rowe L (2006) Debating sexual selection and mating strategies. Science 312:691
Dennett DC (1995) Darwin's dangerous idea. Simon and Shuster, New York
Eberhard WG (1996) Female control: sexual selection by cryptic female choice. Princeton University Press, Princeton (NJ)
Fisher RA (1930) The genetical theory of natural selection. Clarendon Press, Oxford
Furlow B, Kimball RT, Marshall MC (1998) Are rooster crows honest signals of fighting ability? Auk 115:763–766
Ghiselin MT (2006) Debating sexual selection and mating strategies. Science 312:691–692
Gotmark F (1993) An experimental study of the importance of plumage coloration in breeding males of the white-crowned sparrow. Ornis Scandinavica 24:149–154
Gowaty PA (1981) The aggression of breeding Eastern Bluebirds *Sialia sialis* toward each other and intra- and inter-specific intruders. Anim Behav 29:1013–1027
Gowaty PA (1992) Evolutionary biology and feminism. Hum Nat 3(3):217–249
Gowaty P A (1996) Battles of the sexes and origins of monogamy. In: Black JL (ed) Partnerships in Birds. Oxford University Press, Oxford
Gowaty PA (1997a) Principles of females' perspectives in avian behavioral ecology. J Avian Biol 28:95–102
Gowaty PA (1997b) Sexual dialectics, sexual selection, and variation in mating behavior. In: Gowaty PA (ed) Feminism and evolutionary biology: boundaries, intersections, and frontiers. New York, Chapman Hall
Gowaty PA (2003a) Power asymmetries between the sexes, mate preferences, and components of fitness. In: Travis C (ed) Women, evolution, and rape. Cambridge, MIT Press
Gowaty PA (2003b) Sex roles, contests for the control of reproduction, and sexual selection. In: Kappeler PM, Van Schaik CP (eds) Sexual selection in primates: new and comparative perspectives. Cambridge University Press, Cambridge
Gowaty PA, Anderson WW, Bluhm CK, Drickamer LC, Kim Y-K, Moore AJ (2007) The hypothesis of reproductive compensation and its assumptions about mate preferences and offspring viability. Proc Natl Acad Sci U S A 104(38):15023–15027
Gowaty PA (2008) Reproductive compensation. J Evol Biol 21:1189–1200
Gowaty PA (2011) What is sexual selection and the short herstory of female trait variation. Behav Ecol 22:1146–1147
Gowaty PA, Buschhaus N (1998) Ultimate causation of aggressive and forced copulation in birds: Female resistance, the CODE hypothesis, and social monogamy. Am Zool 38:207–225
Gowaty PA, Plissner JH, Williams TG (1989) Behavioral correlates of uncertian parentage: mate guarding and nest guarding by eastern bluebirds, *Sialia sialis*. Anim Behav 38:272–284
Hagelin JC (2002) The kinds of traits involved in male-male competition: a comparison of plumage, behavior, and body size in quail. Behav Ecol 13:32–41
Hamilton WD (1963) The evolution of altruistic behavior. Am Nat 97:354–356
Hamilton WD (1964) The genetical evolution of social behaviour. I. J Theor Biol 7(1):1–16
Hrdy SB (1981) The Woman that never evolved. Harvard University Press, Cambridge
Hurd PL (2006) Debating sexual selection and mating strategies. Science 312:692–693
Johnson LK, Hubbell SP (1984) Male choice: experimental demonstration in a Brentid weevil. Behav Ecol Sociobiol 15:183–188
Lessells CM, Bennett ATD, Birkhead TR, Colegrave N, Dall SRX, Harvey PH, Hatchwell B, Hosken DJ, Hunt, J, Moore AJ, Parker GA, Pitnick S, Pizzari T, Radwan J, Ritchie M, Sheldon BC, Shuker DM, Simmons LW, Stockley P, Tregenza T, Zuk M (2006) Debating sexual selection and mating strategies. Science 312:689–690
Lewontin RC (1970) The units of selection. Ann Rev Ecol Syst 1:1–18
Mateos C, Carranza J (1997) The role of bright plumage in male-male interactions in the ring-necked pheasant. Anim Behav 54:1205–1214

Matos RJ, Mcgregor PK (2002) The effect of the sex of an audience on male-male displays of siamese fighting fish (*Betta splendens*). Behaviour 139:1211–1221

Mayr E (1977) Darwin and Natural Selection: How Darwin may have discovered his highly unconventional theory. Am Sci 65:321–327

Miller G (2006) Debating sexual selection and mating strategies. Science 312(5774):693–693

Parker TH, Ligon JD (2002) Dominant male red junglefowl (Gallus gallus) test the dominance status of other males. Behav Ecol Sociobiol 53:20–24

Pizzari T, Birkhead TR, Blows MW, Brooks R, Buchanan KL, Clutton-Brock TH, Harvey PH, Hosken DJ, Jennions MD, Kokko H, Kotiaho JS, Lessells CM, Macias-Garcia C, Moore AJ, Parker GA, Partridge L, Pitnick S, Radwan J, Ritchie M, Sheldon BC, Simmons LW, Snook RR, Stockley P, Zuk M (2006) Debating sexual selection and mating strategies. Science 312(5774):690–690

Platt John R (1964) Strong inference. Science 146:347–353

Roughgarden J (2004) Evolution's rainbow: diversity, gender, and sexuality in animals and people. University of California Press, California

Roughgarden J, Akcay E, Oishi M (2006a) Debating sexual selection and mating strategies—reponse. Science 312(5774):694–697

Roughgarden J, Oishi M, Akcay E (2006b) Reproductive social behavior: cooperative games to replace sexual selection. Science 311(5763):965–969

Roughgarden J (2007) Challenging Darwin's theory of sexual selection. Daedalus 136(2):23–36

Sherfey MJ (1966) The nature and evolution of female sexuality. Random House, New York

Shuker DM, Tregenza T (2006) Debating sexual selection and mating strategies. Science 312(5774):693–694

Smuts BB, Smuts RW (1993) Male-Aggression and sexual coercion of females in nonhuman-primates and other mammals: Evidence and theoretical implications. Adv Stud Behav 22:1–63

Stewart J (2006) Debating sexual selection and mating strategies. Science 312:694–694

Wallace AR (1889) Darwinism: an exposition of the theory of natural selection, with some of its applications. Macmillian, London

West-Eberhard MJ (1979) Sexual selection, social competition, and evolution. Proc Amer Phil 51:222–234

West-Eberhard MJ (1983) Sexual selection, social competition, and speciation. Q Rev Biol 58:155–183

West-Eberhard MJ (1984) Sexual selection, competitive communication and species-specific signals in insects. In: Trevor L (ed) Insect Communication: Proc. 12th Symposium Royal Entomol. Soc. London. Academic, London

Chapter 7
Sexual Selection: The Logical Imperative

Geoff A. Parker and Tommaso Pizzari

Abstract Modern sexual selection theory, developed from Darwin's original intuition, is a cornerstone of evolutionary theory and represents the most parsimonious and robust explanation for a bewildering array of evolutionary patterns and diversity. Here we first outline the principles of modern sexual selection theory and discuss their heuristic value. Second, we review empirical demonstrations of the operation of sexual selection through the case study of the yellow dung fly. Finally, we propose that a sequence of evolutionary events flows inevitably from the early evolution of sexual recombination and gametes, to anisogamy and in dioecious organisms, to the unity sex ratio via Fisher's principle. As Darwin and Bateman predicted, it was the primary sexual difference—anisogamy—that became an almost obligatory, irreversible transition favouring socio-ecological conditions that ultimately generated secondary differentiation of sexual strategies between the sexes, and typically plays a strong part in their maintenance (though sex roles can, rather rarely, be reversed). When considered within the broader context, sexual selection emerges deductively as the logical consequence of this evolutionary succession. We conclude by highlighting aspects integral to sexual selection theory that are currently the focus of on-going discussion.

Keywords Anisogamy · Sex roles · Sperm competition · Sexual conflict · Yellow dung fly

7.1 Introduction

Darwin (1874) defined sexual selection as competition between individuals of one sex (usually males) to gain matings with the limiting sex (usually females), and interpreted male and female sex roles and much behaviour and morphology in terms of

G. A. Parker (✉)
Institute of Integrative Biology, University of Liverpool, Liverpool L69 7ZB, UK
e-mail: gap@liverpool.ac.uk

T. Pizzari
Edward Grey Institute, Department of Zoology, University of Oxford, Oxford OX1 3PS, UK

T. Hoquet (ed.), *Current Perspectives on Sexual Selection,* History, Philosophy and Theory of the Life Sciences 9, DOI 10.1007/978-94-017-9585-2_7

what is now termed pre-copulatory sexual selection. Bateman (1948) in a classic paper elucidated the mechanism behind this form of selection, and explained Darwin's claim that sexual selection is typically stronger in males in light of factors arising ultimately from the primary sexual difference between the sexes due to anisogamy.

We argue here that there is a remarkable logical beauty in the sequence of events (the 'sexual cascade'; Parker 2014) that flows inevitably from the early evolution of sexual recombination and gamete formation, to result in sexual selection and ultimately the differentiation of sexual strategies in males and females. The evolution of sexual recombination and gametes led (in multicellular organisms and some unicells) directly to the evolution of anisogamy, the primary sexual differentiation underlying the two sexes, males and females—which in turn typically generated the unity sex ratio via Fisher's sex ratio principle in organisms with separate sexes. Ancestrally, sexual selection would have operated entirely by sperm competition, until enhanced mobility allowed higher fertilisation gains via female-targeted sperm release and pre-mating competition, leading (as an economic consequence) to testes reduction, the rise of pre-copulatory sexual selection and enhanced sexual conflict, and to high degrees of secondary sexual differentiation as Darwin (1874) and later Bateman (1948) predicted.

Despite the power of the logic supporting these sequential evolutionary steps and their consequence, the Darwin-Bateman Paradigm (DBP; Dewsbury 2005) of typical male and female sex roles, sexual selection has been controversial. The first critique came from Alfred Russell Wallace (see e.g. Prum 2012), followed by a period where even male-male competition was doubted, ignored, or at best grudgingly accepted (e.g. Huxley in the 1930s). In spite of one or two notable exceptions (e.g. Bateman 1948; Jacobs 1955) these doldrums persisted until the behavioural ecology revolution of the 1970s, after which sexual selection experienced an explosion of interest and accumulation of theory and supporting evidence. In the past few years, however, Darwinian sexual selection has been attacked as a flawed and unsubstantiated concept, and the DBP characterised as a misinterpretation due to gender bias (see Sect. 7.5). We argue that both these attacks are misguided, since the evidence for sexual selection is overwhelming, and its conceptual basis (and that of the DBP) logically sound (see also Parker and Birkhead 2013).

We first introduce the fundamentals of sexual selection theory and argue that this theory represents a powerful heuristic tool and the most parsimonious explanation for a very wide range of biological patterns. Next, we illustrate how the Darwinian concept of sexual selection is the logical consequence of evolutionary transitions originating from the evolution of sex and recombination. Finally, we focus our discussion on some issues that have contributed to ignite recent debate over sexual selection theory.

7.2 Sexual Selection Theory is a Powerful Heuristic Tool

Darwin (1874) viewed sexual selection as a process targeting variation in reproductive success among individuals of the same sex and species solely due to intrasexual competition over access to reproductive opportunities. This definition was later formalized in the concept of the opportunity of sexual selection, I_T, which is the standardized variance in intra-sexual reproductive success

$$I_T = \frac{\sigma^2}{\overline{T}^2},$$

where $\overline{T}$ is the average total reproductive success for an individual and σ^2 is the variance in T across individuals of the same sex (Arnold and Wade 1984; Shuster and Wade 2003; Jones 2009). I_T represents a useful empirical measure of the potential of sexual selection that can operate in a given population (but see Sect. 7.5.3). A trait explains a significant portion of I_T when a standardised increment in the trait causes a standardised change in individual reproductive success, and the strength of sexual selection on the trait is measured by the slope of the regression of individual reproductive success against standardised trait expression (the sexual selection gradient, β, Arnold and Wade 1984). There is an elegant simplicity in this view, which makes it broadly relevant (i.e. applicable to all sexually-selected organisms, from unicellular to human) because it does not depend on assumptions about the proximate (i.e. cognitive, physiological or morphological) mechanisms underpinning variance in T.

Crucially although Darwin developed sexual selection theory to explain sexual dimorphism, he was aware that intra-sexual competition occurred in both males and females, and his framework is equally applicable to both sexes, making no *a priori* assumptions about sex-specific patterns of sexual selection. Instead, sexual dimorphism and 'sex roles' emerge as properties of sex differences in the opportunity of sexual selection. Darwin recognised that intra-sexual variance in T depends on the number and 'quality' of reproductive mates secured. The number of partners corresponds to individual mating success (M). Mate 'quality' (Q) captures a number of ways through which the contribution of an individual to a reproductive event affects the fitness of its mate, so that the total reproductive success of an individual can be expressed as:

$$T = (M \cdot Q) + \varepsilon$$

where ε is an error term with zero mean. Female quality for example, would include clutch size, or the number of eggs produced by a female in a given event, but clearly maternal investment in the eggs and zygotes, as well the expression of maternal genes in the descendants would also play an important role in quality. Similarly, male quality would include paternal investment and the genetic contribution to the offspring.

7.2.1 Sex Roles and the Darwin-Bateman Paradigm

In a series of experiments on *Drosophila melanogaster* published in 1948, Bateman presented empirical evidence suggesting that the regression of T on M was steeper in male rather than female flies, suggesting that sexual selection is more intense in males than in females. Bateman (1948) generalised these results and proposed that the sex experiencing more intense sexual selection has higher standardised variance in both T and M (i.e. I_T and I_M respectively) and a steeper slope β of the regression of T on M:

$$T = (\beta \cdot M) + \varepsilon.$$

Therefore, β measures the gradient of sexual selection on mating success and is often referred to as the Bateman gradient. Bateman (1948) argued that because of anisogamy (see below) males typically produce orders of magnitude more sperm than there are eggs available for fertilisation; their reproductive success is essentially limited by their ability to access mating opportunities, resulting in higher male I_T and I_M, and a steeper Bateman gradient than females. The larger investment in individual gametes would constrain (but not necessarily eliminate) the Bateman gradient on females. For $M \geq 1$, lower (or zero, or even negative) Bateman gradients in females are therefore expected, though positive female gradients can arise for many reasons, including increased fertility, male nuptial gifts, or may partly reflect the proportion of female I_T that is explained by variation in male Q. Anisogamy therefore creates a fundamental difference in the way Darwinian selection operates on adult males and females: sexual selection pushing primarily males to compete with each other over access to mating opportunities, and females to compete more strongly for resources to produce young and to discriminate amongst prospective mates.

Bateman's intuition therefore provided a conceptual framework for Darwin's original prediction that sexual selection favours the exaggeration of male traits that result in higher M and/or Q by conveying a competitive advantage in intra-sexual selection and/or by matching more closely female mating preferences (inter-sexual selection). Darwin gave much evidence for patterns of male-male competition for females, which have subsequently been amplified considerably. A sex role pattern of high observable male-male competition when male parental investment (PI; Trivers 1972) is zero or very low is undeniable, this is correlational only but fits explanations based on the primary sex role divergence due to anisogamy (e.g. see Kokko et al. 2013; Schärer et al. 2012)—as we later discuss. This view, the Darwin-Bateman paradigm (DBP), was further developed and nuanced to become the backbone of modern sexual selection theory (Trivers 1972; Arnold and Duvall 1994).

7.2.2 Sperm Competition

A significant advance in the DBP and developments in modern sexual selection theory has been the realisation of an additional source of variance in T giving rise to episodes of sexual selection unexplored by Darwin. Parker (1970a) reviewed female mating behaviour in insects and proposed that whenever a female mates with more than one male in a way that their sperm co-occur at the time of fertilisation of a set of eggs, these ejaculates compete for fertilisation opportunities, a process that became known as sperm competition. Sperm competition therefore represents a form of intra-sexual selection, which occurs under some degree of female polyandry, and can account for many sexual adaptations (Parker 1970a). Despite scientific history, in evolutionary terms sperm competition long predates Darwinian pre-copulatory sexual selection, as we later argue.

The possibility that a male might not fertilise the entire clutch of his mate(s) introduces variability in the proportion of the eggs fertilised by each of the males that mated with the same female within the relevant time window (i.e. in time for their sperm to have a non-zero probability of fertilisation). Under sperm competition the reproductive success of a male therefore becomes:

$$T = (M \cdot Q \cdot P) + \varepsilon,$$

where P represents the paternity share within a clutch across M females (Webster et al. 1995; Collet et al. 2012).

It has been proposed that polyandry can also create potential for females (and/or their ova) to exert a systematic bias of the outcome of sperm competition in favour of the ejaculates of certain male phenotypes or genotypes, a process known as cryptic female choice (Eberhard 1996). Mechanisms of cryptic female choice represent an episode of inter-sexual selection generated by polyandry. In internally fertilising organisms sperm competition and cryptic female choice follow necessarily episodes of male competition and mate choice (pre-copulatory sexual selection). In some external fertilisers (e.g. corals, echinoderms) however, sperm competition and cryptic female choice might to a large extent replace pre-copulatory sexual selection.

As for pre-copulatory episodes of sexual selection, episodes of post-copulatory sexual selection are not necessarily restricted to males. Under some extreme conditions of male sperm limitation or sex-role reversal, the ova of different females can compete for access to sufficient sperm supplies to guarantee fertilisation. Similarly, like females, males can also perform 'cryptic' mechanisms of mate choice, by strategically adjusting their ejaculate expenditure based on the phenotype or genotype of their mates (see below).

7.2.3 Evidence of Sexual Selection

Over the last 40 years the study of sexual selection has exploded and represents one of the most dynamic and topical areas of evolutionary biology. Reviewing such vast empirical effort is well beyond the scope of this chapter, and rather than give a long list of the successes, in Sect. 7.3 we showcase detailed studies carried out on one species as an example of how sexual selection theory can explain adaptation. The reader is encouraged to refer to a number of excellent reviews on this subject (e.g. Andersson 1994; Andersson and Simmons 2006; Kokko et al. 2006; Jones and Ratterman 2009). Suffice it to note here that overwhelming empirical evidence has accumulated that exaggerated traits, which convey no viability benefits (i.e. the traits that inspired Darwin to think about sexual selection theory in the first place), can deliver significant advantages in intra-sexual competition by increasing *M* and *Q*, as hypothesised by Darwin (1874). In most cases, sexual selection remains the only parsimonious explanation to account for the evolution and maintenance of such traits. There is also robust evidence that sexual selection is determined by direct intra-sexual competition and by differential patterns of mating responses and mate discrimination in the opposite sex, consistent with mate choice. While sexually-selected male traits to increase *M* are intuitive, the adaptive nature and functional significance of mate choice (in other words, the mechanisms through which females evolved preference for certain mate types) is less clear, as one would expect given the following elements: (a) the multidimensional and context-dependent nature of *Q*, (b) the different inter-sexual co-evolutionary trajectories that mould preference, and (c) the often weaker (compared to *M*) effect of *Q* on *T*. It is important to note here that sexual selection theory predicts some degree of inter-sexual selection through forms of mating preferences or discrimination, but does not require specific assumptions about the evolutionary mechanisms underpinning such preferences.

In conclusion, sexual selection theory typically provides the only parsimonious and robust explanation for variation in sexual dimorphism and the evolution of a class of traits, exaggerated ornaments and armaments. One could argue that this is reassuring but hardly surprising considering that exaggerated male ornaments and armaments are the traits that originally inspired Darwin's theory of sexual selection. Crucially however, the heuristic power of sexual selection theory transcends the explanation of such traits, and has been successfully applied to explain biological patterns that were unknown to Darwin. Below, we consider two such triumphs of sexual selection theory.

7.2.4 De Novo Evolution of Sexually-Selected Traits

The yeast, *Saccharomyces cerevisiae*, reproduces sexually through disassortative fusion between sex cells of two types, MATa and MATα, which attract each other through the production of, and attraction to, a-pheromone and α-pheromone. Normally, the diploid organism undergoes meiosis and produces a tetrad of four haploid sex cells (two of each type) and self-fertilization occurs within the tetrad between MATa and MATα cells. This reproductive mode prevents sexual selection by removing sexual competition between individuals. If however, haploid sex cells from different individuals are mixed, potential for competition arises between individuals to attract and fuse with cells of the opposite type. Rogers and Greig (2009) have used this experimental construct to artificially create sexual selection in this system. The authors created six replicate (isogenic) populations in which the ratio of mating types was experimentally biased in favour of MATα. The artificial excess of MATα creates opportunity for sexual selection by forcing competition within this cell type to attract MATa cells. In each of the populations, the authors introduced a rare allele coding for an increased production of α-pheromone in MATα. The study monitored the spread of this strong signalling allele in the replicates evolving under sexual selection and six additional control populations evolving in the absence of sexual selection (i.e. equal ratio of MATa and MATα) for 13 generations. In all sexually-selected populations the strong signalling allele increased rapidly in frequency and had approached fixation in five out of six by the 13th generation. In stark contrast, in five out of the six the control populations there was no appreciable increment in frequency and only a modest increase in the sixth. The study provides an elegant experimental demonstration of the central axiom of sexual selection theory, intra-sexual competition can favour the spread of a trait that confers a competitive advantage.

This is but one example of many 'experimental evolution' studies in which the predictions of sexual selection theory have been validated. Many such investigations have compared populations after several generations of breeding under 'enforced monogamy' (in which sexual selection is absent) with those breeding under increased male-male competition or 'enforced polyandry' (e.g. see Sect. 7.3.2(iv)).

7.2.5 Sperm Competition and Ejaculate Expenditure

As noted above, Darwin's original theory of sexual selection was largely limited to pre-mating events. The intuition of sperm competition and post-copulatory sexual selection came about a century later (Parker 1970a), and led to the discovery of a remarkable diversity of traits and mechanisms—largely unsuspected by Darwin—mediating the outcome of sperm competition. Yet, the application of the general principles of sexual selection theory has enabled evolutionary biologists to understand the operation of post-copulatory events just as successfully as we can predict the operation of pre-copulatory episodes. A large body of theoretical work, ejaculate economic theory, has been developed to predict the way sperm competition drives the evolution of male ejaculate expenditure (Parker and Pizzari 2010). Qualitative and sometimes quantitative support for many of these predictions has been accumulated by a plethora of empirical studies investigating patterns of male ejaculate expenditure in terms of the percentage of body mass devoted to gonads (i.e. the gonadosomatic index or GSI = 100[gonad mass/total mass]), and in terms of number of sperm allocated to individual copulations (Parker and Pizzari 2010; Kelly and Jennions 2011), demonstrating that ejaculate economic theory can be a powerful heuristic to explain variation in ejaculate expenditure across species, across males within species and even plastic changes within individual males (Parker and Pizzari 2010). More recently, ejaculate economic theory has been extended to consider widespread cases where non-sperm ejaculate compounds have gonadotropic effects on females boosting their clutch size or oviposition rate (Alonzo and Pizzari 2010). These models predict that males evolve dynamically strategies of differential ejaculate allocation by preferentially investing in sperm and less in gonadotrophic compounds when they mate with a female after another male. This strategy would enable the second male to invest in sperm competition while simultaneously exploiting the fecundity investment made by the first male in a female. Patterns of strategic ejaculate exploitation consistent with these theoretical expectations have been experimentally demonstrated in *D. melanogaster*, where males preferentially reduce their investment in the gonadotropic accessory gland compound ovulin when they mate second with a female (Sirot et al. 2011).

7.3 The Yellow Dung Fly as a Case Study

Research on the common yellow dung fly, *Scatophaga* (=*Scathophaga*) *stercoraria* L., carried out over many years, has resulted in probably the most extensive data available on sexual selection in a given species. Studies published in the 1970s (reviewed by Parker 1978a) began in 1965 specifically for the purpose of testing Darwin's theory of sexual selection at a time when it was largely ignored, using a hypothetico-deductive approach of testing observations against predictions generated by optimality modelling. This has the aim, not to show that animals behave optimally, but to provide evidence that the selection pressures used in the model are those that have moulded the adaptations under consideration (Parker and Maynard Smith 1990). The early models for the dung fly mating system were constructed on the assumption that sexual selection acts to maximise a male's overall fertilisation rate in competition with other males. An evolutionarily stable strategy (ESS) approach was used to determine the male competitive optima for several traits. Studies of sexual selection in this species continue to present day, and by now have covered a wide range of sexually-selected adaptations (both pre-and post-copulatory). From the outset they have shown that sexual selection can generate very fine-grained adaptive optima (Parker 1978b).

The mating system of dung flies was described by Hammer (1941) and Parker (1978b). Males arrive swiftly around fresh cattle droppings to await the arrival of gravid females, who typically lay all their mature eggs as a batch in a single dropping. Although generally already containing sperm from previous matings, a gravid female copulates at each visit to the dung to oviposit. Struggles between males for the possession of females are common. After copulation the male does not dismount but releases genital contact and then guards the female from other males until she has finished laying her mature eggs, which she signals by side-to-side movements. The male then dismounts and the female flies away immediately, returning only when her next egg batch is mature, when she mates again before laying the next batch, and so on.

This pattern poses the question of why females are polyandrous (the average number of ejaculates stored is around 3 per female; Demont et al. 2011). Tregenza et al. (2003) found no simple benefits or costs of double versus single mating for females, but Hosken et al. (2002) found that females mated once survived longer than those mated three times, suggesting that longevity costs are associated with multiple mating. There are physical costs of mating to females (Hammer 1941; Parker 1978a; Demont et al. 2011). There are also obvious time costs of supernumerary matings (Parker 1970b). All this suggests that significant female benefits must accrue to polyandry to offset its costs. A number of possible advantages have been found or proposed. Polyandrous mating at each return to the dropping for oviposition can result a large saving in time at the dropping to a female, which arises from the benefits of gaining a guarding male for oviposition (Parker 1970b). Also, males that were more successful in sperm competition sired offspring that developed faster (Hosken et al. 2003). Another possible benefit relates to cryptic female choice (see also Sect. 7.3.2(v)). For example, elegant field experiments, in which all reproducing parents and progeny arising from artificial droppings were genotyped, showed that for females the total number of offspring and proportion of offspring emerging increased with the degree of polyandry (Demont et al. 2012).

The maximisation criterion used in optimality models of male-male competition in dung flies is eggs fertilised per minute of reproductive activity. The expected value of time at the dropping to a male (0.23 eggs/min; the mean for over a hundred droppings) was calculated as: total eggs oviposited into the dropping by all females divided by total time spent by all males at that dropping (Parker 1970c).

7.3.1 Pre-Copulatory Adaptations

7.3.1.1 Competitive Mate Searching by Males Matches Ideal Free Predictions

The numbers of each sex at a dropping shows a rise to a peak, then a gradual decay as the dropping ages; the male peak is much earlier than the female peak (Parker 1970c). From the average time each female spends at the dropping, the rate of arrival of females was calculated to be a decay curve, with the highest female arrival rate, $F(t)$, immediately after dropping deposition at time $t=0$. Knowing the number of males present through time t, males present at the earliest times could be shown to experience highest fertilisation rates. Thus males arriving instantly and remaining for a very short "stay time" would appear to be at an advantage. However, such males would experience high fertilisation losses due to excessive times spent searching for new droppings (the average time taken to find a new dropping is c. 4 min). The ESS consists of a distribution of stay times such that all males achieve equal fitness in terms of probability rates of capture of females ($=c$ per min). When travel time between droppings was included, all males were shown to experience similar gain rates, whatever their stay time at the dropping (Parker 1970c). Their behaviour matched the ESS, which is defined by all males arriving as quickly as possible to a given new dropping, then showing a phased departure so that the number of males present at time t, $m(t)$, balances the rate of arrival of new females: $m(t) = c^{-1} \cdot F(t)$, i.e. they should obey 'input matching' (Parker 1978a), a temporal version of the ideal free theory distribution (Fretwell 1972). Later, the claim of equal male gain rates was criticised by Curtsinger (1986) on various grounds, including the fact that differences in stay time had not been tested statistically. However, when examined, no statistical difference from the input matching prediction could be found, and other criticisms were also refuted (Parker and Maynard Smith 1987). The evidence that male dung flies show input matching during mate searching, as predicted by sexual selection, remains strong.

In addition to this temporal ideal free evidence, there is also good evidence that males obey ideal free searching in space around the dropping (Parker 1974b). There are three sources of gain for a male: newly-arriving, gravid females, take-overs of females from copulating pairs, and take-overs of females from guarding males while the female is laying her eggs (after a take-over, the successful male immediately mates with the female and then fertilises over 80 % of the subsequent egg batch). During the first 20 min after deposition, male search strategy is geared towards newly-arriving females (often encountered in the grass round the dropping); the proportions of males searching in each of a series of concentric zones on and around the dropping matched ideal free expectation (Parker 1974b, 1978b). But later, gains from take-overs (especially of ovipositing females) become significant, drawing more males to the dropping surface rather than the surrounding grass. Parker (1974a) predicted y = the ESS proportion of the total searching males expected on the dropping surface in relation to two variables: x = time after dropping deposition, and z = the total number of searching males. This predicted three-dimensional profile of $y(x, z)$ showed a good fit to the observed profile, suggesting that males respond to both (i) time after dropping deposition and (ii) number of competitors in determining where to search for females (surrounding grass versus dropping surface).

7.3.1.2 Males Show Intense Struggles for the Possession of Females

Dung fly males show specialised guarding behaviour and extreme contest behaviour; both are directed exclusively to gaining or retaining females (Parker 1970d)—there can be no doubt that they have evolved through Darwinian intra-sexual selection. The male mounts as soon as a gravid female is encountered, copulates, and then guards the female during oviposition. However, especially when a female flies directly to the dropping with a high density of searching males, two or more males may contact the female simultaneously and a protracted struggle then develops between males for possession of the female. If a searching male approaches a copulating or ovipositing pair, the paired male shows specialised behaviours that deflect the attacker away from the female. But should a second attacker approach while the paired male is deflecting the first attacker, the second attacker may manage to grasp the female and insert himself between the paired male and the female, resulting in a struggle. The probability of take-over is higher during oviposition than during copulation (Parker 1970d), but irrespective of when it occurs, after a take-over the new male copulates and guards the female while she lays her remaining eggs.

Parker and Thompson (1980) examined the time distribution of dung fly struggles, again taking male fitness as expected number of eggs fertilised per min, but modifying the struggle time to take account of the energetic costs of struggling relative to searching for an alternative female. One second spent struggling was assumed to cost k seconds searching; likely values for k were deduced from published data on insects. Superficially, struggles between males could plausibly match the predictions of the symmetric war of attrition with linear costs (Maynard Smith and Price 1973) since struggle durations showed a negative exponential distribution with a mean in the expected range. However, this concordance was lost when the data were examined in categories; for example, there was a much higher probability that the holder will win than the attacker.

Dung fly struggles are asymmetric contests, and males with larger body size have a greater probability of winning (Sigurjónsdóttir and Parker 1981). Struggles are probably settled by assessment as information is acquired during the contest about the relative 'resource holding power' (RHP; Parker 1974b) of the two males. The attacking male is typically larger than the guarding male, and as the relative size of the guarding male increases, the persistence duration of the attacker decreases. The RHP of the guarder appears to be influenced by his size relative to both (i) the attacker and (ii) the female he guards. Interestingly, the duration of struggles in which there was no take over (i.e. when the attacker gave up) increased with the number of eggs remaining to be laid by the female, suggesting that the paired male's choice of persistence time increased with the value of the female, as may be expected from contest theory if the paired male 'knows' how many eggs have been laid. However, this was not so when a struggle resulted in a take over (i.e. when the paired male gave up), suggesting that the attacker had no information about the eggs remaining to be laid, which again seems plausible. Sigurjónsdóttir and Snorrason (1995) examined the body size of flies in relation to their spatial distribution around droppings, and found that males guarding ovipositing females were on average similar in size to those copulating on the dung, but larger than males copulating in the grass, which they interpreted as being due to various effects, including the advantage of male size in take-overs. In flies reared under high and low density conditions, Stockley and Seal (2001) found that the propensity to begin struggles increased in relation to body size among males reared at high density, though the opposite trend was found in those reared at low density.

Sexual selection intensity (male mating success) was measured directly in the field by Jan et al. (2000), and conformed to the behavioural observations. As expected, selection intensity increased with male competitor density at a dropping. Though there was some evidence that small males had higher mating success at very low densities, overall, large males had higher mating success. Jan et al. found higher selection intensity for large size in males than females (see also Blanckenhorn 2007), a result consistent with the observed sexual size dimorphism in yellow dung flies, where males are typically considerably larger than females.

7.3.1.3 Pre-Copulatory Female Choice

Though generally agreed to be a predominantly male-controlled mating system, there is some evidence that at low male densities—when females are potentially able to choose between males—they show preference for pairing with larger males; this was argued to relate to the benefits of having a large male guarding during oviposition, and so avoiding struggles, which can be costly and damaging to females (Borgia 1981).

These studies on the average copula duration with gravid females show a small discrepancy between the observed (36 min; Parker 1970f) and the predicted optima (42 min, including meetings with undetected virgins, Parker et al. 1993). More recent studies have sought to evaluate dung fly copula duration in greater detail by examining optima in terms of phenotypic size variation of males and fecundity variation in females. The evidence suggests that copula duration is optimised across all male size phenotypes (i.e. the observed regression of copula duration against male body size matches the optimal regression, holding female size constant at the species average). Further, holding male size constant at the average, the observed regression varying female size also appears to be around the optimum predicted for the male.

Two factors influence the optimum in relation to male size: (i) sperm displacement rate increases with male size, and (ii) time to find and guard a new female decreases with male size, due to a size advantage in gaining take-overs in struggles for females (Parker and Simmons 1994). Charnov and Parker (1995) showed that these two effects interact so that optimal total sperm allocation should remain approximately constant with male size. Hence small males, with lower displacement rates, should copulate for longer time than large males to achieve equal input. As expected, observed copula duration decreases with male size (Ward and Simmons 1991; Parker and Simmons 1994; Simmons et al. 1999). The first calculation of the predicted relationship between copula duration and male size assumed that males displace sperm directly from the female sperm stores (Parker and Simmons 1994). This gave a good fit with the observed relationship, except for small males, where longer copula durations were predicted than were observed. Later, it was found that sperm displacement is indirect; sperm flow from the male aedeagus into the female's bursa, and is then transferred by movements of the female tract to the spermathecae (Hosken 1999; Hosken and Ward 2000; Simmons et al. 1999). When the predicted relationship was remodelled for this indirect transfer method the poor fit for small males disappeared, generating a very good fit between predicted and observed copula durations across all natural male sizes (Parker and Simmons 2000). Thus size-dependent optimal sperm displacement in dung flies can thus be explained by fertilisation rate maximisation in relation to the factors (i) and (ii) above.

Male dung flies vary their sperm allocation in relation to female fecundity: copula duration increases with female egg content (Parker et al. 1999). This observation matches predictions, and the match is again quantitative both for matings with new, fully gravid females arriving at the dropping, and also for matings with females taken over by a new male part way through oviposition. In addition to egg content, a second factor that must be taken into account in optimality models is the fact that a female's reproductive tract dimensions (notably her spermathecal volumes) increase with her size, decreasing the sperm displacement rate (Parker et al. 1999). Fertilisations in future clutches exert only a small effect on predicted copula duration for matings with fully gravid females, but exert an increasing effect as oviposition proceeds. For gravid, newly-arriving females, the number of mature eggs increases linearly with female size (Parker 1970f). The observed copula duration was found to increase with female size in a close quantitative fit with the predicted relation, and males probably assess female size rather than egg content directly

(Parker et al. 1999). For females taken over during oviposition, the eggs remaining to be laid depends on the timing of the takeover, and the observed copula duration decreases as eggs decrease, again fitting the prediction qualitatively except that the latter is slightly steeper than the observed relation. Males successful at take over may assess female egg content by how her much abdomen is distended; distension decreases notably throughout oviposition.

7.3.2.4 Experimental Evolution Produces Changes Predicted by Sexual Selection

Studies of experimental evolution in dung flies have generated the evolutionary responses predicted by sexual selection (Hosken and Ward 2001; Hosken et al. 2001; Martin et al. 2004). These experiments involved lines selected under either enforced polyandry (each female mated with 3 different males before oviposition, enabling post-copulatory sexual selection), or monogamy (each female mated only once, precluding sexual selection). Theory predicts that relative testis size should increase with the mean level of sperm competition in a population (reviewed in Parker and Pizzari 2010), and matches to this prediction have been found in so many comparative studies that relative testis size is now used ubiquitously as an indicator of sperm competition level. Monitored after only 10 generations, a strong divergence in testis size was found between monandrous and polyandrous dung fly lines, with much larger testes in polyandrous lines, where sperm competition was present (Hosken and Ward 2001; Hosken et al. 2001). Females in polyandrous lines evolved larger accessory sex glands, which are argued to increase female ability to influence paternity: males' success as second mates was lower in females in polyandrous lines (Hosken et al. 2001). However, males from polyandrous lines achieved higher paternity under sperm competition, supporting the prediction of increased testis size. A trade off may apply here: increased investment in testis mass appears to correlate with decreased immune function (Hosken 2001). By rearing larvae under high and low density conditions, Stockley and Seal (2001) found that males reared at high density with larger testes were less active in mate-searching, suggesting a trade off between testis investment and mate searching activity; however, the same trend was not found in males reared at low density.

Martin et al. (2004) compared fitness traits (lifetime reproductive success and longevity) of females evolved under enforced monogamy with those evolved under polyandry after each female had a single mating with a male from one of the two selection regimes. Females from polyandrous lines had lower fitness; they died earlier and produced significantly fewer progeny. Martin et al. plausibly argue that these results arise from sexual conflict inherent with the polyandrous selection regime.

7.3.2.5 Studies on Sperm Selection by Females

There have been several attempts to demonstrate female choice aspects of sexual selection in dung flies in terms of cryptic female choice, i.e. post-copulatory sexual selection in which the female selects sperm from alternative ejaculates (Eberhard 1996). Ward (1993) was first to propose that female dung flies bias paternity, and that this may account for some of the (typically high) variation in the proportion of last-male fertilisations (P_2) typically seen in paternity studies (Ward 2000). Hellriegel and Ward (1998) investigated theoretically plausible mechanisms enabling sperm preference with single or multiple sperm stores. For example, females having one store could apply different storage rates for different ejaculates; those with two or more stores could also separate ejaculates across stores. Ability to choose sperm from a given store enables far more effective paternity control, and offers a plausible hypothesis for why females often have more than one sperm store (e.g. Matsuda 1976; Ward 1993; Eberhard 1996; Hellriegel and Ward 1998). Dung fly females typically have three spermathecae and infrequently four.

The success of dung fly eggs depends on the topography and microclimate of the place of oviposition on the dropping; choice of a suitable oviposition site increases female reproductive success (Ward et al. 1999). Ward (1998) raised larvae of different phosphoglucomutase (*pgm*) genotypes in two different dung conditions with the same means for humidity and temperature, but in one set the temperature remained constant and in the other set it was variable. He found that the most successful genotype differed between the two sets. In an experiment in which females were constrained to lay in simulated 'sun' or 'shade' conditions, one of two *pgm* alleles was relatively commoner in eggs laid in 'sun' the other relatively more common 'shade'; differences in hatching or mortality could be discounted from this effect. He suggested that females use sperm selection to lay eggs of different genotypes under different sun/shade conditions, increasing offspring fitness by matching their genotypes to the larval growth conditions. Ward (1993, 1998) also found that in fixed length copulations, females stored more sperm from larger males, though whether this is due to cryptic female choice (Ward 1998) or to the fact that larger males have higher sperm input (and hence displacement) rates (Simmons et al.1996) remains controversial. Ward (2000) also found higher last male paternity if the second of two males to mate was genetically similar to the female at the *pgm* locus, and suggested that this involved cryptic female choice. In the field, *pgm* alleles from eggs were found to be non-randomly distributed between both (i) north and south slopes and (ii) shaded and sunny areas of artificial cow pats (Ward et al. 2002), but whether this effect arose from sperm selection by females or from different behaviour of females of different genotypes could not be determined.

However, two more recent studies generate pessimistic conclusions about the hypothesis of cryptic female choice of *pgm* alleles. Blanckenhorn et al. (2012) performed extensive lab and field investigations on the activity of *pgm* alleles and their effects on larval development times at different temperatures, and on the distribution of alleles in eggs deposited on the warmer southern slopes of droppings compared to those on the north slopes. They concluded that although *pgm* activity differences were apparent, and that *pgm* genotype did differentially affect development time, eggs laid on the north versus south slopes showed no biases in *pgm* composition as indicated from the previous work, removing the basis for cryptic female choice of sperm with different *pgm* genotypes. Further, Demont et al. (2012) performed field experiments in which females could choose to lay eggs in three different dropping microenvironments (south slope, ridge, and north slope), and genotyped both (i) the resulting offspring, and (ii) the sperm remaining in the female sperm stores after oviposition. Although (as expected) females showed a greater preference to oviposit on north slopes as ambient temperature increased, they found no evidence that females biased paternity towards certain male genotypes depending on the offspring's microclimate.

Bussière et al. (2010) used molecular techniques to demonstrate that although the mean proportion of sperm stored in the spermathecae match the published mean average paternity for the last male (the P_2 value), sperm from different males are not stored randomly across the female's sperm stores (see also Otronen et al. 1997; Hellriegel and Bernasconi 2000). The mean number of ejaculates stored also differs across spermathecae (Demont et al. 2011, 2012). Thus while hints are present, and the capacity for it certainly exists, clear evidence for cryptic female choice in dung flies has so far proven elusive.

7.3.3 General Comments on the Dung Fly Sexual Selection Studies

We have reviewed the dung fly studies at length because they represent perhaps the most detailed investigations of a wide range of aspects of sexual selection in a single species. The male-male competition studies have shown many *quantitative* fits between field and lab observations and model predictions across a wide range of male pre- and post-copulatory reproductive activities, providing very strong evidence that this component of sexual selection has indeed been a prime selective force moulding male behaviour in this species. This evidence clearly runs quite counter to the claim that "....There are fundamental problems that universally undercut all applications of sexual selection theory to any species...." (Roughgarden et al. 2006).

However, while the potential for females to exercise cryptic post-copulatory choice has been well established, and fertilisation biases detected, attempts to demonstrate that females select sperm in a manner that yields clear adaptive benefits have not yet been successful.

Thus the large amount of empirical work and modelling on sexual selection in the yellow dung fly reveals a trend that appears to be rehearsed in general for sexual selection studies: while evidence for male-male competition as a major selective force in evolution is widespread and highly supportive, evidence for female choice is less advanced and often controversial.

7.4 The Logical Imperative: Evolutionary Steps in Sexual Strategy

The logical imperative for Darwinian sexual selection is founded upon a predictable sequence of evolutionary events beginning with the evolution of recombination and sexual reproduction. The inevitability of this sequence is remarkable, since each step drives the next in an evolutionary cascade (the 'sexual cascade'; Parker 2014) leading to males and females that coexist commonly as two highly differentiated sexual morphs with internal fertilisation. We outline these events as a series of separate steps; although in general the evolution of one step precedes and creates the selective pressure for the next, some degree of synchronicity in adaptation is likely. A further perspective on the sexual cascade is given in Parker (2014).

We argue that the transitions in sexual strategy are driven initially by gamete competition, and after the evolution of anisogamy, by sperm competition in association with changes in mobility and mode of fertilisation, eventually enabling pre- as well as post-copulatory sexual selection to operate. Figure 7.1 gives a summary to accompany the text.

7.4.1 The Evolution of Sex: Sexual Recombination and Isogamous Gamete Production

Sexual reproduction is a composite phenomenon that can be subdivided into a number of components—fusion, recombination, fission, and the male-female phenomenon—each component being subject to selection (Baker and Parker 1973). Genetic recombination is ubiquitous in living organisms from the simplest to the most complex, and may have evolved in the earliest organisms. Gametic fusion

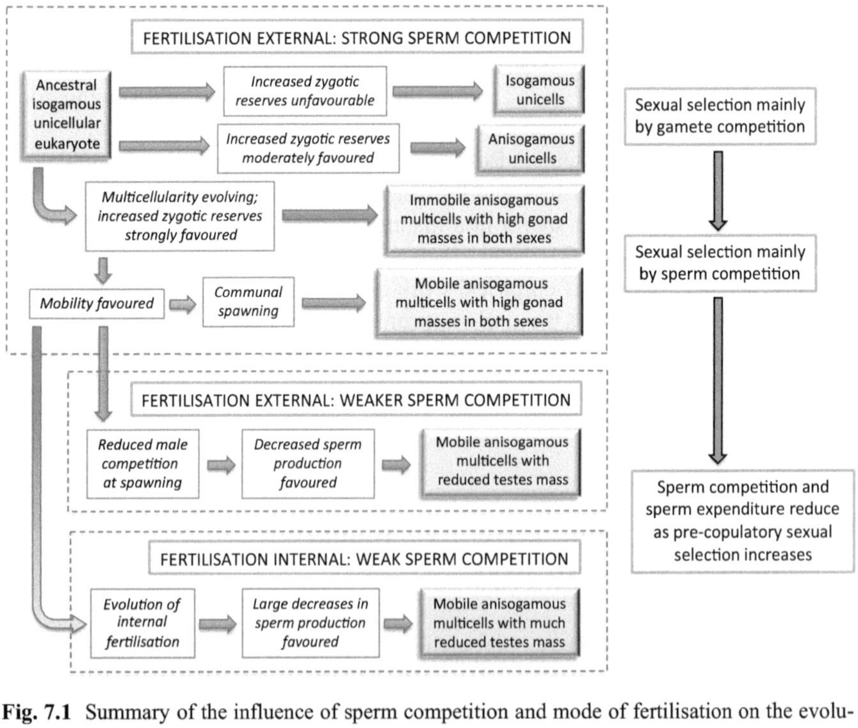

Fig. 7.1 Summary of the influence of sperm competition and mode of fertilisation on the evolution of sexual strategies and sexual selection

(syngamy) and recombination in eukaryotes involve the evolution of meiosis and the haploid-diploid cycle (e.g. see Maynard Smith and Szathmáry 1995). The selective advantage of sexuality over asexuality has been one of the longest and most enduring puzzles for evolutionary biologists. For example, in the nineteenth century, Weismann (1889) proposed that sex functioned to generate genetic variation, while Darwin (1889) favoured an explanation in terms of hybrid vigour. Later, Fisher (1930) proposed that a sexual population could evolve (and hence adapt) faster than an asexual population, and explicitly envisaged that sexual recombination was one of the very few adaptations that relied upon group selection. Muller (1932) noted that individuals in an asexual population irreversibly accumulate deleterious mutations (termed 'Muller's rachet' by Felsenstein 1974), and explained the success of sexual reproduction over asexual reproduction as a means of overcoming this costly accumulation. This benefit may not apply in asexual organisms that have asexual forms of recombination. The start of an avalanche of theoretical research of the past 40 years on the evolution of sexual recombination appears to have begun with Maynard Smith's (1971) classic paper "What use is sex?".

Maynard Smith (1978) noted that the advantage of sex must be sufficiently large to overcome the 'two-fold cost of sex' which arises as the cost of anisogamy (i.e. mainly a cost of producing males, but see Lehtonen et al. 2012): a mutant female able to reproduce parthenogenetically by producing similar females would replicate at twice the rate as a sexual female. Note that this is a requirement for the maintenance of sex in a sexual population rather than a requirement for its origin. Sex is likely to have occurred in an isogamous population (in which parents share the investment in the zygote, allowing each parent to produce twice as many offspring as a female in a sexual population), in which the evolution of sex would be much less costly (e.g. Lehtonen et al. 2012). However, the 'two-fold cost of sex' suggests that the overall advantage of sexual reproduction must be high in order that it is maintained against invasion by asexual mutants.

By now, many different theories have been proposed for the widespread maintenance of sex (see recent review of Hartfield and Keightley 2012), generating a vast literature (including several books). Later in his life, Maynard Smith (pers. comm.) became convinced that no single theory offers a general explanation for sex, but that the many mechanisms taken together may offer a sufficient account. West et al. (1999) have extended this pluralist view, stressing the advantages gained from considering that multiple mechanisms and their interactions operate to maintain the ubiquity of sex.

Whatever the advantage of genetic recombination, sexual fusion (syngamy), with its merging of the cellular investment of one gamete with that of another gamete of different genetic constitution, can set the scene for conflict or cooperation over the investment from each partner, and the evolution of anisogamy.

7.4.2 The Evolution of Two Sexes: Anisogamy

The ancestral state in eukaryotes is likely to have been a unicellular organism with isogamy, i.e. where the fusing gametes are of similar size, and hence contribute equally to the zygote (Fig. 7.1). However, it is clear that under many conditions isogamy is unstable, and in such cases soon after the evolution of gametes and sexual re-combination, selection is likely to have favoured a drive for anisogamy from the ancestral isogamy.

It is convenient to define sexes in terms of the gamete size-morph that an individual produces (Parker 2011). Thus an isogamous population consists of individuals of just one sex; and an anisogamous population consists of two sexes—males (microgamete producers) and females (macrogamete produces). This definition of sexes differs from one defining sexes in terms of mating types, in which one gametic mating type fuses with a dissimilar gametic mating type, which may or may not be the same size. Much less confusion is caused by defining sexes in terms of the gamete size a phenotype transmits (i.e. males convey small gametes, females large gametes, and hermaphrodites, which are male and female in one soma), and by defining mating types in terms of gamete types that can or cannot fuse together. Hermaphroditism (where two sexes coexist in one phenotype) is probably a derived state arising from special conditions (e.g. see Charnov et al. 1976); the initial mutations are likely to have been those affecting the size of gametes produced by given parents, leading to gamete dimorphism with two separate sexes.

There are several theories for the evolution of two sexes, most of which assume an origin from pre-exisiting gametic mating types (e.g. see review of Lessells et al. 2009). Two leading proposals—'gamete limitation' and 'gamete competition'—both focus on: (i) fusions gained, and (ii) zygote survival prospects. The initial theory, gamete limitation, dates back to Kalmus (1932; see also Kalmus and Smith 1960; Scudo 1967; Dusenbery 2000), who showed that when the probability of fusion is limited, a population with anisogamy and union between many micro- and few macro-gametes could achieve more surviving zygotes than an isogamous population with intermediate numbers of gametes. This theory was revitalised in an individual selection context by Cox and Sethian (1984, 1985), and Levitan (e.g. 1996, 1998) who explicitly considered the effect of how gamete size affects collision probability through its effects on 'target' size. More recent demonstrations that the Kalmus effect alone can generate anisogamy under individual selection (e.g. Iyer and Roughgarden 2008) appear marred by bias to demonstrate that cooperation rather than sexual conflict shapes sexual strategies, in line with Roughgarden's 'social selection' hypothesis (see Parker 2011). However, the most advanced recent analysis (Lehtonen and Kokko 2011) clearly and elegantly confirms that Kalmus' gamete limitation hypothesis can account for the evolution of anisogamy by individual selection provided that gamete competition is low or absent. It appears that Darwin had achieved some intuition about this effect, when he wrote: "With lowly-organised aquatic animals, permanently affixed to the same spot and having their sexes separate, the male element is invariably brought to the female; and of this we can see the reason, for even if the ova were detached before fertilisation, and did not require subsequent nourishment or protection, there would yet be greater difficulty in transporting them than the male element, because, being larger than the latter, they are produced in far smaller numbers." (Darwin 1874, p. 222).

The gamete competition theory of Parker et al. (1972) envisaged a large population of ancestral marine unicells with essentially isogamous gametes. Unlike the gamete limitation models, their model (analysed by computer simulation) does not assume that gamete size is linked to mating types; selective fusion is envisaged to evolve later (Parker 1978c). Parents release gametes varying somewhat in size, *m*, fusion is random and most or all gametes fuse, so that the set of gametes produced by each parent compete in the same 'pool' for fusions. The ESS is isogamy or anisogamy, depending on how the viability or success, *f*, of the zygote increases with its size $S = m_i + m_j$, i.e. on the zygote-size fitness function, $f(S)$. Many, but not all, of the subsequent developments of this model start with the assumption of mating types, as does that of Bulmer and Parker (2002) who include both a zygote-size fitness function and a gamete-size fitness function, $g(m)$, to show how these interact to determine which ESS is achieved, anisogamy or isogamy. As the zygote-size fitness function moves further away from the gamete-size fitness function, requiring a larger size before fitness begins to increase steeply, the ESS changes from isogamy to anisogamy. Bulmer and Parker (2002) argued that this change would reflect the transition from uni- to multi-cellularity, as originally proposed by Parker et al. (1972).

Which effect, gamete limitation or gamete competition, has been more important in the origin and evolution of anisogamy? Lehtonen and Kokko (2011) have generated important new insights by showing that both gamete competition and gamete limitation can lead to anisogamy, depending on the conditions. Using a development of Bulmer and Parker's (2002) model in which they included 'consistency' (i.e. average fitness of male and female must be equal if the sex ratio is unity), Lehtonen and Kokko modelled the situation where the number of parents in the local mating group could vary. Thus there is no gamete competition when just two parents of different mating type occur, and gamete competition increases with the number of parents in the group. Their analysis shows that anisogamy could indeed originate through either gamete limitation and gamete competition mechanisms. Even low levels of gamete competition generate anisogamy when gametes can fuse fairly readily, but conditions of gamete limitation and low gamete competition can also generate anisogamy. The isogamy ESS disappears relatively quickly (but not immediately) with the numbers of parents in the mating group.

Given that both gamete limitation and gamete competition can lead to anisogamy (and hence the two sexes), which condition has had the bigger influence on its origin depends on conditions in the ancestral isogamous unicells from which anisogamy evolved. Though some gamete limitation in these organisms seems quite plausible, so does fairly intense gamete competition due to gametes being released into the sea. The latter effect (and recent theory; Parker and Lehtonen 2014) would certainly favour gamete competition as the more potent selective pressure, and reflects our own view (see also Lessells et al. 2009), but we suspect that this question may never yield a definitive answer. Gamete (sperm) competition certainly offers a plausible solution for the maintenance of anisogamy in most current populations (Parker 1982).

Theory suggests strongly that increasing organismal complexity during the evolution of multi-cellularity favours anisogamy because of the need for larger zygotes (Parker et al. 1972; Bulmer and Parker 2002; Lehtonen and Kokko 2011), for which there is also empirical evidence (see review of Parker 2011). Thus, once sexual reproduction and sygamy have evolved, we can readily explain the evolution of two sexes due to increased importance of zygotic reserves associated with the evolution of increased organismal complexity in multi-cellularity. Starting from an ancestral isogamous (probably marine) unicellular eukaryote, if increased zygotic reserves are not favoured by selection, the ancestral isogamous state is retained, but if increased reserves are favoured, anisogamy will develop (Fig. 7.1). The most likely candidate driver of this event, in our view gamete competition, is a primitive form of fertilisation competition analogous to sperm competition under sexual selection.

7.4.3 The Evolution of the Sex Ratio

Darwin (1874) struggled with the problem of the evolution of the sex ratio, which was later solved by Fisher (1930; 'Fisher's principle') in a cryptic verbal argument, first made formal by Shaw and Mohler (1953). Envisage a large, random-mating population, in which anisogamy and selective sperm-egg fusions are established, most eggs are fertilised, and the cost of each male or female offspring is equal. Since each offspring has a mother and a father, the summed fitness of all male individuals must equal the summed fitness of all females (a requirement termed 'Fisher consistent' by Lehtonen and Kokko 2011). Thus in a population with unequal numbers of males and females, individuals of the rarer sex will have higher mean fitness, and genes for production of the rarer sex will increase until the sex ratio at the end of parental care becomes unity. This is an equilibrium, at which (deterministically) the mean fitness of each male equals the mean fitness of each female, and selection on sex ratio genes becomes neutral. Differential mortality of males and females after the end of parental investment does not affect the unity sex ratio, since if one sex suffers greater juvenile mortality it becomes the rarer sex, which compensates for its higher mortality. When selection has produced the unity sex ratio in a population, the expected gain from producing a male or a female offspring becomes equal for the parent. Hamilton's (1967) classic paper established a theoretical basis for 'extraordinary' sex ratios, and sex allocation theory has by now become a large research field (Charnov 1982; West 2009).

Lehtonen and Kokko (2011) note that analyses of the evolution of anisogamy that start by assuming the existence of mating types are not strictly 'Fisher consistent'. However, the analysis of the evolution of anisogamy by Parker et al. (1972) involved a genetics-based computer simulation in which gametes fused randomly (i.e. no mating types). Gamete size was determined by alleles at a 'gamete-size locus'. When anisogamy evolved by disruptive selection against alleles for intermediate gamete sizes, the result was a polymorphic equilibrium in which large and small gamete-size alleles coexist. In conditions generating high degrees of anisogamy, most fusions occurred among the vast numbers of microgametes, which died because they lacked enough reserves to survive as zygotes. What remained was a population consisting of equal numbers of proto-males and proto-females. Thus if J is a dominant allele for micro-gamete producing, and A its recessive allele for macro-gamete producing, the surviving genotypes were JA males and AA females, resembling the XY, XX sex-determining system. Reversing the dominance gave JJ males and JA females, resembling the ZZ female, ZW male system. Had the simulations started from mating types, or had they allowed selective fusion to evolve during the evolution of anisogamy, the unity sex ratio would also have been generated but without wastage of huge numbers of micro- x micro-gamete fusions. Parker et al. (1972) interpreted this unity sex ratio result as being due essentially to Fisher's principle operating in their simulations.

Thus unless special conditions apply (Hamilton 1967), anisogamy typically generates an equal sex ratio, essentially by Fisher's principle. With the drive to multicellularity and increasing complexity, anisogamy can readily be explained and will typically generate equal numbers of males and females.

7.4.4 The Evolution of Copulation and Internal Fertilisation

Primitively, fertilisation is likely to have been external. While the ancestral unicells may have been motile, e.g. through the action of cilia and flagella, in plants multicellular forms are usually sessile, as are many primitive multicellular invertebrate animals. Primitive sessile invertebrates such as sponges, certain coelenterates and echinoderms typically broadcast sperm into the sea, and fertilisation may be either external or occur after contact with ova held within the female (or hermaphrodite) soma. Such marine systems may involve intense sperm competition in which ejaculates from large numbers of different males (or hermaphrodites) compete for ova. Theory predicts that intense sperm competition is likely to result in high male gametic expenditure (Parker and Pizzari 2010), and since females should also maximise gamete production, this should result in similar high gonad expenditures in the two sexes in sessile or weakly mobile forms (Fig. 7.1). In conformity with this prediction, equal male and female body size and similar, high gonad masses are indeed characteristic of broadcast spawning marine invertebrates such as relatively immobile sea urchins, where sexual dimorphisms in body and gonad size are rare and related to special biological circumstances (see Levitan 2005). Gonad expenditures remain high and are often similar for the two sexes (or occasionally higher for males than females), resembling what would probably have been the ancestral state (Table 7.1). Sperm limitation has also been argued to maintain high male gametic expenditure in broadcast spawners (Levitan and Petersen 1995). It is important to remember, however, that sperm competition can apply even when there is sperm limitation—the sperm competition level increases with the number of different ejaculates competing for a given set of ova rather than with decreasing sperm limitation, though there will often be a negative correlation between sperm competition level and increasing sperm limitation.

High male gametic expenditure is also retained in many mobile vertebrates with external fertilisation when spawning is communal, so that high levels of sperm competition are retained, and can be even higher than in relatively immobile broadcast spawning invertebrates (Table 7.1), though in such cases males typically use their mobility to release sperm as close to a spawning female as possible (e.g. many communal spawning fish; Breder and Rosen 1966).

Table 7.1 Examples of gonadosomatic index (GSI) in relation to mode of reproduction across the animal kingdom, in taxa with separate sexes

Taxon	Species	Male GSI (%)	Authority	GSI dimorphism	Authority	Comment	Mode of reproduction
Invertebrates							
Cnidaria							
Anthozoa							
	Paramuricea clavata			Male > female	Gori et al. (2007); Coma et al. (1995)	No GSI value available but gonad volume values per polyp given in Gori et al. (2007) and calculated from data in Coma et al. (1995, Table 7.1), both at reproductive peaks	Sessile marine broadcast spawner, zygotes brooded on surface of female colonies (Coma et al. 1995, Gori et al. 2007)
	Eunicella singularis			Male > female	Gori et al. (2007)	No GSI value available but gonad volume values per polyp given in Gori et al. (2007) at reproductive peaks	Sessile marine broadcast spawner, zygotes brooded inside female polyps (Gori et al. 2007)
Scyphozoa	*Aurelia aurita*	6–9 (dry weight) 9–17 (ash free dry weight)	Lucas and Lawes (1998)	Roughly equal	C. H. Lucas, pers. comm	Male GSI values are deduced from female GSI taken from dry weight at peaks of maturity	Sperm shed externally in proximity to females (Lucas 2001) in spawning aggregations (Hamner et al. 1994); fertilisation internal
	Periphylla periphylla	2.8	C. H. Lucas, unpublished data	Male > female; 2.8/1.4	C. H. Lucas, unpublished data	GSI values from wet weights; significant difference between GSI means ($P<0.003$, df=19) but not bell diameters (C. H. Lucas, unpublished data)	Marine non-synchronous broadcast spawner
Echinoderms				Usually equal	Levitan (2005)	Usually high degree of multiple paternity (Levitan 2005)	Usually marine broadcast spawners

Table 7.1 (continued)

Taxon	Species	Male GSI (%)	Authority	GSI dimorphism	Authority	Comment	Mode of reproduction
Asteroidea	*Odontaster validus*	5.5–11.7	Grange et al. (2007)	Roughly equal	Grange et al. (2007)	Female GSI varied between 4.6-10.4 %; SD overlapping with male GSI (Grange et al. 2007)	Marine broadcast spawner
Ophiuroidea	*Ophiocoma alexandri*	28–43	Benítez-Villalobos et al. (2012)	Male > female	Benítez-Villalobos et al. (2012)	From June, September, October and April values; can be lower at other times of year	Marine broadcast spawner
	Ophiocoma aethiops	17–30	Benítez-Villalobos et al. (2012)	Equal	Benítez-Villalobos et al. (2012)	From June, September, October and April values; can be lower at other times of year	Marine broadcast spawner
	Ophionotus victoriae	0.5–2	Grange et al. (2004)	Equal	Grange et al. (2004)	GSI values are at peak maturity	Marine broadcast spawner
Echinoidea	*Sphaerechinus granularis*	2–8	Martínez-Pita et al. (2008)	Equal	Martínez-Pita et al. (2008)	Much variation between localities	Marine broadcast spawner
Molluscs							
Bivalvia	*Adamussium colbecki*	35	Tyler et al. (2003)	Male > female; 35/24.8	Tyler et al. (2003)	GSI values are at peak maturity; Chiantore et al. (2002) give lower values for female GSI of 7-10 %	Marine broadcast spawner
Gastropoda	*Fissurella maxima*	13	Bretos et al. (1983)	Appear roughly equal	Bretos et al.(1983)	GSI for November peak	Marine broadcast spawner
	Helcion pruinosus	25–34	Henniger (1998)	Male > female;	Henniger (1998)	Peak value; two localities	Marine broadcast spawner

Table 7.1 (continued)

Taxon	Species	Male GSI (%)	Authority	GSI dimorphism	Authority	Comment	Mode of reproduction
Cephalopoda	*Illex coindetii*	4.9	(Rosa et al. 2005)	Male < female; 4.9/11.4	(Rosa et al. 2005)	GSI values are at maturity	Internal fertiliser: male places spermatophore inside the female's mantle during mating. High levels of sperm competition reported in cephalopods (Marian 2012)
	Todaropsis eblanae	5.5	(Rosa et al. 2005)	Appear equal; 5.5/5.7	(Rosa et al. 2005)	GSI values are at maturity	Internal fertiliser: male places spermatophore inside the female's mantle during mating. High levels of sperm competition reported in cephalopods (Marian 2012)
Insects						Male GSI positively correlated with level of sperm competition all 5 groups studied (reviewed by Simmons and Fitzpatrick 2012)	Internal fertilisation, usually with copulation
Drosophilidae		2–11	Pitnick (1996)	Male < female?		Male GSI increased with body mass and with sperm length; high GSI associated with production of few remarkably long sperm (Pitnick 1996)	Internal fertilisation with copulation
Tettigoniidae		1–14	Vahed et al. (2011)	Male < female?		High male GSI associated with high polyandry and low nuptial gift investment (Vahed et al. 2011)	Internal fertilisation with copulation

Table 7.1 (continued)

Taxon	Species	Male GSI (%)	Authority	GSI dimorphism	Authority	Comment	Mode of reproduction
Onthophagus species		1.6–3.6	Simmons et al. (2007)	Male < female?		High male GSI ssociated with high frequency of minor males (Simmons et al. 2007)	Internal fertilisation with copulation
Photinus species		0.6–3.2	Demary and Lewis (2007)	Male < female?		Male GSI positively correlated with level of sperm competition which can be high (Demary and Lewis 2007); GSI value is dry weights of testes and seminal vesicles, increases to 0.7–14 if accessory glands added	Internal fertilisation with copulation
Vertebrates							
Fish		0–12	Stockley et al. (1997)	Usually male < female?		Male GSI positively correlated with level of sperm competition (Stockley et al. 1997), highest in communal spawners, lowest in species with internal fertilisation (see also Simmons and Fitzpatrick 2012)	Mostly external fertilisers, range from communal spawning to pair spawning, occasionally internal fertilisation
Amphibia							
Anura		0–8	Jennions and Passmore (1993)			Male GSI positively correlated with level of sperm competition (reviewed by Simmons and Fitzpatrick 2012) and other factors (Emerson 1997)	Mostly external fertilisation with amplexus, sometimes in aggregations
Anura	*Fejervarya limnocharis*	0.3	Othman et al. (2011)	Male < female; 3/12.4	Othman et al. (2011)	GSI for March—April peaks	External fertilisation with amplexus

Table 7.1 (continued)

Taxon	Species	Male GSI (%)	Authority	GSI dimorphism	Authority	Comment	Mode of reproduction
	Rana leptoglossa	0.5	Saha and Gupta (2011)	Male < female; 0.5/6.4	Saha and Gupta (2011)	GSI peak values	Males hold territories; courtship followed by amplexus with external fertilisation (Saha and Gupta 2011)
Reptiles							
Birds		0–10 mean = 1	Pitcher et al. (2005)	Parental care, often by both sexes		Reviewed by Simmons and Fitzpatrick (2012)	Internal fertilisation
Mammals		0–7 64% < 1 6% > 3	Kenagy and Trombulak (1986)	Parental care, usually by female only		Male GSI positively correlated with level of sperm competition in most groups (reviewed in Dixson 2009; Simmons and Fitzpatrick 2012)	Internal fertilisation

The evolution of mobility, involving various advanced modes of locomotion, has had important consequences: it has enabled release of sperm close to spawning females. Thus in many invertebrates and most vertebrates, higher fertilisation benefits have become available through mate searching and female-targeted sperm release than through sheer expenditure on sperm production. Enhanced mobility and female-targeted sperm release, coupled with the fact that Fisher's principle maintains the sex ratio at unity, favoured dramatic reductions in testes mass, associated with the trend towards either pair spawings with external fertilisation or, eventually, to copulation and internal fertilisation (Parker 1970a, 1984 2014; Levitan 1998). Targeted forms of sperm release yield both increased fertility benefits and immediate sperm competition benefits compared to untargeted broadcast spawning, and once expenditure on mate searching and female-targeted sperm release yields higher marginal fertilisation gains than numerical sperm production, reduction in sperm expenditure is predicted (Parker 2014). The ultimate extrapolation of this trend in sexual selection ends in copulation with internal fertilisation, and then the evolution of specialised male intromittent organs, driven perhaps predominantly by the benefits of reduced sperm competition (Parker 1970a). The reduction in testis expenditure creates the 'resource space' for the expansion in pre-copulatory male-male competition (Fig. 7.1)—it 'enables' Darwinian pre-copulatory sexual selection. Thus pre-copulatory sexual selection must be regarded as evolving secondarily to gamete and sperm competition.

The view that internal fertilisation has evolved primarily by sexual selection contrasts with classical views of its evolution solely by natural selection to increase fertilisation efficiency and gamete and zygote survival. While the latter benefits are highly likely to have played a significant part in reinforcing the drive towards internal fertilisation, we see them as being secondary to sexual selection as the main motive force. Copulation, followed by the evolution of specialised male intromittent organs, probably represents the final stages of sexual selection and reduced sperm competition. An alternative scenario could be that they were driven predominantly by female choice (Eberhard 1996), though we envisage that where it occurs, this has evolved later.

There is little doubt that internal fertilisation typically (but need not necessarily) results in much lower sperm competition than external fertilisation with communal spawning. Polyandry, and the evolution of female sperm stores, however, can maintain sperm competition as a powerful selective force. The range of sperm competition levels seen even in one taxon is often so great as to favour a vast array of relative testes sizes, from tiny to huge (e.g. fish; see Stockley et al. 1997).

While anisogamy, with tiny sperm and large ova, is clearly favoured under conditions of high sperm competition, the question remains as to why so many tiny sperm are still produced in internal fertilising species when sperm competition has reached very low levels (Parker 1982). One may imagine intuitively that with low sperm competition, it would pay males to provision sperm so that they could then contribute to the reserves of the zygote, which would challenge the stability of anisogamy. Provided that sperm competition risk is not trivially small, there is a very good reason why anisogamy is not lost due to sperm contributing to zygotic investment (Parker 1982, 2011; Lehtonen and Kokko 2011). This is because extra provisioning would need to be provided to each and every sperm, which would generally have large costs (e.g. under sperm competition) that outweigh the marginal benefits. For instance, it will not pay to add extra provisioning to each sperm to aid with zygote provisioning if the probability that a female will mate twice (generating sperm competition) rather than once (i.e. no sperm competition) is greater than $2/A$, where A = ovum size/sperm size (Parker 1982, 2011). This is an extremely robust condition, since A is usually likely to be a very large number, and for species with maternal care A should include the parental investment as well as the ovum costs, making the condition even more robust. Therefore, anisogamy appears to represent an irreversible evolutionary transition in most multicells. Typically, only when the probability of sperm competition falls towards zero can anisogamy be threatened, and in such circumstances males appear to increase parental care rather than sperm contribution to the zygote (e.g. sea horses, see Parker 2011 p. 49).

In summary, the evolution of enhanced mobility allows a trend towards female-targetted sperm release through higher fertilisation gains with increased proximity of sperm to the ova (for further details, see Parker 2014). Under communal spawning, male gametic investment remains high (often equalling that of females) since sperm competition remains high, but less communal spawning favours reduced sperm expenditure and an increase in expenditure on pre-copulatory male-male competition. Gains from female-targeted sperm release may account for the evolution of internal fertilisation and the evolution of male intromittent organs.

7.4.5 The Consequences: Pre-Copulatory Sexual Selection and High Secondary Sexual Differentiation

Sperm allocation theory predicts ESS expenditure on testes to increase with mean sperm competition level across populations (Parker and Pizzari 2010), and a comprehensive recent review (Simmons and Fitzpatrick 2012) of the many studies available have shown that this expectation is generally met: relative testes size usually (but not always) increases with sperm competition level across many animal taxa, and is now commonly used as an index of sperm competition level in comparative studies. A complication is that sperm competition level is associated with polyandry level (and hence the mating rate of both sexes), which affects sperm demand, though mating rate is more likely to affect the investment per ejaculate than testes investment (Parker and Ball 2005; Vahed and Parker 2012).

Theoretical models assume a fixed resource budget for reproduction, so that pre-copulatory expenditure on gaining matings trades off against post-copulatory expenditure on ejaculates, resulting in a negative relation between pre- and post-copulatory expenditures. While most theory assumes pre-copulatory male-male competition to be some form of scramble competition in which the number of matings increases linearly with pre-mating expenditure (e.g. competitive mate-searching), Parker et al. (2013) have outlined how different forms of male-male pre-copulatory competition can affect the ESS balance between pre- and post-copulatory expenditures, depending on the mean level of sperm competition experienced by the population. This does not appear to affect the general prediction that post-copulatory expenditure (i.e. on testes and ejaculate production) increases with sperm competition level.

Table 7.1 shows some examples of how GSI varies in relation to the mode of reproduction and expected sperm competition level across the animal kingdom, for taxa with separate sexes, in reproductive condition. We stress that (i) GSI is typically allometric, particularly in males, so the ranges shown must bear this in mind, and (ii) the maximum GSI level possible will vary considerably in taxa depending on somatic requirements, so that it is often more informative to consider how male GSI relates to female GSI in discussing the trends in Fig. 7.1. We anticipate that selection will generally push females towards maximum expenditure on GSI, while this will not be so for males when sperm competition is relaxed and pre-mating competition possible.

Marine broadcast spawning invertebrates show much variation in GSI just before spawning, depending on locality (and presumably feeding resources). They typically show (i) no obvious sexual size dimorphism, (ii) either similar GSI in males and females (published measures are often not separated for males and females) or higher male GSI, and (iii) much more male-biased GSI dimorphism than internal fertilising taxa (though exceptions can occur, e.g. the ophiuroid, *Ophionotus victoria*). Thus for cnidarians and echinoderms, selection typically maintains body size at similar levels in the sexes, and we anticipate that in both sexes virtually all reproductive investment is directed towards gametes. In Scyphozoa, the best index of GSI is probably ash free dry weight (see *Aurelia aurita*, Table 7.1). GSI values for *Periphylla periphylla* are expressed in wet weight; converting to dry weight would increase GSI because the percent dry weight of gonads, which are fairly organic-rich, is much higher than whole tissue which is predominantly watery (>95 %) mesoglea (Dr. C. H. Lucas, pers. comm.).

The same trends appear to apply for broadcast spawning marine molluscs, with a reduction GSI coinciding with internal fertilisation in cephalopods. Though they have specialised sperm stores and are therefore candidates for raised sperm competition, insects typically show fairly low male GSI (e.g. *Onthophagus* beetles, Table 7.1) unless special features intervene (e.g. *Drosophila*, Tettigonids, Table 7.1). They appear to show associations between sperm competition level and male GSI.

For vertebrates, there is evidence for many of the major taxa that relative testes size correlates positively with sperm competition level. Marine communal spawning fish retain high GSI, and have similar characteristics to marine broadcast spawning invertebrates (see above), despite their high mobility and the fact that males are often competitive in their attempts to ejaculate close to spawning females. Their high GSI and low sexual dimorphism is probably maintained by the high sperm demand due to the high sperm competition prevalent in communal spawns. In amphibians, though anurans usually have external fertilisation this is achieved in pair spawnings in amplexus (mating embrace), which results in reduced male GSI (Table 7.1) and also reduced male body size (Arak 1988). For land animals, internal fertilisation is almost obligatory (though not copulation; e.g. male thysanurans and collembolans deposit spermatophores on the substrate, which are picked up by females). It is quite possible that internal fertilisation first arose in aquatic ancestors to increase fertility and to reduce sperm competition, and served as a preadaptation to land colonisation. Thus birds and mammals have male GSI typically below 1 %, though it can rise to 7–10 %, and there is much evidence that relative testes size increases with sperm competition in these groups. Female GSI is not included in Table 7.1 since birds and mammals typically show high levels of parental care, which forms the large part of the female budget for reproduction.

Also in line with expectations, there is empirical evidence that reductions in relative testis size through reduced sperm competition are associated with increased expenditure on adaptations to pre-copulatory sexual selection, such as male armament, mate-searching and mate-guarding, etc. (e.g. see Poulin and Morand 2000; Parker et al. 2013). As expected, this can generate high levels of secondary sexual dimorphism.

Sexual size dimorphism is usually explained in terms of a different balance for the two sexes between the benefits of larger size through enhanced reproduction and the increased costs of juvenile mortality risk through delaying sexual maturation (e.g. see review of Blanckenhorn 2000). For females, fecundity typically increases with size, while pre-copulatory male-male competition is usually seen as the major selective pressure favouring increased male body size. Male-biased sexual size dimorphism is characteristic of species with high male-male contests for females (see chapters in Fairbairn et al. 2007). When males compete by sperm production alone rather than contests, selection on male size can occur in order to maintain large testes (Parker 1992). While high levels of sperm competition under communal spawning can prevent male size dropping below female size, sperm competition alone (without contest competition) cannot easily push male size above female size, and if sperm competition is very low, small or dwarf males are predicted (Parker 1992).

The notion of gradual evolutionary transition from sexual selection mainly by sperm competition, to a mixture of both pre- and post-copulatory sexual selection with the evolution of mobility and copulation as sperm competition reduces (Fig. 7.1) was, in fact, foreshadowed rather cryptically in Darwin's original treatise. Darwin (1874, pp. 260–265) dismissed sexual selection (i.e. pre-copulatory sexual selection) in "the lower classes of the animal kingdom" on the grounds that they are sometimes hermaphrodite, or sessile (precluding male-male competition: "the one cannot search or struggle for the other"), or because they "have too imperfect senses and much too low mental powers to appreciate each other's beauty or other attractions, or to feel rivalry"). The number of pages he devotes on evidence for (pre-copulatory) sexual selection increases roughly in inverse proportion to relative testis size of the taxa, with most pages devoted to insects, birds and mammals. That pre-copulatory sexual selection (i.e. as envisaged by Darwin) is essentially the province of species with relatively low testis expenditure (viewed across the entire animal kingdom) is hard to dispute.

Thus anisogamy, mobility and internal fertilisation resulted in reduced sperm expenditure and allowed higher male pre-copulatory competition (Fig. 7.1). Coupled with the constraint of the unity sex ratio, the scene was set for the ecological asymmetry between the two sexes leading to the consequences of pre-copulatory sexual selection in terms of stereotypical sex roles, i.e. the Darwin-Bateman Paradigm (DBP). Thus DBP relies on (i) the unity sex ratio, and (ii) ejaculates being cheap and male parental care negligible relative to the cost of ova and any female parental care, i.e. on the rise of pre-copulatory male-male competition at the expense of expenditure on sperm, as outlined above (see Fig. 7.1). DBP, and its many causal interpretations, i.e. in terms of relative parental investment (PI; Trivers 1972), operational sex ratio (OSR; Emlen and Oring 1977), potential reproductive rate (PRR; Clutton-Brock and Vincent 1991), or the relative 'time in' and 'time out' of each sex's availability for mating during adulthood (Clutton-Brock and Parker 1992) all rely on this asymmetry, and are extensions of Darwin's original insight. These three measures of 'sexual selection intensity' are all closely related mathematically (Parker and Birkhead 2013). As such, DBP serves as a null model for mobile species with reduced relative expenditure on testes, and negligible or low male parental care. Of course, there are exceptions to such a proviso, but DBP nevertheless covers most animal species.

Our evolutionary arguments have so far not included the origin of parental care, or parental investment other than that in the gametes. It is too seldom stressed that across the animal kingdom as a whole, parental care by either sex is relatively rare in invertebrates, and the rule only in two taxa, mammals and birds. Parental care has evolved later, and is a complex problem that should not be (but often is) confused with the events described in Fig. 7.1 leading initially to the generality of the DBP. Paradoxically, the male-biased OSR predicted under DBP generates frequency-dependent selection, analogous to Fisherian sex ratio selection, that favours increased parental investment by males (Kokko and Jennions 2008). Undoubtedly, the primary asymmetry of anisogamy and the mode of fertilisation have influenced the subsequent evolution of parental care (Maynard Smith 1977), which is highly biased towards females. Kokko and Jennions (2008) suggest that the predominance in conventional sex roles in species with parental care are maintained by sexual selection on males, reduced paternity through female multiple mating or group spawning, and higher male mortality generating female-biased adult sex ratios.

However, cases of bi-parental care and male-only care have evolved in some taxa (notably fishes and birds). In a few species (notably birds), male-only care can lead to sex role reversal. Though counter to sex roles predicted initially by DBP, such cases can sometimes be explained by a reversal in the ecological asymmetry from that predicted simply by anisogamy and reduced sperm expenditure (e.g. Simmons 1992). Note that when ecological conditions promote higher male PI, this will typically lead to the evolution of different forms of paternal investment, but for the reasons outlined above anisogamy will remain.

As a final evolutionary consequence, sexual selection almost inevitably generates sexual conflict, i.e. an evolutionary conflict of interest between some males and females (Parker 1979; Arnqvist and Rowe 2005), though this is the subject of recent controversy (see Sect. 7.5.2).

7.5 Some Current Controversies

7.5.1 Criticisms of the Darwin-Bateman Paradigm (DBP)

In view of the logical imperative for sexual selection, DBP remains a satisfactory first expectation for species with zero male care and internal fertilisation (i.e. the vast majority of species). As such, it fulfils a similar heuristic purpose as does, say, the Hardy-Weinberg equilibrium for neutral selection on two alleles—i.e. when we find deviations from it, we need to examine why these occur. They are likely to be due to special biological features, which, however interesting, do not negate the validity of DBP as a general rule for the majority of cases (Parker and Birkhead 2013). The current criticisms of DBP relate partly to the fact that deviations from DBP in species with relatively high male parental care are (unsurprisingly) not that uncommon; moreover, there are various other reasons why DBP expectations may not be met (see e.g. Klug et al. 2010).

Attacks on sexual selection and/or DBP have recently arisen from two related but rather different sources. First, Roughgarden et al. (2006) have claimed that the entire concept is flawed and that solutions to male and female sexual adaptation should be sought in terms of 'social selection'—the principle that mating and associated reproductive activities between the sexes will be cooperative. This critique is based on erroneous claims relating to the quality of the evidence for Darwinian sexual selection and has attracted much criticism (see the multiple responses in *Science*, 2006, vol. 312, 689–694). Further, while the notion that reproduction can involve cooperation is certainly not novel (e.g. we have long known that sexual cooperation can occur, for instance in animals such as birds with biparental care), this itself involves sexual conflict, which must be fully considered in understanding its evolution and stability. Further, while mutual benefits to each sex may arise from a given reproductive adaptation, these may offer a weak or negligible selective force compared to those arising directly through sexual selection. For example, the guarding phase of male dung flies was originally (Foster 1967) seen as co-operation with the female to ensure more efficient oviposition, but there is strong evidence to suggest that it has arisen through sexual selection for paternity guarding (Parker 1970e).

A second recent critique is the attack on the DBP paradigm, resulting in a 'gender role' controversy, i.e. whether DBP—and ultimately the primary sexual difference of anisogamy—does offer an explanation of male and female sex roles and behaviour (see review of Parker and Birkhead 2013). This has also been strongly argued to be misguided (Schärer et al. 2012; Kokko et al. 2013; Parker and Birkhead 2013). However, aspects of this critique are possibly not entirely unrelated to what must be regarded as an excellent and growing development in sexual selection studies, namely an increasing focus on the female perspective, and on female interests in multiple mating, i.e. polyandry (e.g. see the recent theme issue in *Phil. Trans. R. Soc. Lond. B* 2013, vol. 368 on polyandry).

7.5.2 Conflict and Co-operation in Sexual Dynamics

One of the areas of current debate is the extent to which sexual selection generates an evolutionary conflict of fitness interests between individual males and females. The debate has been strongly polarised: while some biologists have proposed that sexual selection necessarily coincides with sexual conflict (e.g. Arnqvist and Rowe 2005), others have called for sexual interactions can only be understood in the light of social cooperation (Roughgarden 2006). The biological reality is likely to be more complex.

First, it is undeniable that sexual selection implies a tension between the evolutionary interests of some males and females. Sexual conflict can occur over a number of reproductive events, from mating to parental allocation, and through different mechanisms (Parker 1979). The primordial sexual conflict probably began during the evolution of anisogamy and sperm-ovum fusions (Parker 1978c, 2011), but as divergence in the two sexual phenotypes becomes more exaggerated through the evolution of enhanced mobility, reduced sperm expenditure, and increased male-male mating competition, so does the potential for sexual conflict. For example, the very concept of mate choice necessarily creates a conflict of interests between the chooser and those members of the opposite sex that are less preferred. More generally, conflict will occur whenever an individual gains by differentially allocating reproductive resources to reproduction with individual partners. In the blue head wrasse, *Thalassoma bifasciatus*, the most successful males invest their reproductive resources to attract and mate guard a large number of females. This investment however limits the number of sperm that a male is able to allocate to the eggs spawned by each female, leaving about 7 % of their eggs not fertilised (Warner et al. 1995). Therefore, while this strategy yields a larger number of eggs fertilised by a male across all the females attracted, it imposes fertility costs on individual females (see Ball and Parker 1996 for other predictions on 'adaptive infertility' in external fertilisers). Sexual selection can also promote traits that convey an advantage in intra-sexual competition while imposing a fitness cost on mating partners. These costs are likely to represent collateral side-effects in the majority of cases (e.g. Siva-Jothy 2006), however, in principle it is also possible that sexual selection might favour a male trait precisely for the costs that it imposes on females (e.g. Johnstone and Keller 2000; Lessells 2005). A wide range of such traits has been documented mostly in males. Therefore an element of conflict is unavoidable whenever alternative reproductive opportunities are available to an interacting male and female.

Second, despite the near-ubiquitous potential for sexual conflict, sexual selection does not eliminate potential for inter-sexual cooperation over a number of reproductive decisions. One such example is conflict over female re-mating decisions. Clearly, by mating with a second male a female can reduce the reproductive success of the first male, which may lose paternity through sperm competition. Therefore, whenever females benefit by re-mating (e.g. Arnqvist and Nilsson 2000), sexual conflict is expected between the female and the first male. However, in many species males can stimulate female fecundity, for example through the gonadotropic effect of ejaculate compounds (see above). Alonzo and Pizzari (2010) have shown that when a mating more than doubles female fecundity, as has been documented in a number of taxa, two males actually gain by mating with the same female as double mating yields a higher number of eggs available for fertilisation than would be available to either male mating exclusively with a female. This generates a scenario of inter- and intra-sexual cooperation over female re-mating decisions. A similar example, analysed long ago by Maynard Smith and Ridpath (1972), is that of wife sharing in the Tasmanian Native Hen, *Tribonyx* (= *Gallinula*) *mortierii.* When two males share a female, the number of progeny that can be produced is increased. Conditions favouring wife sharing are more permissive, requiring (in the simplest case) only an increase in progeny of 67 % since the two males are brothers. Note that clearly in both cases, conflict remains over who gets to fertilise the eggs.

Finally, much of the current debate over conflict and cooperation in sexual dynamics is narrowly focused on direct consequences that a certain sexual trait or behaviour has on the fitness of the actor and recipient. However, sexual interactions like many other social traits, may also influence the fitness of third parties, creating potential for indirect effects to contribute to the evolution of sexually-selected traits. Indirect effects are the relatedness-weighted effects on the inclusive fitness of social partners (Hamilton 1964; Pizzari and Gardner 2012). Inclusive fitness effects expand the diversity of evolutionary outcomes of sexual interactions by adding the possibility of altruism and spite to conflict and mutualism driven by selfishness and direct effects. Pizzari and Gardner (2012) identify two conditions under which indirect benefits can arise: "(i) the recipient is related to the actor; or (ii) the actor is related to a third party, who will at some point also interact with the recipient." The former condition (i) represents the case of inbreeding. Because of anisogamy and sex differences in parental investment we expect males to gain from inbreeding in situations in which females would lose from inbreeding, and we expect this potential for conflict to expand as opportunity costs associated with mating are progressively reduced (Parker 1979, 2006). Parker (1979, 2006) had already demonstrated how indirect effects –through kin selection—can modulate sexual conflict over inbreeding (see also Kokko and Ots 2006). The latter condition is more broadly relevant but so far has received little consideration. However, recently it has been shown that when male competition occurs locally amongst rivals that are more related to each other than the population average, indirect effects are likely to buffer sexual selection for male traits that harm females (Rankin 2011; Wild et al. 2011), thus reducing the intensity of sexual conflict ('virulence' *sensu* Pizzari and Gardner 2012).

An alternative mechanism through which indirect effects might modulate sexual dynamics is through potential 'greenbeard' effects (Pizzari and Gardner 2012). 'Greenbeards' are genes that allow their carriers to increase each other's fitness through mutual recognition and differential interactions (West and Gardner 2010). The preference and ornament genes in sexual signalling can be thought of as an inter-sexual green beard, and the rapid coevolution of exaggerated ornament and preferences envisaged by Fisher is clearly modulated by green beard indirect effects (Pizzari and Gardner 2012).

These considerations illustrate that potential for sexual conflict should not be assumed but carefully measured for individual reproductive decisions including both direct and indirect fitness effects. It is important to note however, that current debate on conflict and cooperation reflects a development rather than a limitation of sexual selection theory.

7.5.3 Intensity of Sexual Selection

OSR and I_T (and related indices) have long been proposed and used as measures of the intensity of sexual selection. Recently, Klug et al. (2010) have strongly criticised their use on the grounds that they only accurately predict sexual selection under a limited set of circumstances, and more specifically, only when mate monopolization is extremely strong. However, their analysis has been seen as pessimistic by Parker and Birkhead (2013), mainly because it ignored the direct effect of the likely relation between OSR and male time out of the mating pool, which when included, shows that OSR and I_T can indeed be reasonable measures of sexual selection intensity. Independently, a detailed study of how and when male time out and its relation to OSR can allow OSR and I_T to give fair measures of the intensity of sexual selection has been given by Kokko et al. (2012). While measures of the intensity of sexual selection are sometimes useful (e.g. particularly for comparative analyses) many would agree with Klug et al. (2010) that ideally—and provided that one knows *a priori* what traits are currently targeted by sexual selection in a given species—one would measure selection directly on the phenotypic trait of interest, a point originally stressed by Grafen (1987).

7.6 Concluding Comments

We conclude that sexual selection theory is a powerful heuristic tool providing the most parsimonious explanation for a vast diversity of traits, across sexually-reproducing organisms, from unicellular taxa to primates. Overwhelmingly strong qualitative and quantitative evidence has accumulated over the last decades vindicating Darwin's original insight. Crucially, sexual selection theory has also been successfully applied as predictive tool to explain biological phenomena that were unknown to—or not considered by—Darwin. Sexual selection has a strong underlying deductively logical imperative that follows from the predictable sequence of evolutionary events arising after the evolution of sexual recombination and fusion.

It is fallacious to argue that because (actually rather low frequency) differences from DBP occur that DBP does not offer a general explanation. For the vast majority of species where there is no male parental care, it tells us what to expect when special features of biology do not act to change that expectation.

Sexual selection theory certainly does not need to replaced as has been advocated recently (Roughgarden et al. 2006); rather, it represents one of the major triumphs of adaptive explanation. However, we argue that current debates may be more effectively resolved by bringing sexual selection theory more firmly within the framework of social evolution (e.g. Rankin 2011; Pizzari and Gardner 2012) and by a more dynamic integration of theory with the ecological and physiological details of sexual interactions.

Acknowledgements We thank Thierry Houquet for inviting us to write this chapter, for inviting us to the workshop he organised on sexual selection, which GAP was unfortunately unable to attend. We are especially indebted to Dr. Cathy H. Lucas for information on Cnidaria and for kindly allowing us to include her unpublished results on *Periphylla periphylla*, and to Prof. Paul A. Tyler for much helpful advice and information on GSI in marine invertebrates.

The notion of sexual selection as a logical imperative arising through the sequence of evolutionary events leading to highly differentiated males and females was to have been the central theme of a book entitled *The Evolution of Sexual Strategy* by GAP during the 1970s. This project was 70 % accomplished, mostly during a year (1978–79) in the Research Centre of King's College, Cambridge, but was never completed after his return to Liverpool University in September 1979. GAP has often regretted this failure, but wishes to thank King's College, Cambridge for the opportunity to work in the Research Centre, which nevertheless proved most stimulating.

References

Alonzo SH, Pizzari T (2010) Male fecundity stimulation: conflict and cooperation within and between the sexes: model analyses and coevolutionary dynamics. Am Nat 175:174–185

Andersson M (1994) Sexual selection. Princeton University Press, Princeton

Andersson M, Simmons LW (2006) Sexual selection and mate choice. Trends Ecol Evol 21:296–302

Arak A (1988) Sexual dimorphism in body size: a model and a test. Evolution 42:820–825

Arnold SJ, Duvall D (1994) Animal mating systems: a synthesis based on selection theory. Am Nat 143:317–348

Arnold SJ, Wade MJ (1984) On the measurement of natural and sexual selection: theory. Evolution 38:709–719

Arnqvist G, Nilsson T (2000) The evolution of polyandry: multiple mating and female fitness in insects. Anim Behav 60:145–164

Arnqvist G, Rowe L (2005) Sexual conflict. Princeton University Press, Princeton

Baker RR, Parker GA (1973) The origin and evolution of sexual reproduction up to the evolution of the male-female phenomenon. Acta Biotheoretica 22:49–77

Ball MA, Parker GA (1996) Sperm competition games: external fertilization and 'adaptive' infertility. J theor Biol 180:141–150

Bateman AJ (1948) Intra-sexual selection in Drosophila. Heredity 2:349–368

Benítez-Villalobos F, Aguilar-Duarte C, Avila-Poveda OH (2012) Reproductive biology of *Ophiocoma aethiops* and *O. alexandri* (Echinodermata: Ophiuroidea) from Estacahuite Bay, Oaxaca, Mexico. Aquat Biol 17:119–128

Birkhead TR, Møller AP (1998) Sperm competition and sexual selection. Academic Press, London

Blanckenhorn WU (2000) The evolution of body size: what keeps organisms small? Q Rev Biol 75:385–407

Blanckenhorn WU (2007) Case studies of the differential-equilibrium hypothesis of sexual size dimorphism in two dung fly species. In: Fairbairn DJ, Blanckenhorn WU, Székely T (eds) Sex, size, and gender roles: evolutionary studies of sexual size dimorphism. Oxford University Press, Oxford, pp 106–114

Blanckenhorn WU, Briegel U, Choffat Y, Demont M, Gautier R, Pemberton KL, Roschitzki-Voser H, Willi Y, Ward PI (2012) Temperature-mediated microhabitat choice and development time based on the *pgm* locus in the yellow dung fly *Scathophaga stercoraria*. Biol J Linn Soc 107:686–696

Borgia G (1981) Mate selection in the fly *Scatophaga stercoraria*: female choice in a male-controlled system. Anim Behav 29:71–80.

Breder CM, Rosen DE (1966) Modes of reproduction in fishes. Natural History Press, Garden City

Bretos M, Tesorieri I, Alvarez L (1983) The biology of *Fissurella maxima* Sowerby (Mollusca: Archaeogastropoda) in northern Chile. 2. Notes on its reproduction. Biol Bull 165:559–568

Bulmer MG, Parker GA (2002) The evolution of anisogamy: a game-theoretic approach. Proc R Soc Lond B 269:2381–2388

Bussière LF, Demont M, Pemberton AJ, Hall MD, Ward PI (2010) The assessment of insemination success in yellow dung flies using competitive PCR. Mol Ecol Res 10:292–303

Charnov EL (1982) The theory of sex allocation. Princeton University Press, Princeton

Charnov EL, Parker GA (1995) Dimensionless invariants from foraging theory's marginal value theorem. Proc Natl Acad Sci USA 92:1446–1450

Charnov EL, Maynard Smith J, Bull JJ (1976) Why be an hermaphrodite? Nature 263:125–126

Chiantore M, Cattaneo-Vietti R, Elia L, Guidetti M, Antonini M (2002) Reproduction and condition of the scallop *Adamussium colbecki* (Smith 1902), the sea-urchin *Sterechinus neumayeri* (Meissner 1900) and the sea-star *Odontaster validus* (Koehler 1911) at Terra Nova Bay (Ross Sea): different strategies related to inter-annual variations in food availability. Polar Biol 25:251–255

Clutton-Brock TH, Parker GA (1992) Potential reproductive rates and the operation of sexual selection. Q Rev Biol 67:437–456

Clutton-Brock TH, Vincent ACJ (1991) Sexual selection and the potential reproductive rates of males and females. Nature 351:58–60

Collet J, Richardson DS, Worley K, Pizzari T (2012) Sexual selection and the differential effect of polyandry. PNAS 109:8641–8645

Coma R, Ribes M, Zabala M, Gili JM (1995) Reproduction and cycle of gonadal development in the Mediterranean gorgonian Paramuricea clavata. Mar Ecol Prog Ser 117:173–183

Cox PA, Sethian JA (1984) Search, encounter rates, and the evolution of anisogamy. PNAS 81:6078–6079

Cox PA, Sethian JA (1985) Gamete motion, search and the evolution of anisogamy, oogamy, and chemotaxis. Am Nat 125:74–101

Curtsinger JW (1986) Stay times in *Scatophaga* and the theory of evolutionarily stable strategies. Am Nat 128:130–136

Darwin C (1874) The descent of man and selection in relation to sex, 2nd edn. John Murray, London

Darwin C (1889) The effects of cross and self fertilisation in the vegetable kingdom. Chapter XII. General results. D. Appleton and Company, New York, pp 436–463

Demary KC, Lewis SM (2007) Male reproductive allocation in fireflies (*Photinus* spp.). Invertebr Biol 126:74–80

Demont M, Buser CC, Martin OY, Bussiére LF (2011) Natural levels of polyandry: differential sperm storage and temporal changes in sperm competition intensity in wild yellow dung flies. Funct Ecol 25:1079–1090

Demont M, Martin OY, Bussiére LF (2012) Wild yellow dung fly females may not select sperm based on dung pat microclimate but could nevertheless benefit from polyandry. Evol Ecol 26:715–731

Dewsbury DA (2005) The Darwin—Bateman Paradigm in historical context. Integr Comp Biol 45:831–837

Dixson AF (2009) Sexual selection and the origins of human mating systems. Oxford University Press, Oxford

Dusenbery DB (2000) Selection for high gamete encounter rates explains the success of male and female mating types. J Theoretical Biol 202:1–10

Eberhard WG (1996) Female control: sexual selection by cryptic female choice. Princeton University Press, Princeton

Emerson SB (1997) Testis size variation in frogs: testing the alternatives. Behav Ecol Sociobiol 41:227–235

Emlen ST, Oring LW (1977) Ecology, sexual selection, and the evolution of mating systems. Science 197:215–223

Fairbairn DJ, Blanckenhorn WU, Székely T (2007) Sex, size and gender roles: evolutionary studies of sexual size dimorphism. Oxford University Press, Oxford

Felsenstein J (1974) The evolutionary advantage of recombination. Genetics 78(2):737–756. (PMC 1213231)

Fisher RA (1930) The genetical theory of natural selection. Clarendon Press, Oxford

Foster WA (1967) Co-operation by male protection of ovipositing female in the Diptera. Nature 214:1035–1036

Fretwell SD (1972) Populations in a seasonal environment. Princeton University Press, Princeton

Gori A, Linares C, Rossi S, Coma R, Gili J-M (2007) Spatial variability in reproductive cycle of the gorgonians Paramuricea clavata and Eunicella singularis (Anthozoa, Octocorallia) in the Western Mediterranean Sea. Mar Biol 151:1571–1584

Grafen A (1987) Measuring sexual selection: why bother? In: Bradbury JW, Andersson MB (eds) Sexual selection: testing alternatives. Wiley, Chichester, pp 221–223

Grange LJ, Tyler PA, Peck LS, Cornelius N (2004) Long-term inter- annual cycles of the gametogenic ecology of the Antarctic brittle star Ophionotus victoriae. Mar Ecol Prog Ser 278:141–155

Grange LJ, Tyler PA, Peck LS (2007) Multi-year observations on the gametogenic ecology of the Antarctic seastar Odontaster validus. Mar Biol 153:15–23

Hamilton WD (1964) Genetical evolution of social behaviour. I. J Theor Biol 7:1–16

Hamilton WD (1967) Extraordinary sex ratios. Science 156:477–488

Hammer O (1941) Biological and ecological investigations on flies associated with pasturing cattle and their excrement. Vidensk Meddr dansk naturh Foren 105:1–257

Hamner WM, Hamner PP, Strand SW (1994) Sun-compass migration by Aurelia aurita (Scyphozoa): population retention and reproduction in Saanich Inlet, British Columbia. Mar Biol 119:347–356

Hartfield M, Keightley PD (2012) Current hypotheses for the evolution of sex and recombination. Integr Zool 7:192–209

Hellriegel B, Bernasconi G (2000) Female-mediated differential sperm stor- age in a fly with complex spermathecae, *Scatophaga stercoraria*. Anim Behav 59:311–317

Hellriegel B, Ward PI (1998) Complex female reproductive tract morphol- ogy: its possible use in postcopulatory female choice. J Theor Biol 190:179–186

Henninger TO (1998) Aspects of the ecology and reproductive biology of the limpet, Helcion pruinosus (Gastropoda: prosobranchia). Masters thesis, Rhodes University

Hosken DJ (1999) Sperm displacement in yellow dung flies: a role for females. Trends in Ecol Evol 14:251–252

Hosken DJ (2001) Sex and death: microevolutionary trade-offs between reproductive and immune investment in dung flies. Curr Biol 11:R379–R380

Hosken DJ, Ward PI (2000) Copula in yellow dung flies (*Scathophaga stercoraria*): investigating sperm competition models by histological observation. J Insect Physiol 46:1355–1363

Hosken DJ, Ward PI (2001) Experimental evidence for testis size evolution via sperm competition. Ecol Lett 4:10–13

Hosken DJ, Garner TWJ, Ward PI (2001) Sexual conflict selects for male and female reproductive characters. Curr Biol 11:489–493

Hosken DJ, Uhía E, Ward PI (2002) The function of female accessory reproductive gland secretion and a cost to polyandry in the yellow dung fly. Physiol Entomol 27:87–91

Hosken DJ, Garner TWJ, Tregenza T, Wedell N, Ward PI (2003) Superior sperm competitors sire higher-quality young. Proc R Soc Lond B 270:1933–1938

Iyer P, Roughgarden J (2008) Gametic conflict versus contact in the evolution of anisogamy. Theor Popul Biol 73:461–72

Jacobs ME (1955) Studies on territorialism and sexual selection in dragonflies. Ecology 36:566–86
Jann P, Blanckenhorn WU, Ward PI (2000) Temporal and microspatial variation in the intensities of natural and sexual selection in the yellow dung fly *Scathophaga stercoraria*. J Evol Biol 13:927–938
Jennions MD, Passmore NI (1993) Sperm competition in frogs: testis size and a "sterile male'" experiment on *Chiromantis xerampelina* (Rhacophoridae). Biol J Linn Soc 50:211–220
Johnstone RA, Keller L (2000) How males can gain by harming their mates: sexual conflict, seminal toxins, and the cost of mating. Am Nat 156:368–377
Jones AG (2009) On the opportunity for sexual selection, the Bateman gradient and the maximum intensity of sexual selection. Evolution 63:1673–1684
Jones AG, Ratterman NL (2009) Mate choice and sexual selection: What have we learned since Darwin? PNAS 106:10001–10008
Kalmus H (1932). Über den Erhaltungswet den phenotypishen (morphologishen) Anisogamie und die Entstehung der ersten Geshlectsuntershiede. Biol Zent 52:716
Kalmus H, Smith CAB (1960) Evolutionary origin of sexual differentiation and the sex-ratio. Nature 186:1004–1006
Kelly CD, Jennions MD (2011) Sexual selection and sperm quantity: meta-analyses of strategic. Biol Rev 86:863–884
Kenagy GJ, Trombulak SC (1986) Size and function of mammalian testes in relation to body size. J Mammal 67:1–22
Klug H, Heuschele J, Jennions MD, Kokko H (2010) The mismeasurement of sexual selection. J Evol Biol 23:447–462
Kokko H, Jennions M (2008) Parental investment, sexual selection and sex ratios. J Evol Biol 21:919–948
Kokko H, Ots I (2006) When not to avoid inbreeding. Evolution 60:467–475
Kokko H, Jennions MD, Brooks R (2006) Unifying and testing models of sexual selection. Ann Rev Ecol Evol Syst 37:43–66
Kokko H, Klug H, Jennions MD (2012) Unifying cornerstones of sexual selection: operational sex ratio, Bateman gradient and the scope for competitive investment. Ecol Lett 15:1340–1351
Kokko H, Booksmythe I, Jennions MD (2013) Causality and sex roles: prejudice against patterns? A reply to Ah-King. Trends in Ecol Evol 28:2–4
Lehtonen J, Kokko H (2011) Two roads to two sexes: unifying gamete competition and gamete limitation in a single model of anisogamy evolution. Behav Ecol Sociobiol 65:445–459
Lehtonen J, Jennions MD, Kokko H (2012) The many costs of sex. Trends in Ecol Evol 27:172–178
Lessells CM (2005) Why are males bad for females? Models for the evolution of damaging male mating behaviour. Am Nat 165:46–63
Lessells CM, Snook RR, Hosken DJ (2009) The evolutionary origin and maintenance of sperm: selection for a small, motile gamete mating type. In: Birkhead TR, Hosken D, Pitnick S (eds) Sperm biology: an evolutionary perspective. Academic Press, London, pp 43–67
Levitan DR (1996) Effects of gamete traits on fertilization in the sea and the evolution of sexual dimorphism. Nature 382:153–155
Levitan DR (1998) Sperm limitation, sperm competition and sexual selection in external fertilizers. In: Birkhead T, Møller A (eds) Sperm competition and sexual selection. Academic Press, San Diego, pp 173–215
Levitan DR (2005) The distribution of male and female reproductive success in a broadcast spawning marine invertebrate. Integr Comp Integr Comp Biol 45:848–855
Levitan DR, Petersen C (1995) Sperm limitation in the sea. Trends Ecol Evol 10:228–231
Lucas CH (2001) Reproduction and life history strategies of the common jellyfish, *Aurelia aurita*, in relation to its ambient environment. Hydrobiologia 451:229–246
Lucas CH, Lawes S (1998) Sexual reproduction of the scyphomedusa *Aurelia aurita* in relation to temperature and variable food supply. Mar Biol 131:629–638
Marian JE (2012) A model to explain spermatophore implantation in *cephalopods (M*ollusca: Cephalopoda) and a discussion on its evolutionary origins and significance. Biol J Linn Soc 105:711–726

Martin OY, Hosken DJ, Ward PI (2004) Post-copulatory sexual selection and female fitness in *Scathophaga stercoraria*. Proc R Soc Lond B 271:353–359

Martínez-Pita I, Sánchez-España AI, García FJ (2008) Gonadal growth and reproduction in the sea urchin *Sphaerechinus granularis* (Lamarck 1816) (Echinodermata: Echinoidea) in southern Spain. Scient Mar 72:603–611

Matsuda R (1976) Morphology and evolution of the insect abdomen. Pergamon, Oxford.

Maynard Smith J (1971) What use is sex? J Theor Biol 30:319–335

Maynard Smith J (1977) Parental investment: a prospective analysis. Anim Behav 25:1–9

Maynard Smith J, Ridpath MG (1972) Wife sharing in Tasmanian native hen, Tribonyx mortierii: A case of kin selection? Am Nat 106:447–452

Maynard Smith J (1978) The evolution of sex. Cambridge University Press, Cambridge

Maynard Smith J, Price GR (1973) The logic of animal conflicts. Nature 246:15–18

Maynard Smith J, Szathmáry E (1995) The major transitions in evolution. W.H. Freeman Spektrum, Oxford

Muller HJ (1932) Some genetic aspects of sex. Am Nat 66(703):118–138

Othman MS, Khonsue W, Kitana J, Thirakhupt K, Robson MG, Kitana N (2011) Reproductive mode of *Fejervarya limnocharis* (Anura: Ranidae) caught from Mae Sot, Thailand based on its gonadosomatic indices. Asian Herpetol Res 2:41–45

Otronen M, Reguera P, Ward PI (1997) Sperm storage in the yellow dung fly *Scathophaga stercoraria*: identifying the sperm of competing males in separate female spermathecae. Ethology 103:844–854

Parker GA (1970a) Sperm competition and its evolutionary consequences in insects. Biol Rev 45:525–567

Parker GA (1970b) The reproductive behaviour and the nature of sexual selection in *Scatophaga stercoraria* L. (Diptera: Scatophagidae). V. The female's behaviour at the oviposition site. Behaviour 37:140–168

Parker GA (1970c) The reproductive behaviour and the nature of sexual selection in *Scatophaga stercoraria* L. (Diptera: Scatophagidae). II. The fertilization rate and the spatial and temporal relationships of each sex around the site of mating and oviposition. J Anim Ecol 39:205–228

Parker GA (1970d) The reproductive behaviour and the nature of sexual selection in *Scatophaga stercoraria* L. (Diptera: Scatophagidae). IV. Epigamic recognition and competition between males for the possession of females. Behaviour 37:113–139

Parker GA (1970e) The reproductive behaviour and the nature of sexual selection in *Scatophaga stercoraria* L. (Diptera: Scatophagidae). VII. The origin and evolution of the passive phase. Evolution 24:774–788

Parker GA (1970f) Sperm competition and its evolutionary effect on copula duration in the fly *Scatophaga stercoraria*. J Insect Physiol 16:1301–1328

Parker GA (1971) The reproductive behaviour and the nature of sexual selection in *Scatophaga stercoraria* L. (Diptera: Scatophagidae). VI. The adaptive significance of emigration from oviposition site during the phase of genital contact. J Anim Ecol 40:215–233

Parker GA (1974a) The reproductive behaviour and the nature of sexual selection in *Scatophaga stercoraria* L. (Diptera: Scatophagidae). IX. Spatial distribution of fertilization rates and evolution of male search strategy within the reproductive area. Evolution 28:93–108

Parker GA (1974b) Assessment strategy and the evolution of fighting behaviour. J Theor Biol 47:223–243

Parker GA (1978a) Searching for mates. In: Krebs JR, Davies NB (eds) Behavioural ecology: an evolutionary approach. Blackwell, Oxford, pp 214–244

Parker GA (1978b) Selfish genes, evolutionary games, and the adaptiveness of behaviour. Nature 274:849–855

Parker GA (1978c) Selection on non-random fusion of gametes during the evolution of anisogamy. J Theor Biol 73:1–28

Parker GA (1979) Sexual selection and sexual conflict. In: Blum MS, Blum NA (eds) Sexual selection and reproductive competition in insects. Academic Press, New York, pp 123–166

Parker GA (1982) Why are there so many tiny sperm? Sperm competition and the maintenance of two sexes. J Theor Biol 96:281–294
Parker GA (1984) Sperm competition and the evolution of animal mating strategies. In: Smith RL (ed) Sperm competition and the evolution of animal mating systems. Academic Press, Orlando, pp 1–60
Parker GA (1992) The evolution of sexual size dimorphism in fish. J Fish Biol 41(Suppl. B):1–20
Parker GA (2006) Sexual conflict over mating and fertilization: an overview. Phil Trans R Soc B 361:235–259
Parker GA (2011) The origin and maintenance of two sexes (anisogamy), and their gamete sizes by gamete competition. In: Togashi T, Cox PA (eds) The evolution of anisogamy: a fundamental phenomenon underlying sexual selection. Cambridge University Press, Cambridge, pp 17–74
Parker GA (2014) The sexual cascade and the rise of Darwinian sexual selection, sex roles and sexual conflict. In: Rice WR, Gavrilets S (eds) Sexual conflict, in press. Cold Spring Harbor Laboratory Press, New York
Parker GA, Ball MA (2005) Sperm competition, mating rate and the evolution of testis and ejaculate sizes: a population model. Biol Lett 1:235–238
Parker GA, Birkhead TR (2013) Polyandry: the history of a revolution. Phil Trans R Soc B 368:20120335. http://dx.doi.org/10.1098/rstb.20102.0335. Accessed 21 Jan 2013
Parker GA, Lehtonen J. (2014) Gamete evolution and sperm numbers: sperm competition versus sperm limitation. Proc R Soc Lond B 281: 20140836.
Parker GA, Maynard Smith J (1987) The distribution of stay times in *Scatophaga*: reply to Curtsinger. Am Nat 129:621–628
Parker GA, Maynard Smith J (1990) Optimality theory in evolutionary biology. Nature 348:27–33
Parker GA, Stuart RA (1976) Animal behaviour as a strategy optimizer: evolution of resource assessment strategies and optimal emigration thresholds. Am Nat 110:1055–1076
Parker GA, Simmons LW (1994) Evolution of phenotypic optima and copula duration in dung fies. Nature 370:53–56
Parker GA, Simmons LW (2000) Optimal copula duration in yellow dungflies: ejaculatory duct dimensions and size-dependent sperm displacement. Evolution 54:924–935
Parker GA, Pizzari T (2010) Sperm competition and ejaculate economics. Biol Rev 85:897–934
Parker GA, Thompson EA (1980) Dung fly struggles: a test of the war of attrition. Behav Ecol Sociobiol 7:37–44
Parker GA, Baker RR, Smith VGF (1972) The origin and evolution of gamete dimorphism and the male-female phenomenon. J theor Biol 36:529–553
Parker GA, Simmons LW, Ward PI (1993) Optimal copula duration in dung flies—effects of frequency-dependence and female mating status. Behav Ecol Sociobiol 32:157–166
Parker GA, Simmons LW, Stockley P, McChristie DM, Charnov EL (1999) Optimal copula duration in yellow dung flies: effects of female size and egg content. Anim Behav 57:795–805
Parker GA, Lessells CM, Simmons LW (2013) Sperm competition games: a general model for precopulatory male-male competition. Evolution 67:95–109
Pitcher TE, Dunn PO, Whittingham LA (2005) Sperm competition and the evolution of testes size in birds. J Evol Biol 18:557–567
Pitnick, S. (1996) Investment in testes and the cost of making long sperm in *Drosophila*. Am Nat 148:57–80
Pizzari T, Gardner A (2012) The sociobiology of sex: inclusive fitness consequences of intersexual interactions. Phil Trans Roy Soc B 367:2314–2323
Poulin R, Morand S (2000) Testes size, body size and male–male competition in acanthocephalan parasites. J Zool 250:551–558
Prum RO (2012) Aesthetic evolution by mate choice: Darwin's really dangerous idea. Phil Trans R Soc B 367:2253–2265
Rankin DJ (2011) Kin selection and the evolution of sexual conflict. J Evol Biol 24:71–81
Rogers DW, Greig D (2009) Experimental evolution of a sexually selected display. Proc R Soc Lond B Biol Sci 276:543–549

Rosa R, Costa PR, Nandarra N, Nunes ML (2005) Changes in tissue biochemical composition and energy reserves associated with sexual maturation in the ommastrephid squids *Illex coindetii* and *Todaropsis eblanae*. Biol Bull 208:100–113

Roughgarden J, Oishi M, Akçay E (2006) Reproductive social behaviour: cooperative games to replace sexual selection. Science 311:965–970

Saha BK, Gupta BBP (2011) Studies on annual activity cycle of gonads and breeding behavior of the endangered frog, *Rana leptoglossa* (Cope, 1868). Int J Sci Nat 2:683–691

Schärer L, Rowe L, Arnqvist G (2012) Anisogamy, chance and the evolution of sex roles. Trends Ecol Evol 27:260–264

Scudo FM (1967) The adaptive value of sexual dimorphism: I. Anisogamy. Evolution 21:285–291

Shaw RF, Mohler, JD (1953) The selective advantage of the sex ratio. Am Nat 87:337–342.

Shuster SM, Wade MJ (2003) Mating systems and strategies. Princeton University Press, Princeton

Sigurjónsdóttir H, Parker GA (1981) Dung fly struggles: evidence for assessment strategy. Behav Ecol Sociobiol 8:219–230

Sigurjónsdóttir H, Snorrason SS (1995) Distribution of male yellow dung flies around oviposition sites: the effect of body size. Ecol Entomol 20:84–90

Simmons LW (1992) Quantification of role reversal in relative parental investment in a bush-cricket. Nature 358:61–63

Simmons LW, Fitzpatrick JL (2012) Sperm wars and the evolution of male fertility. Reproduction 144:519–534

Simmons LW, Stockley P, Jackson RL, Parker GA (1996) Sperm competition or sperm selection: no evidence for female influence over paternity in yellow dung flies, *Scatophaga stercoraria*. Behav Ecol Socioblol 38:199–206

Simmons LW, Parker GA, Stockley P (1999) Sperm displacement in the yellow dung fly, *Scatophaga stercoraria*: an investigation of male and female processes. Am Nat 153:302–314

Simmons LW, Emlen DJ, Tomkins JL (2007) Sperm competition games between sneaks and guards: a comparative analysis using dimorphic male beetles. Evolution 61:2684–2692

Sirot LK, Wolfner MF, Wigby S (2011) Protein-specific manipulation of ejaculate composition in response to female mating status in *Drosophila melanogaster*. PNAS 108:9922–9926

Siva-Jothy MT (2006) Trauma, disease and collateral damage: conflict in cimicids. Philos Trans R Soc Lond B Biol Sci 361: 269–275

Stockley P, Seal NJ (2001) Plasticity in reproductive effort of male dung flies (*Scatophaga stercoraria*) as a response to larval density. Funct Ecol 15:96–102

Stockley P, Gage MJG, Parker GA, Møller AP (1997) Sperm competition in fishes: the evolution of testis size and ejaculate characteristics. Am Nat 149:933–954

Tregenza T, Wedell N, Hosken DJ, Ward PI (2003) Maternal effects on offspring depend on female mating pattern and offspring environment in yellow dung flies. Evolution 57:297–304

Trivers RL (1972) Parental investment and sexual selection. In: Campbell B (ed) Sexual selection and the descent of man 1871–1971. Aldine, Chicago, pp 136–179

Tyler PA, Reeves S, Peck L, Clarke A, Powell D (2003) Seasonal variation in the gametogenic ecology of the Antarctic scallop *Adamussium colbecki*. Polar Biol 26:727–733

Vahed K, Parker DJ (2012) The evolution of large testes: sperm competition or male mating rate? Ethology 118:107–117

Vahed K, Parker DJ, Gilbert JDJ (2011) Larger testes are associated with a higher level of polyandry, but a smaller ejaculate volume, across bushcricket species (Tettigoniidae). Biol Lett 7:261–264

Ward PI (1993) Females influence sperm storage and use in the yellow dung fly *Scathophaga stercoraria* (L.). Behav Ecol Sociobiol 32:313–319

Ward PI (1998) A possible explanation for cryptic female choice in the yellow dung fly (L.), *Scathophaga stercoraria*. Ethology 104:97–110

Ward PI (2000) Cryptic female choice in the yellow dung fly *Scathophaga stercoraria* (L.). Evolution 54:1680–1686

Ward PI, Simmons LW (1991) Copula duration and testes size in the yellow dung fly, *Scathophaga stercoraria* (L.): the effects of diet, body size, and mating history. Behav Ecol Sociobiol 29:77–85
Ward PI, Foglia M, Blanckenhorn WU (1999) Oviposition site choice in the yellow dung fly *Scathophaga stercoraria*. Ethology 105:423–430
Ward PI, Vonwil J, Scholte EJ, Knop E (2002) Field experiments on the distributions of eggs of different phosphoglucomutase (PGM) genotypes in the yellow dung fly *Scathophaga stercoraria* (L.). Mol Ecol 11:1781–1785
Warner RR, Shapiro DY, Marcanato A, Petersen CW (1995) Sexual conflict: males with highest mating success convey the lowest fertilization benefits to females. Proc R Soc Lond B 262:135–139
Webster MS, Pruett-Jones S, Westneat DF, Arnold SJ (1995) Measuring the effects of pairing success, extra-pair copulations and mate quality on the opportunity for sexual selection. Evolution 49:1147–1157
Weismann A (1889) Essays upon heredity and kindred biological problems. Clarendon Press, Oxford
West SA (2009) Sex allocation. Princeton University Press, Princeton
West SA, Gardner A (2010) Altruism, spite and greenbeards. Science 327:1341–1344
West SA, Lively CM, Read AF (1999) A pluralist approach to sex and recombination. J Evol Biol 12:1003–1012
Wild G, Pizzari T, West SA (2011) Sexual conflict in viscous populations: the effect of the timing of dispersal. Theor Popul Biol 80:298–316

Chapter 8
Selfish Genetic Elements and Sexual Selection

Nina Wedell and Tom A. R. Price

Abstract One aspect of sexual selection that Darwin was completely unaware of was the role of Selfish Genetic Elements (SGEs). SGEs are genes, organelles or microorganisms present within the genome or cell of an organism that spread through populations by subverting normal patterns of inheritance in ways that increase their representation in the next generation. SGEs are ubiquitous in living organisms, have a dramatic ability to manipulate host reproduction, including the frequent reduction in male fertility and sperm competitive ability, yet their impact on sexual selection remains little explored. Here we discuss the pervasiveness and power of SGEs as an agent of sexual selection and show they can have remarkably wide-ranging impacts on male and female reproduction and therefore in shaping mating systems, even when present at low frequencies.

Keywords Intralocus sexual conflict · Male killing endosymbionts · Meiotic drive · Sex ratio distortion · Segregation distortion

8.1 Introduction

8.1.1 Levels of Selection

This review will deal with how the selfish action of genes, chromosomes, organelles and cells can drive conflict, and how this impacts on sexual selection. Key to this is the concept of levels of selection. Biological systems are organised hierarchically. Genes make up chromosomes, which are the major component of the genome of an

N. Wedell (✉)
Biosciences, University of Exeter, Cornwall Campus, Penryn TR10 9FE, UK
e-mail: N.Wedell@Exeter.ac.uk

T. A. R. Price
Institute of Integrative Biology, University of Liverpool, Biosciences Building, Crown Street, Liverpool L69 7ZB, UK

T. Hoquet (ed.), *Current Perspectives on Sexual Selection,* History, Philosophy and Theory of the Life Sciences 9, DOI 10.1007/978-94-017-9585-2_8

individual. Individuals can be part of families, colonies, and populations, and form a species. Another biological hierarchy is cells forming individuals. Selection can potentially work at the level of any hierarchical unit as long as it possesses heritable variation (Keller 1999). For example, imagine a hypothetical disaster removed the vast majority of oxygen from the atmosphere. All organisms that rely on oxygen for respiration (aerobes) would rapidly be eliminated. Organisms that respire using chemicals other than oxygen (obligate anaerobes), such as *Clostridium botulinum*, and many of the archaea microorganisms that live in hot springs, the deep ocean, and the low oxygen mud at the bottom of lakes, might survive. In this situation, selection would occur at the level of the species or higher phylogenetic groups, as there is simply no significant within species genetic variation in ability to survive without oxygen. All dogs would die, but all strains of *C. botulinum* would be likely to survive the atmospheric change, at least until subsequent impacts of the ecosystem collapse began to affect them. However, some species are facultative anaerobes: they use oxygen to respire, but in the absence of oxygen will respire using other chemicals. In these species there is likely to be genetic variation between individuals in their ability to anaerobically respire, and hence in their ability to survive. In these species selection would be likely to occur at the level of the gene or individual.

Conflict often occurs through selection at two levels working in opposite directions (Keller 1999) (Box 1). Prions are proteins that naturally occur in organisms, but which have mis-folded and changed into a new shape that catalyses the transformation of similar proteins to the same configuration (Johnson 2005). The catalysis allows the prions to spread through the individual in which they arise, and also to potentially infect other individuals through contact or consumption. Hence they can be transmitted between individuals. However, the loss of a functional protein due to transformation to the prion form can have serious consequences for the individual. In humans, prions are responsible for Creutzfeld-Jakob disease (CJD, "mad cow disease"), an incurable and fatal brain disease. Clearly the prions that cause CJD increase their transmission within individuals through converting proteins, and so are positively selected at the molecular level. However, the mortality they cause to humans is a form of counter selection at the individual level. In CJD there is a lack of transfer from infected humans back to cattle, making human infection a dead end for the prion. This is not always the case. Kuru is an infective prion disease found in New Guinea that causes death within a few months (Gajdusek and Zigas 1957). The disease was passed between individuals during bereavement rituals in which the dead relative is partially eaten and flesh rubbed on the skin, allowing transmission through consumption and through skin lesions. The decline in these rituals has led to a rapid decline in cases (Alpers 2008). The success and spread of prions such as these will be determined by both selection at the level of individuals, and their ability to spread within individuals.

Box 1. Levels of Selection: Cancer

A cancer cell possesses heritable genetic differences to the other cells in the body, allowing it to rapidly clonally reproduce, while avoiding the control mechanisms that usually eliminate uncontrolled rapidly reproducing cells. This usually occurs due to mutations in somatic cells that prevent the operation of normal cell death. Cancer can be viewed as a conflict between two levels of selection (Keller 1999). The mutations that allow cancer cells to multiply increase the fitness of those cells relative to normal cells, and allow the cancer to spread within the organism. This method of thinking about cancer can be useful in explaining why cancers can become more rapid growing with time, as cancer cell lines with mutations that increase their rate of proliferation outcompete older lineages that grow more slowly. However, selection at the level of the individual counteracts the success at the level of cells, as individuals carrying cancer frequently die and fail to reproduce.

Cancer cells could avoid this fate if they were able to transmit between individuals. The biochemical self recognition systems in most organisms make this extremely difficult, as cells that are genetically distinct from the host are typically attacked and rapidly eliminated. However, there are three known examples where cancer has been able to avoid the costs of killing individuals by becoming capable of transmission between individuals. The best known of these is canine transmissible venereal cancer (CTVC), which affects dogs and wolves. This cancer occurs on the genitals of dogs and wolves, but is not derived from the cells of the host (Murgia et al. 2006). Instead it is an ancient cell line of cancer that is transmitted between individuals, usually through sex. First described over a 140 years ago (Novinski 1876), analysis of the gene sequence of CTVC suggests that it has existed for at least 6000 years (Rebbeck et al. 2009). CTVC is usually not fatal, instead undergoing relapse 3 months after initial infection. This relatively low level of harm to the hosts might explain why this transformation from a multicellular sexual animal to a unicellular parasite has been so successful. The conflict between levels of selection is far more obvious in a second transmissible cancer. Devil facial tumor disease (DFTD) is a cancer affecting Tasmanian devils (*Sarcophilus harrisii*). First identified in 1996, this cancer occurs on the faces of Tasmanian devils, and is transferred between individuals during fights or when feeding on the same food. The disease is a major conservation problem, which may lead to extinction of the species in the wild (Miller et al. 2011). The cancer seems to be able to transmit between individuals because the population is highly inbred, with such low genetic diversity that immune systems of most individuals fail to recognise the cancer cells as "non-self" (Siddle et al. 2007).

8.1.2 What are Selfish Genetic Elements?

Selfish genetic elements (SGEs) are genes that are present in the genome or cell of an organism that ensure they are passed on at a higher frequency than the rest of the genome to subsequent generations. This 'selfish' nature of SGEs ensures they will accumulate within the genome (Burt and Trivers 2006). However, the subversion of the normal pattern of inheritance by SGEs generates a conflict with the rest of the genome (Burt and Trivers 2006; Hurst and Werren 2001; Werren 2011). Hence there are opposing selective forces operating: selection on the SGE to increase its transmission, opposing selection on the rest of the genome to suppress the selfish action of the SGE, and finally selection on the individual carrying the genome (harbouring the SGE) to increase its fitness. SGEs are ubiquitous in living organisms and can make up a large proportion of the genome (Burt and Trivers 2006; Hurst and Werren 2001; Werren 2011). There are a variety of different types (Table 8.1).

Most multicellular organisms are made up of two types of cells, the somatic cells involved in the building and maintenance of the body of an organism and the sex cells—the gametes—that transmit the genetic material from one generation to the next. SGEs frequently target the sex cells, as this is an effective way to increase their own transmission at the expense of the rest of the genes within the genome. This generates conflict between the SGEs and the rest of the genome. The transmission bias of SGEs generates strong selection for suppression of SGEs to ensure a more equitable inheritance of genes. This is particularly true for SGEs that distort the population sex ratio, as this creates strong selection to restore the sex ratio to

Table 8.1 Different classes and examples of selfish genetic elements and some of their, following Hurst and Werren (2001); Burt and Trivers (2006); Werren (2011)

Class	Example	Impacts
Autonomous replicators	*Transposable elements*	Affect genome size and gene function, cause mutations
	Retrotransposons	Affect gene regulation and gene function, cause mutations
Converting elements	*Homing endonucleases*	Chromosome breakage and insertion
	Prions	Affect brain structure and other neural tissues
Segregation distorters	*Driving chromosomes*	Can cause sex ratio distortion
	B chromosomes	Increase crossing over and recombination
Post segregation distorters	*Endosymbionts*	Reproductive disruption, sex ratio distortion
	Medea	Killing of offspring that do not inherit the selfish gene from their mother

unity. Such genetic suppression of SGEs can in turn have far reaching impacts, and may have given rise to key evolutionary innovations. For example, it is suggested that methylation arose to silence SGEs, making them key to the evolution of genomic imprinting and the placenta in mammals (Haig 2012). Similarly, it is suggested that genetic suppression of SGEs has given rise to a variety of RNA interference mechanisms as a means to silence SGE expression (Vagin et al. 2006). There is also evidence that SGEs can, instead of being silenced, be domesticated and take on new and beneficial functions. For example, the vertebrate immune system may have domesticated transposable elements as mutators to generate new variants of T-cells to protect against novel pathogens (Kidwell and Lisch 2000).

8.1.2.1 Autonomous Replicators and Converting Elements

Autonomous replicators and converting elements are by far the most common type of SGEs. They can have dramatic impacts on genome structure by regulating the size and function of the genome. For example, in *Zea mays* it is estimated that $>50\,\%$ of the genome is comprised of transposable elements (TE), and that TE insertions have caused an almost doubling of the size of the genome in the last few million years (SanMiguel et al. 1998). Similarly, $\sim 45\,\%$ of the human genome is made up of TEs, many of which are inactive and ancient (Lander et al. 2001). Transposons are TEs that move around the chromosome, so called "Jumping genes". They are DNA sequences that encode enzymes that catalyze their own movement within the genome, can cause mutations and change the amount of DNA in the genome. They may have had their origins in DNA repair systems, or a viral origin. Homing endonucleases work in a similar way, producing enzymes that cut DNA at a particular site and insert a copy of a copy of the homing endonuclease. They can multiply by moving from one chromosome location to another. Transposable elements are a potent force in shaping the structure and function of the genome. During an insertion event they frequently disrupt the function of a gene, often causing additional mutation events. For example, it is estimated that in *Drosophila melanogaster* flies, around 50 % of all mutations are caused by TE insertions (Charlesworth et al. 2004). In addition, they frequently increase the gene content of a genome by giving rise to repetitive DNA sequences that we refer to as 'junk DNA'. They can also cause inversions and translocation of large genome segments that can have dramatic impact on the structure of chromosomes. This is in part due to the misalignment during DNA repair after TE insertions. This ability of TEs to change location within the genome can result in new functional activities (Alzohairy et al. 2013).

8.1.2.2 Segregation Distorters

Segregation distorters ensure that after meiosis they are present $>50\,\%$ of the offspring. Meiotic drive chromosomes are well-studied segregation distorters, which alter the meiotic process so that the driving chromosome is present in more than

50% of the gametes. Most animals and plants are diploid: their non-sexual cells (somatic cells) carry chromosomes organised into pairs. The process that transforms diploid somatic cells into haploid gametes is called meiosis. Complex molecular mechanisms ensure that meiosis is generally fair: each chromosome has a 50% chance of entering each gamete. However, it is possible for alleles to "cheat" (Burt and Trivers 2006). Any allele that manipulates meiosis to increase its success is referred to as a "meiotic driver" (Jaenike 2001). Eventually such an allele might spread until every copy of that chromosome in the population carried the allele. At this point, it would no longer have any effect, as all chromosomes carry the allele, and all will be equally good at being passed on to the next generation, Meiotic drive can occur both through the ova or though sperm, although sperm drive appears to be more common. An allele that increases the success of the sperm that carry it relative to sperm that do not carry the allele in the same male's ejaculate should be able to fertilise more ova than rival alleles, and so increase in frequency. The simplest way for an allele to increase its success in sperm competition against other sperm in the same ejaculate is to sabotage the sperm that do not carry the allele. This is exactly what we see in "post-meiotic" meiotic drivers. These drivers damage, and in many cases completely eliminate, sperm produced by the same male that do not carry the driving allele (Jaenike 2001).

If the meiotic drive chromosome is sex linked this gives rise to sex ratio distortion. In a population with a female biased sex ratio, each male will on average mate with more than one female, which means that males have higher success than females, thus making sons more valuable than daughters. Hence, carrying a sex ratio distorter imposes a cost on the rest of the genome of the males that carry it, as these alleles will only be passed on to daughters. This cost generates selection for resistance to the meiotic driver. This can result in the evolution of suppressor alleles that prevent the Y-chromosome sperm being eliminated. For example, in the fruit fly *D. paramelanica* there is a meiotic driving X-chromosome, and two Y-chromosome forms. One of the Y-chromosomes is able to resist the driving X, and males carrying a driving X-chromosome and a resistant Y produce broods with equal number of sons and daughters (Stalker 1961). Resistance to drive can also evolve on the autosomes, as these will also benefit from being passed on to a higher frequency of sons (Tao et al. 2007). Another common form of segregation distorters is B-chromosomes, which are parasitic nonessential heterochromatic chromosomes, and are widespread in eukaryotes.

8.1.2.3 Post-Segregation Distorters

Post-segregation distorters chromosomes, and are after fertilization and development has commenced to ensure they are present in >50% offspring. There are several different types of post-segregation distorters including Medea and endosymbionts. Medea is a selfish killer allele that is passed on from mother to offspring, killing all offspring that do not inherit the Medea allele from their mother, allowing it to rapidly spread. Endosymbionts are organisms that live inside the cell of another organism. They are very common, and are either beneficial to their host by providing key nutrients, or selfishly serve their own interest at the expense of reduced host fitness. In this latter role they act as post-segregation distorters by causing a variety of reproductive manipulations of the host that can result in reproductive failure and sex ratio distortion. The reason for these reproductive manipulations is that endosymbionts are typically vertically transmitted through the cytoplasm in the egg, meaning they are predominantly inherited from mothers to offspring. Therefore from the endosymbionts point of view, males represent a dead end. Endosymbionts induce a variety of reproductive manipulations to increase their transmission by avoiding sons. The most well-characterised endosymbiont is the bacterium *Wolbachia pipientis* that is estimated to infect between 30 and 70 % of all arthropods, potentially making is one of the more common organisms on Earth. *Wolbachia* increases its transmission by inducing parthenogenesis and abolishing the production of males altogether (e.g. several species of wasp), by the killing of sons and thereby benefiting their sisters through reduced competition, risk of inbreeding, or simply by providing additional nutrients (e.g. flies, ladybirds and butterflies), by feminization of genetic males into functional females (e.g. isopods, moths and butterflies), or by inducing reproductive incompatibilities in crosses between infected males and uninfected females—termed Cytoplasmic Incompatibility (CI). This benefits *Wolbachia* as it result in the production of more infected than uninfected offspring, as infected females are compatible with both infected and uninfected males (producing infected young), whereas uninfected females are only compatible with uninfected males. CI is the most common and widespread form of post-segregation reproductive manipulation. There are several different strains of *Wolbachia* that are often not compatible with each other, whereas some show partial bi-directional compatibility between strains. The mechanisms of CI induction is not clear, although it involves modification of the infected males' sperm that is thought to require a 'rescue' factor present in the cytoplasm of infected females' eggs that ensure embryo development progress normally, resulting in the production of infected offspring. When an infected males' sperm fertilises the egg of an uninfected female, the entry of the male pro-nucleus is delayed whereas the maternal chromosomes segregate normally resulting in the production of haploid embryos that die. Interestingly, there is large variation in the severity of CI in crosses between infected males and uninfected females, suggesting either that *Wolbachia* does not modify all sperm in infected males and/or that the manipulation is not as severe.

8.1.3 Sexual Selection

A little over 150 years ago Darwin revolutionised biology with his theory of natural selection (Darwin 1859). He convincingly argued that species carry heritable variation in their ability to survive, and that only a subset of individuals each generation are able to survive and reproduce. This selection of beneficial variation each generation will, over the generations, "fit" organisms to their environment. Assuming that new variation occurs each generation, which we now know is true, this continuing process is capable of developing entirely new traits and creating new species, thereby explaining the diversity of life we see around us. The simplest part to this is natural selection; the importance of the ability to survive. It is obvious that a hare that is better able to withstand cold might be better able to survive harsh winters. An antelope faster than its siblings might be better able to escape from predators. However, Darwin also put forward another key hypothesis in *The Origin Of Species*, that individuals need not only to survive but also to successfully reproduce. This theory of sexual selection was expanded in *The Descent of Man, and Selection in Relation to Sex* (Darwin 1871), which laid the foundation for all subsequent work on sexual selection.

Sexual selection is the competition within a sex to secure mates and produce offspring (Anderson 1994). This may occur through conflict between members of that sex, such as when bull elephant seals fight to control access to a mating beach and the females on it. It can also occur through mate choice. In most species females require more time between reproductive bouts than males and hence are frequently the limiting sex (Bateman 1948). Males cannot coerce females to mate in the majority of species, and females tend to choose between abundant males. This mate choice has often favoured evolution of extreme traits in males, such as bright plumage and elongated tail feathers in many male birds. However, because of the cost of reproduction (Trivers 1972), the reproductive interests of male and female often do not coincide, resulting in sexual conflict over parental investment and female mating frequency (Parker 1979). Selection will favour exploitation of mating partners to make a larger investment in reproduction and, if a male gains by a female mating only with him whereas a female gains by mating with several different mates, then there is also potential for conflict over female receptivity (Arnqvist and Rowe 2005). This conflict can at times even lead to traits that harm mating partners, such as male dungflies occasionally drowning females in their efforts to coerce them into mating, or female praying mantises eating their mates.

Sexual selection and sexual conflict are powerful evolutionary forces. Sexually selected and in particular sexual conflict traits typically evolve faster than other classes of traits, except those involved in resistance to fast evolving parasites and diseases (Swanson and Vacquier 2002). This is because sexual conflict typically involves adaptations in one sex aimed at manipulating the other to increase its fitness, despite the cost it may generate in the other sex. This promotes counter selection, favouring adaptations in the other sex to ameliorate costs incurred from such manipulations and regain reproductive fitness. This in turn generates selection on the other sex, and so on. The result is sexually antagonistic co-evolution that is thought to be responsible for the rapid divergence in reproductive traits between populations, and is even suggested to promote reproductive isolation (Arnqvist and Rowe 2005).

8.1.4 Aim of Review

In this review we discuss how selfish genetic elements can be major drivers of sexual selection. Selfish genetic elements are ubiquitous and arise due to the inevitable conflicts that occur within individuals between alleles, chromosomes and cell lines. We show that SGEs can have major impacts on the reproductive success of individuals that carry them, on the individuals that interact with carriers of SGEs during reproduction, and on the populations in which they occur. We discuss the various ways in which the impact of SGEs can drive sexual selection by focusing on a few recent case studies, highlighting the far-reaching effect of different SGEs on mating systems.

8.2 Why do SGEs Affect Sexual Selection?

So how might SGEs affect sexual selection? Because SGEs are associated with a variety of costs such as reduced male fertility (see below) and production of the more common sex in the case of sex ratio distorters, non-carrying individuals are expected to avoid mating with SGE-carrying mates. There is some evidence of mate discrimination against SGE-carriers prior to mating in some species. However, as discussed below, such cases are remarkably rare and tend to involve suppressed recombination between the SGE and the ornament used in mate choice. Instead SGE seems to play a greater role in post-copulatory sexual selection. Moreover, the scarcity of males caused by sex ratio distorting SGEs can have dramatic effects on mating systems (Werren 2011), as discussed below. Finally, there is strong selection for suppression of SGEs to ensure a more equitable inheritance and to restore the sex ratio to unity that can directly affect sexual selection.

8.2.1 SGEs and Pre-Copulatory Mate Choice

Many SGEs reduce the fitness of the individuals that carry them. An obvious prediction from this is that individuals should be selected to avoid mating with SGE carriers. However, examples of this are surprisingly rare (Price and Wedell 2008). Perhaps the best example comes from the stalk-eyed flies of the Diopsidae family. Their most noticeable trait is their huge eyestalks (Fig. 8.1), which are far larger in males than females. In some species, large males can have eyestalks wider than their body is long. In many species males arrange themselves in a "lek", an area where males congregate. Females visit and assess males, preferring to mate with males with wider eye spans (Rogers et al. 2008). Eye span correlates with body size, and is also used by males to assess rivals in fights for the best locations in the lek. In normal males, males of low genetic quality have higher costs of producing wide eye-spans, and so cannot produce the extreme eye-spans of the high genetic

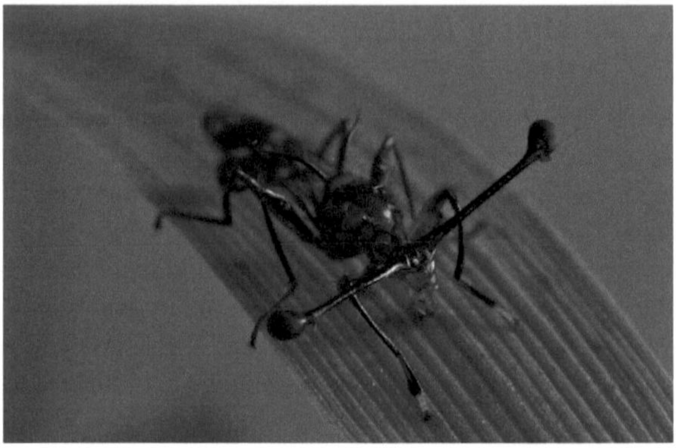

Fig. 8.1 A male stalk-eyed fly (*Teleopsis dalmanni*) showing the enormous eye stalks. Females prefer to mate with males with wider eye spans, which correlates both with body size and resistance to meiotic drive. (this photo is a creative commons by attribution photo, by Rob Knell, CC by-SA 2.5 (http://creative-commons.org/licenses/by-sa/2.5/))

fitness males (Baker et al. 2003). Hence, females choosing to mate with wider eye-span males will tend to be mating with males carrying good genes, and should have higher fitness offspring as a result (David et al. 2000).

Several species are known to carry sex ratio distorting X-chromosome meiotic drive. In the two best-studied species, *Teleopsis dalmanni* and *T. whitei*, there are multiple strains of driving X, some of which are transmitted to more than 90 % of a male carrier's offspring (Presgraves et al. 1997). These drivers are common in natural populations, often reaching frequencies of 35 % (Wilkinson et al. 1998). As a result, populations consist of more females than males and an average male is expected to mate with several females. Hence producing a higher proportion of sons is associated with higher fitness. Females that avoid mating with meiotic drive carrying males would increase their fitness due to increased production of sons and their offspring not inheriting the driving X. In *T. dalmanni* about 30 % of heritable variation in male eye-span is found on the X-chromosome, and meiotic drive X-chromosomes are associated with shorter eye-spans (Johns et al. 2005). Furthermore, there are Y-chromosomes that are able to resist the driving effect and are associated with wider eye-spans (Wilkinson et al. 1998). By preferring mates female with wider eye-spans, females are able to avoid mating with costs meiotic drive carrying males.

A second example is found in mice. Several mouse species carry autosomal meiotic drivers called *t* haplotypes (Artz et al. 1982; Lenington 1991). In the house mouse, *Mus musculus*, there are several *t* haplotypes, and these can be found worldwide at frequencies of up to 25 % (Ardlie and Silver 1998). If a male carries a *t* haplotype, the development of sperm that do not carry that *t* haplotype is suppressed. There is considerable variation in *t* haplotypes, with some strains eliminating 80 % of sperm carrying rival chromosomes, and thereby being passed on to 90 % of the male's offspring (Lenington 1991). These *t* haplotypes kill homozygotes while they are foetuses. Males that carry two different *t* haplotypes are completely sterile, whereas females with two different *t* haplotypes develop normally. However, if a heterozygote female mates with a heterozygote male, 45 % of their offspring will be homozygous and die, 50 % will inherit a *t* haplotype, and only 5 % will not inherit a *t* haplotype. As predicted, there is evidence that females do indeed prefer to avoid mating with *t* haplotype carrying males. Moreover, this preference is much stronger in females that carry a *t* haplotype than females that do not. However, the majority of this evidence comes from studies of scent, where females prefer to nest in the bedding of males that do not carry *t* haplotypes, or to spend time in chambers near to them, but is not based on actual mating data (Lenington et al. 1992). Moreover, females strongly prefer to mate with dominant males, which can mask effects. Furthermore, there is an effect of the reproductive cycle of females, with females in oestrus showing stronger preferences (Williams and Lenington 1993). Nevertheless, there is good evidence that heterozygote females do indeed prefer to avoid mating with *t* haplotype carrying males, at least in some populations.

There is also evidence that some endosymbionts can drive the evolution of mate choice (Goodacre and Martin 2012). The best evidence comes from spider mites (*Tetranychus urticae*). In this species, females that do not carry *Wolbachia* will lose fitness if they mate with an infected male due to the death of offspring through CI. Uninfected females prefer to mate with uninfected males, whereas infected females show no preference (Vala et al. 2004). In the fruitfly *Drosophila paulistorum*, subspecies are associated with different strains of *Wolbachia* that are important for ensuring successful reproduction. Females strongly prefer to mate with males carrying their own strain (Miller et al. 2010). However, when *Wolbachia* levels are reduced by antibiotics the preference for males harbouring the same strain is dramatically lower. In *D. melanogaster*, curing strains of *Wolbachia* also alters mate preference, although in unpredictable ways (Markov et al. 2009).

So why are mate preferences against carriers of SGEs so rare despite large fitness cost of not discriminating? Female choice against SGE carrying males is expected to be vulnerable to a breakup of the linkage between the SGE and the detectable trait (Nichols and Butlin 1989; Pomiankowski and Hurst 1999). For example, if a recombination in stalk eyed flies resulted in a meiotic driving X that was associated with wider eye-spans, it would be likely to spread through the population to such a high frequency that eye-span stopped being a useful signal of drive status. Any choice system against a SGE is likely to require that recombination between the driver and signal genes is extremely rare. In both stalk eyed flies and mice this does seem to be the case. The association between drive and small eye-span alleles seems to be due to very tight genetic linkage between the two (Johns et al. 2005). In mice, the *t* haplotype and scent genes are tightly linked in an area of very low recombination (Lenington et al. 1992). *Wolbachia*-based mate preferences may involve changes to the odour profiles used in mate choice.

8.2.2 SGEs and Post-Copulatory Sexual Selection

The limited evidence for the importance of SGEs in promoting pre-copulatory mate choice suggests that post-copulatory influences may be more important, potentially favouring female multiple mating (Zeh and Zeh 1996). Many SGEs target male gametes during spermatogenesis in order to increase their transmission rate and get passed on to the next generation. This manipulation can be achieved by modifying sperm during development, as is the case for many endosymbionts (e.g. Snook et al. 2000; Lewis et al. 2011a). Other SGEs such as meiotic drivers, some B-chromosomes and some transposons destroy sperm that do not pass the selfish gene to subsequent generations (e.g. Policansky and Ellison 1970). This can result in dramatically reduced numbers of sperm produced by male carriers compared to non-carrying males (Price and Wedell 2008). In addition, the method of sperm killing can itself have a detrimental side effect on the surviving sperm that carry the SGE that compromises their sperm performance (e.g. meiotic drive, Price et al. 2008a). As a consequence of this sperm manipulation and sperm-killing, male carriers frequently suffer reduced fertility and sperm competitive ability compared to non-carrying males (Price and Wedell 2008; Price et al. 2008a). There are numerous examples of this fertility reduction of SGE-carrying males, and this effect ranges from a slight drop in fertility to a dramatic reduction of 50 % or more in some species (Price and Wedell 2008). This is important since a reduction of a minimum of 50 % is required to stabilize the transmission advantage of an SGE that kills all non-carrier sperm (Haig and Bergstrom 1995). The reduced sperm production of these low fertility SGE-carrying males translates into compromised sperm competitive ability, as relatively higher sperm numbers are advantageous in sperm competition (Simmons 2001; Lewis et al. 2008). In *D. melanogaster*, for example, the fertility reduction of a segregation distorter (SD) is greater than expected purely from by the elimination of non-carrying sperm (Hartl et al. 1967), and several other studies have also shown that a higher degree of drive in males is associated with a greater fertility reduction than expected based on sperm numbers (Fry and Wilkinson 2004; Price et al. 2008a).

The reduced fertility and sperm competitive ability of SGE-carrying males provides an important link between the presence of SGEs and female mating frequency. This link is predicted to promote polyandry as a female strategy to bias paternity against low-fertility SGE-carrying males and avoid passing on the SGEs to offspring (Zeh and Zeh 1996; Price and Wedell 2008; Wedell 2013). As discussed in detail below, there is evidence that this is indeed the case in several taxa. The advantage to polyandrous females is a reduced risk of only mating to SGE-carrying males, although this bet-hedging strategy will only work under limited circumstances (Yasui 1998). A potentially more important aspect is the reduced sperm competitive ability of carrier males, which means that polyandrous females may be able to swamp the sperm of SGE-carrying males with that of normal males. In addition, polyandry is also hypothesized to decrease the population frequency of any SGEs that reduce the sperm competitive ability of males, and so polyandry might protect against SGEs at the population level (Haig and Bergstrom 1995), by undermining their transmission advantage (Price et al. 2010a). Hence there is a dynamic link between polyandry and the frequency of SGEs that may be important in favouring female multiple mating in general (Wedell 2013).

8.3 Case Studies

Here we focus on a few recent case studies to illustrate the potency of SGEs in affecting sexual selection, highlighting the diverse and far-reaching impact they have on animal mating systems.

8.3.1 Sex Ratio Distortion and Coevolution Between the Sexes in a Fruit Fly

One of the best-studied sex-ratio drive systems is that in the fruit fly *Drosophila pseudoobscura* a small fly found in wooded areas from Canada to Guatemala (Dobzhansky and Epling 1944). These flies are diploid, with females carrying two X-chromosomes, and males an X and Y. In 1936 some males were found that produced almost entirely female broods (Sturtevant and Dobzhansky 1936). Further work showed that this was due to these males carrying an X-chromosome with a meiotic driver, which was named "*sex-ratio*", or *SR*. The Y-chromosome bearing sperm of males carrying *SR* fails to develop properly during spermatogenesis, so all functional sperm of the *SR* males carry the *SR* X-chromosome (Policansky and Ellison 1970). As a result, all the functional offspring of an *SR* male are daughters that inherit a copy of the *SR* X-chromosome. In contrast, the X-chromosome of a non-*SR* male will be passed on to half his offspring. Hence, if *SR* and non-*SR* males have the same number of offspring on average, then *SR* should rapidly spread through the population, outcompeting both the non-driving X-chromosomes and the Y-chromosome. If *SR* continued to spread through a population, eventually it should reach a high enough frequency that no males are produced and the population should go extinct (Hamilton 1967). However, surveys over the past 70 years have shown no evidence of an increase in the frequency of *SR* (Dobzhansky 1958; Powell 1997). Instead, there seems to be a stable distribution, with *SR* most common in the southern USA and northern Mexico where 30% of all X-chromosomes are *SR* chromosomes. From the US/Mexican border there is a gradual decrease to the north, and probably a similar decrease to the south. The long-term stability of *SR* in natural populations is a mystery and the reason for this is poorly understood. Nevertheless, populations where *SR* is common tend to have female biased sex ratios (Bryant et al. 1982).

Remarkably, there is no genetic resistance against *SR* (Policansky and Dempsey 1978). This should select females to avoid mating with *SR* males. However, there is no evidence that females can distinguish between *SR* and non-*SR* males prior to mating (Wu 1983; Price et al. 2012). However, the elimination of Y-chromosome sperm greatly reduces the number of sperm an *SR* male produces (Price et al. 2008a). Moreover, there is some evidence that the killing of Y-sperm also damages the X-chromosome sperm. As a result, *SR* males produce fewer sperm than non-*SR* males that may be of low quality (Wu 1983). This can result in females that mate with *SR* males running out of sperm and being unable to lay fertile eggs.

However, even *SR* male *D. pseudoobscura* transfer thousands of sperm to females (Price et al. 2008a), and so it is unlikely that lack of sperm is ever important in nature. But importantly, the reduced sperm production of *SR* males means they are less successful than non-*SR* males in sperm competition. When a female mates with a non-*SR* male, followed by a *SR* male, on average the *SR* male fathers less than 20 % of the offspring. A non-*SR* male in a similar sperm competitive scenario is expected to father 70 % of the offspring (Price et al. 2008a). Hence, even though females cannot tell which males carry *SR*, they can reduce the chance of their offspring being fathered by an *SR* male simply by mating with multiple males (Haig and Bergstrom 1995). Indeed, when laboratory experimental evolution populations were set up (see Box 2) where females risked mating with *SR* males, females rapidly evolved higher remating rates (Price et al. 2008b). This higher mating rate is likely to reduce their risk of siring sex ratio-biased broods, and this only evolved in populations where *SR* was present.

Furthermore, the non-*SR* males also evolved in response to the presence of the *SR* males. This was due to the increased female remating frequency in these populations, providing direct evidence of co-evolution between males and females in the experimental *SR* evolution populations. The increased remating rates in the *SR* populations meant a higher level of sperm competition was experienced by males. Males responded to the higher risk of sperm competition and evolved to transfer bigger ejaculates containing more sperm. Higher sperm number is favoured in sperm competition. In addition, these males were also better at suppressing female receptivity (Fig. 8.2; Price et al. 2010b), indicating evolution of more 'potent' ejaculates. In *D. melanogaster* the male ejaculate contains a cocktail of accessory gland proteins that are known to affect female reproductive physiology, including suppressing her tendency to remate (Ravi Ram and Wolfner 2007). Furthermore, there was evidence of a tight relationship between the level of remating in females and the ability of the males in the very same populations to suppress female receptivity. This indicates rapid co-evolution between the sexes with regards to female remating rate. In populations where females evolved the highest level of remating, the non-*SR* males were best able to suppress female remating, whereas in populations where females evolved the least increase in remating rate, males were least able to prevent the female from remating. This relationship was remarkably tight and evolved after only 11 generations of experimental evolution, by which point the frequency of *SR* was $<5\%$ (Price et al. 2010b). This indicates that even at a low frequency, *SR* has a dramatic effect on the mating system of this fly species: it promoted the evolution of increased female remating rates, which in turn favoured the evolution of bigger and more 'potent' male ejaculates in populations with elevated levels of polyandry. This is a powerful illustration of the power of an SGE to generate sexual selection, even when present at a low frequency by causing a chain of adaptations and counter-adaptations in females and males.

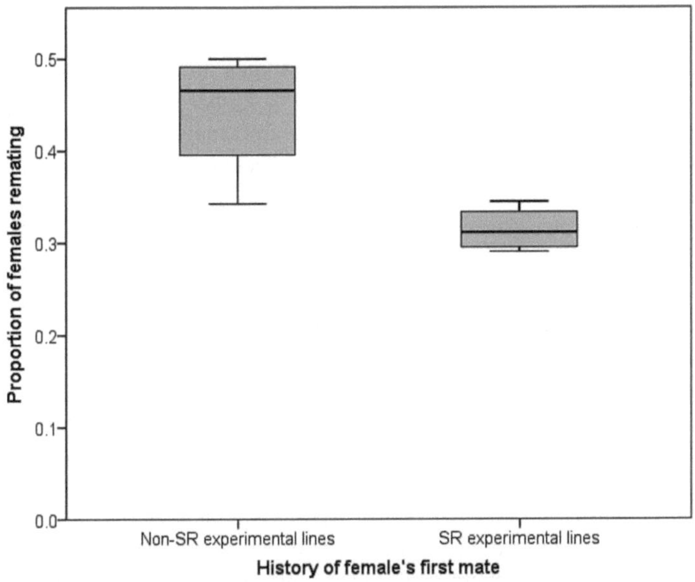

Fig. 8.2 Box plot of the proportion of female *Drosophila pseudoobscura* that remated when presented with a second male following initial mating to a male from an experimental evolution population where *SR* was present or where *SR* was absent (Price et al. 2010b). Female remating tendency was suppressed by mating to a male from a population evolving in the presence of *SR* (Mann-Whitney U test: $N=8$, $U=0.5$, $P=0.029$), due to males in these populations being faced with higher female remating rates. Median, interquartile range, and range are shown.

Box 2. Experimental Evolution

Evolution is difficult to study in nature. Many organisms may show very little change over vast periods of time (Hull and Norris 2009). For example, pelicans today appear very similar to those that appeared 30 million years ago (Louchart et al. 2011). On the other hand, evolution can be extremely rapid, when changes to the environment cause strong selection. For example, DDT resistance in insects has spread extremely rapidly after the use of pesticides due to strong directional selection (see also 3.5 below). If we want to examine evolution, the lack of examples of strong directional selection in nature reduces our ability to determine what evolution is capable of. A second problem with studying evolution in natural populations is that many selective forces will act simultaneously, making it difficult to determine which factors cause the evolutionary change in a population and the relative importance of each factor. A third problem is that it limits us to study the selection that occurs in nature. To understand biological systems, we need to understand what could happen, not just what actually happens in the natural populations we are observing. For example, there is a clear need to understand how species will respond to increasing global temperatures. But it is likely that many species will not experience these conditions until it is too late for the information to be useful (Gienapp et al. 2008). A final problem is the lack of replication in nature. Replication is essential to be sure that an evolutionary response is genuinely linked to a particular selective pressure, and not occur merely chance. Unfortunately, in nature, it is almost impossible to be certain that populations are separate, and hence evolving in parallel. Even geographically isolated populations, on islands for example, could be linked by rare gene flow and hence not represent true replication. It is only possible to be certain that populations are separate in controlled environments.

Experimental evolution, the creation of a specific set of conditions to which a suite of traits may evolve, provides a solution to these problems (Kawecki et al. 2012). For this technique a laboratory population is split into several genetically similar laboratory populations. A group of populations can then be subjected to a potential selective factor, with a second group not exposed and used as a control. Unlike a traditional experiment, the members of each subject population are allowed to reproduce after selection. The treatment is then repeated for multiple generations. This allows the population to respond to the selection by evolving. Experimental evolution is similar to artificial selection where selection is directly applied to the organism (Edward et al. 2010). Examples are the traditional breeding of domestic animals for particular traits such as size, appearance or productivity. A frequently used approach is to only allow the largest (or smallest) 20 % of individuals each generation to breed (Mackay et al. 2009). This typically produces rapid evolution in the trait selected for. Experimental evolution differs from artificial selection because the traits(s) and response are not chosen by the researcher. A population directly selected for size is unlikely to respond in any way other than size, whereas a population exposed to a competitor species could respond in a wide variety of ways. However, there is no clear distinction between experimental evolution and artificial selection, they are better considered as ends of a continuum of the level of direct manipulation by the researcher.

There are limitations to experimental evolution. Firstly, experimental evolution is always limited in scope compared to real populations. The size of the population that can be used is limited by practical considerations. As laboratory populations are typically very small compared to natural populations, novel beneficial mutations are likely to be extremely rare. As a result, the ability of a laboratory population to respond to selection will be limited by the amount of genetic variation contained within that population when the experiment starts. In other words, if the population has little variation at the start of the experiment, either due to small population size, inbreeding, or chance, then there is unlikely to be an evolutionary response. The amount of selection will also be limited by the number of generations that can be used in the experiment. A long-lived species that becomes sexually mature after a period of years, such as elephants or cetaceans would be highly unsuitable for experimental evolution. These considerations mean that many species are completely unsuitable for use in experimental evolution, due to size, inability to survive in the laboratory, or long generation times. Nonetheless, experimental evolution is one of the most powerful techniques available to evolutionary biologists, and has provided insights into areas as diverse as the mechanisms under-lying social evolution (Buckling et al. 2009), life history, toxin tolerance (Mackay et al. 2009), the spread of new mutations (Atwood et al. 1951), and resistance to parasites (Lenski 1988). It provides an opportunity to examine evolution in real time, in a replicated fashion, which is all too rare in natural studies.

8.3.2 Male-Killing Endosymbionts in Butterflies

There is additional evidence that the population sex ratio caused by segregating distorting SGEs can directly influence the level of sexual selection. In the butterfly *Hypolimnas bolina* for example, many populations harbour a male-killing strain of *Wolbachia* (Dyson and Hurst 2004). Throughout the South Pacific there is large variation in the frequency of male-killers between islands that is associated with differences in the degree of female sex ratio bias, with a higher frequency of male killers associated with a more severe female biased population sex ratio (Charlat et al. 2007). As a consequence, the mating system differs depending on the frequency of male killers. In high prevalence populations, males provide smaller sperm packets than in low prevalence populations, which is likely due to their higher mating frequency in these high prevalence populations as a consequence of the pronounced female sex ratio bias. This directly influences the degree of female multiple mating. In populations with a high frequency of male-killers and a severe female-biased sex ratio, females engage in higher levels of multiple mating. This is likely due to the increased severity of sperm limitation experienced by females in these high prevalence populations. This in turn means a higher male mating rate that creates a cycle of male fatigue with a concomitant female sperm shortage that will further promote female multiple mating. At extreme levels of male killing, female mating frequency becomes directly limited by access to males (Charlat et al. 2007). Multiple mating is favoured in this species because of the limited supply of sperm due to the short-age and exhaustion of the few males because of the female biased sex ratio that in some populations can be as high as 100 females to one male (Dyson and Hurst 2004).

So what consequences might this have for sexual selection? Increased levels of female multiple mating will affect the risk of sperm competition, but this risk is likely to be reduced in high prevalence populations since even though females are willing to mate many times, the number of different males they mate with is reduced. It is even possible that a male may mate with a female whom he has already inseminated. In such a situation there is clearly a low risk of sperm competition, and males should instead discriminate against already inseminated females and prefer novel females to mate with. Hence, we may expect male mate choice to be more prevalent under such situations. In addition, it may also favour changes in female behaviour to increase the likelihood of successful insemination. In populations of the butterfly *Acrea encedon* harbouring male-killing *Wolbachia*, females engage in lekking behaviour (mating aggregations believed to increase visibility), presumably in an attempt to advertise their presence to the rare males and increase their mating success (Jiggins et al. 2000). In populations without male-killers, lekking is normally only performed by males. This illustrates the potency of sex ratio distorters in affecting both female mating rate, but also mating behaviour and mate choice.

8.3.3 Increased Male Mating Rates in Wolbachia-Infected Drosophila

The presence of *Wolbachia* can also directly affect male mating rate and mating strategies. Infected male *D. melanogaster* and *D. simulans* harbouring CI-inducing *Wolbachia*, show significantly higher mating rates than uninfected males (Champion de Crespigny et al. 2006). On average, infected *D. simulans* males mate almost 50 % more frequently than uninfected males, whereas, infected *D. melanogaster* males mate about 16 % more frequently than uninfected males. Interestingly, this difference in male mating rate covaries with the level of CI. It is more severe in *D. simulans* where the *wRi* strain causes >95 % of crosses between uninfected females and infected males to fail to hatch, compared to <30 % in similar crosses in *D. melanogaster*. There is evidence that the level of CI declines with both male age and male mating frequency, although the mechanism is not known (Karr et al. 1998). It is therefore possible that the higher level of mating by infected male *Drosophila* may be a male strategy to regain reproductive compatibility with all females, including uninfected females thereby increasing their reproductive success (Champion de Crespigny et al. 2006). Alternatively, the infection may be beneficial to males and enabling them to maintain a higher mating rate than uninfected males. There are documented benefits of *Wolbachia*-infection in terms of conferring protection against RNA viruses in both fly species (Hedges et al. 2008; Teixeira et al. 2008), but this benefit is enjoyed by both sexes and cannot explain why infected males should mate at a higher rate than uninfected males (and endosymbiont protection against viruses is not universal, i.e. [Longdon et al. 2012]). In addition, infected *D. simulans* males suffer reduced sperm production (Snook et al. 2000) and sperm competitive ability (Champion de Crespigny and Wedell 2006) relative to uninfected males, indicating that *Wolbachia*-infection is not associated with a reproductive advantage to infected males. This suggest that the higher mating frequency of *Wolbachia*-infected males is a strategy to reduce the induction of CI, that will restore the reproductive compatibility with uninfected females and therefore increase male fitness. This is turn will affect the overall level of multiple mating in *Wolbachia*-infected fly populations increasing the scope for both sexual selection, but also sexual conflict over female remating rate.

8.3.4 Sex Ratio-Distorting SGEs and Sex Determination

There is evidence that sex ratio distorters such as male killers and feminizers can promote counter-selection and evolution of new sex determination pathways to restore sex ratio to unity and that this process can be rapid. For example, in the isopod *Armadillium vulgare*, chromosomal sex-determination follows female heterogamety (ZZ males and ZW females). However, *Wolbachia*-infected females produce female-biased broods, due to feminization of ZZ males that in turn will also produce infected female-biased broods. The consequence of this on-going feminization is the loss of the W female chromosome, since feminized ZZ-individuals can produce females without the W-chromosome. Here sex-determination is now controlled by *Wolbachia* infection and not by the sex chromosomes, with infected individuals becoming females and uninfected individuals remaining male. Over time, there can be evolution of resistance genes to *Wolbachia*-feminization, with infected females producing male-biased broods that resist *Wolbachia* and transmit the resistance to their offspring. Here selection on host nuclear genes has promoted evolution of resistance to feminization. The resistance genes restore males by reducing the rate of *Wolbachia* infection and directly impact on sex determination with individuals carrying the resistance genes now becoming male (Cordaux et al. 2011).

There is also evidence that multiple SGEs can interfere with sex determination. In the haplodiploid wasp *Trichogramma kaykai* (females diploid, males haploid) for example, *Wolbachia* induce parthenogenesis in infected females. The frequency of infected females is <30 % and kept at a low frequency by the presence of a selfish sex-ratio distorting B-chromosome (paternal sex-ratio, PSR), that is paternally inherited and converts diploid fertilized female-destined eggs into males by destroying the paternal chromosome (although not itself) and making it haploid (van Vugt et al. 2009). Thus the selfish PSR chromosome restores males despite the presence of feminizing *Wolbachia* keeping it at a low frequency and ensuring sexual reproduction and therefore also allowing sexual selection. As yet it is not known to what extent these opposing SGEs affect either pre- or post-copulatory aspects of sexual selection such as mate choice and fertility in sperm competition. It is even suggested that haplodiploidy itself may have evolved as a consequence of male-killing endosymbionts (Normark 2004). The idea is that male-killing is achieved by destroying the paternal chromosomes in diploid males (as the endosymbiont is only passed on through females), thereby turning them into lethal haploids destined to die. This is turn will favour host genes that evolve to 'rescue' these haploid embryos and convert them into viable males. Again this will restore sexual reproduction and therefore the potential for sexual selection to operate.

8.3.5 DDT Resistance in Drosophila melanogaster

It has recently been realised that SGEs can have different effects when expressed in males and females, potentially causing sexual conflict, as it can have the opposite fitness effect in the two sexes. A recent example of this is DDT-resistance alleles

(DDT-R) in *D. melanogaster* flies. Resistance is caused by a retrotransposon inserted into the promoter region of a detoxification gene (*Cyp6g1*) that upregulates the gene in both sexes. Remarkably, DDT-resistant females are more fecund and have offspring of higher fitness than susceptible females (McCart et al. 2005). Despite this fitness advantage to females, DDT-R did not spread before the use of pesticides, implying a cost to males that balances the benefits to females. This is the hallmark of a sexually antagonistic allele—a gene with opposite fitness effects when expressed in the two sexes (see Box 3). Indeed, this is the case in *D. melanogaster*, in which DDT-R males are less likely to obtain matings when competing for females against susceptible males (Smith et al. 2011). This cost (relative fitness reduction to DDT-R males −0.28) almost perfectly balances the fitness benefit to females (+0.25), when DDT-R is expressed in the same genetic background (Fig. 8.3). This implies that the sexual conflict it causes has the potential to affect the spread of the resistance alleles. It is possible that other SGEs may function as sexually antagonistic alleles with opposite fitness effects when expressed in males and females. If this is the case, it indicates that SGEs not only generate sexual selection by favouring polyandry, but may also generate sexual conflict by acting as a sexually antagonistic allele with dramatic sex-specific fitness effects. To date, it is not known the extent to which SGEs may have such sex specific fitness effects, but considering their prevalence throughout the genome and their direct involvement in gene regulation, it is likely to be an overlooked possibility.

Box 3. Intralocus Sexual Conflict

Males and females are designed to do different things, but they share the same genome, and this combination causes inter-sexual conflict. This is because shared traits are encoded by alleles at the same locus in both sexes, but have different optima when expressed in males and females (Arnqvist and Rowe 2005). Such intra-locus conflict traits are selected in different directions and this may prevent males and females from reaching their sex-specific phenotypic optima (Lande 1980). In particular, shared life-history traits are especially likely to be subject to intralocus sexual conflict because they have a strong relationship to fitness and frequently exhibit different optima in the two sexes (Wedell et al. 2006; Lewis et al. 2011b). As a consequence, we predict that the genetic correlation for fitness between the sexes should be negative during the acute and attenuated phases of intra-locus sexual conflict (Bonduriansky and Chenoweth 2009). This prediction is corroborated in a number of studies showing a negative genetic correlation for fitness between the sexes (e.g. Arnqvist and Rowe 2005; Pischedda and Chippindale 2006; Foerster et al. 2007; Mainguy et al. 2009). The potential implications of intra-locus sexual conflict are widespread: it may help preserve genetic variation in a population (Foerster et al. 2007), diminish the benefits of sexual selection (Pischedda and Chippindale 2006), increase the risk of population extinction (Kokko and Brooks

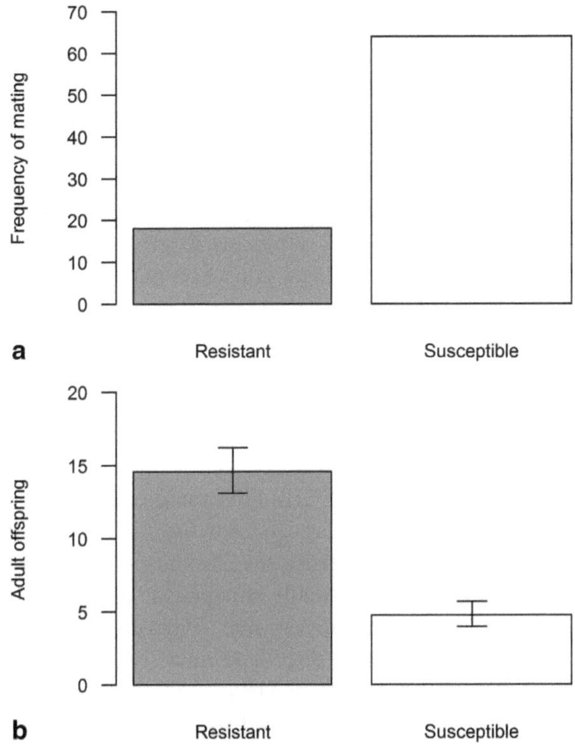

Fig. 8.3 Male competitive mating success and female fecundity of DDT resistant (RR) and DDT susceptible (SS) males of the *Canton-S* genetic background showing that the DDT-R allele has opposing fitness effects when expressed in males and females. This is the hallmark of a sexually antagonistic allele. DDT-R males have reduced mating success in competition with DDT-S males (**a**), whereas DDT-R females enjoy higher fecundity than DDT-S females (**b**). The relative fitness benefit to DDT-R females almost perfectly matched the fitness cost to DDT-R males. After McCart et al. (2005), and Smith et al. (2011).

2003) and/or drive speciation (Parker and Partridge 1998). For example, in the collared flycatcher (*Ficedula albicollis*) there is sexually antagonistic natural selection on body size (Merilä et al. 1997). Adult locomotory activity in *Drosophila melanogaster*, an important component of a male's life-history strategy for locating potential mates, also shows opposing selection in the sexes (Long and Rice 2007). Similarly, in the green-veined white butterfly (*Pieris napi*) male sperm production and female sperm storage, which determines female mating rate, are antagonistically selected (many sperm are good for males but bad for females), and there is a genetic correlation for sperm production in males and sperm storage in females (Wedell et al. 2009).

8.4 Concluding Remarks

Sexual selection and sexual conflict are powerful selective forces that have shaped the diversity of reproductive traits and behaviours in males and females. There are many well-documented factors fuelling these processes, but one that has been largely overlooked is the role of SGEs. This is despite SGEs being ubiquitous in eukaryotes and frequently associated with reduced fertility and fertilization efficiency of male carriers. Coupled with the risk of passing on genes that are associated with reduced fitness, they can exert a dramatic impact on mating decisions and mating patterns of animals. Here we have reviewed several examples highlighting the potency of a variety of SGEs in affecting the mating strategies of males and females, and have provided evidence that SGEs can have remarkably wide-ranging influences on sexual selection, even when present at low frequencies. Further research will reveal the extent to which other SGEs influences reproductive behaviours of animals. In addition, other aspects of sexual selection that may not first be apparent could potentially also be affected by SGEs. For example, the potential for SGEs to promote new sex determination systems can have direct impact on alleles with sexually antagonistic effects. This is because sex chromosomes and sex-linked genes are believed to be the hot spot for sexually antagonistic alleles as they are subject to strong sex-specific selection. In addition, since SGEs can affect the level of gene expression they have the potential to directly regulate sex-specific genes that will determine the level of sexual dimorphism. Hence it is possible that SGEs also play an overlooked role at the transcriptomic level by influencing expression levels of sex-linked genes and alleles with sexually antagonistic effects. Further molecular advances in combination with studies exploring the impact of SGEs on mating decisions are needed to evaluate the impact of SGEs on mating systems and resulting gene expression patterns in the two sexes more generally, but it is clear SGEs ultimately contribute to sexual selection.

References

Alpers MP (2008) The epidemiology of kuru: monitoring the epidemic from its peak to its end. Philos Trans R Soc Lond B Biol Sci 363:3707–3713

Alzohairy AM, Gyulai G, Jansen RK, Bahieldin A (2013) Transposable elements domesticated and neofunctionalized by eukaryotic genomes. Plasmid 69:1–15

Anderson M (1994) Sexual selection. Princeton University Press, Princeton

Ardlie KG, Silver LM (1998) Low frequency of *t* haplotypes in natural populations of house mice (*Mus musculus domesticus*). Evolution 52:1185–1196

Arnqvist G, Rowe L (2005) Sexual Conflict. Princeton University Press, Princeton

Artz K, Shin H-S, Bennet D (1982) Gene mapping within the *T/t* complex of the mouse. II. Anomalous position of the H-2 complex in *t* haplotypes. Cell 28:471–476

Atwood KC, Schneider LK, Ryan FJ (1951) Periodic selection in *Escherichia coli*. Proc Natl Acad Sci U S A 37:146–155

Baker RH, Denniff M, Futerman P, Fowler K, Pomiankowski A, Chapman T (2003) Accessory gland size influences time to sexual maturity and mating frequency in the stalk-eyed fly, *Cyrtodiopsis dalmanni*. Behav Ecol 14:607–611

Bateman AJ (1948) Intra-sexual selection in *Drosophila*. Heredity 2:349–368
Bonduriansky R, Chenoweth SF (2009) Intralocus sexual conflict. Trends Ecol Evol 24:280–288
Buckling A, Maclean RC, Brockhurst MA, Colegrave N (2009) The beagle in a bottle. Nat Rev Genet 457:824–829
Burt A, Trivers R. (2006) Genes in conflict: the biology of selfish genetic elements. Harvard University Press, Harvard
Bryant SH, Beckenbach AT, Cobbs G (1982) "Sex-ratio" trait, sex composition, and relative abundance in *Drosophila pseudoobscura*. Evolution 36:27–34
Champion de Crespigny FE, Wedell N (2006) *Wolbachia* infection reduces sperm competitive ability in an insect. Proc R Soc Lond Ser B 273:1455–1458
Champion de Crespigny FE, Pitt T, Wedell N (2006) Increased male mating rate in *Drosophila* is associated with *Wolbachia* infection. J Evol Biol 19:1964–1972
Charlat S, Reuter M, Dyson EA, Hornett EA, Duplouy AMR, Davies N, Roderick GK, Wedell N, Hurst GDD (2007) Male killing bacteria trigger a cycle of increasing male fatigue and female promiscuity. Curr Biol 17:273–277
Charlesworth B, Borthwick H, Bartolome C, Pignatelli P (2004) Estimates of the genomic mutation rate for detrimental alleles in *Drosophila melanogaster*. Genetics 167:815–826
Cordaux R, Bouchon D, Grève P (2011) The impact of endosymbionts on the evolution of host sex-determination mechanisms. Trends Genet 27:332–341
Darwin C (1859) On the origin of species by means of natural selection, or the preservation of favoured races in the struggle for life. John Murray, London
Darwin C (1871) The descent of man, and selection in relation to sex. John Murray, London
David P, Bjorksten T, Fowler K, Pomiankowski A (2000) Condition-dependent signalling of genetic variation in stalk-eyed flies. Nature 406:186–188
Dobzhansky T (1958) The genetics of natural populations. XXVII. The genetic changes in populations of *Drosophila pseudoobscura* in the American Southwest. Evolution 12:385–401
Dobzhansky T, Epling C (1944) Contributions to the genetics, taxonomy, and ecology of *Drosophila pseudoobscura* and its relatives. Carnegie Institution of Washington, Washington, D.C
Dyson EA, Hurst GDD (2004) Persistence of an extreme sex-ratio bias in a natural population. Proc Natl Acad Sci U S A 101:6520–6523
Edward DA, Fricke C, Chapman T (2010) Adaptations to sexual selection and sexual conflict: insights from experimental evolution and artificial selection. Philos Trans R Soc Lond B Biol Sci 365:2541–2548
Foerster K, Coulson T, Sheldon BC, Pemberton JM, Clutton-Brock T, Kruuk LEB (2007) Sexually antagonistic genetic variation for fitness in red deer. Nature 447:1107–1111
Fry CL, Wilkinson GS (2004) Sperm survival in female stalk-eyed flies depends on seminal fluid and meiotic drive. Evolution 58:1622–1626
Gajdusek DC, Zigas V (1957) Degenerative disease of the nervous system in New Guinea: the epidemic occurrence of "kuru" in the native population. N Engl J Med 257:974–978
Gienapp P, Teplitsky C, Alho JS, Mills JA, Merilä J (2008) Climate change and evolution: disentangling environmental and genetic responses. Mol Ecol 17:167–178
Goodacre SL, Martin OY (2012) Modification of insect and arachnid behaviours by vertically transmitted endosymbionts: infections as drivers of behavioural change and evolutionary novelty. Insects 3:246–261
Haig D (2012) Retroviruses and the placenta. Curr Biol 22:R609–R613
Haig D, Bergstrom C (1995) Multiple mating, sperm competition and meiotic drive. J Evol Biol 8:265–282
Hamilton WD (1967) Extraordinary sex-ratios. Science 156:477–488
Hartl DL, Hiraizumi Y, Crow JF (1967) Evidence for sperm dysfunction as the mechanism of segregation distortion in *Drosophila melanogaster*. Genetics 58:2240–2245
Hedges LM, Brownlie JC, O'Neill SL, Johnson KN (2008) *Wolbachia* and virus protection in insects. Science 322:702
Hull PM, Norris RD (2009) Evidence for abrupt speciation in a classic case of gradual evolution. Proc Natl Acad Sci U S A 106:21224–21229

Hurst GDD, Werren JH (2001) The role of selfish genetic elements in eukaryotic evolution. Nat Rev Genet 2:597–606

Jaenike J (2001) Sex chromosome meiotic drive. Annu Rev Ecol Syst 32:25–49

Jiggins FM, Hurst GDD, Majerus MEN (2000) Sex-ratio-distorting *Wolbachia* causes sex-role reversal in its butterfly host. Proc Roy Soc Lond Ser B 267:69–73

Johns P, Wolfenbarger L, Wilkinson G (2005) Genetic linkage between a sexually selected trait and X chromosome meiotic drive. Proc R Soc Lond Ser B 272:2097–2103

Johnson RT (2005) Prion diseases. Lancet Neurol 4:635–642

Karr TL, Yang W, Feder ME (1998) Overcoming cytoplasmic incompatibility in *Drosophila*. Proc R Soc Lond Ser B 265:391–395

Kawecki TJ, Lenski RE, Ebert D, Hollis B, Olivieri I, Whitlock MC (2012) Experimental evolution. Trends Ecol Evol 27:547–560

Keller L (1999) Levels of selection in evolution. Princeton University Press, Princeton

Kidwell MG, Lisch DR (2000) Transposable elements and host genome evolution. Trends Ecol Evol 15:95–99

Kokko H, Brooks R (2003) Sexy to die for? Sexual selection and the risk of extinction. Ann Zool Fenn 40:207–219

Lande R (1980) Sexual dimorphism, sexual selection, and adaptation in polygenic characters. Evolution 34:292–305

Lander ES, Linton LM, Birren B, Nusbaum C, Zody MC et al (2001) Initial sequencing and analysis of the human genome. Nature 409:860–921

Lenington S (1991) The *t*-complex: a story of genes, behavior, and populations. Adv Study Behav 20:51–86

Lenington S, Coopersmith C, Williams J (1992) Genetic basis of mating preferences in wild house mice. Am Zool 32:40–47

Lenski RE (1988) Experimental studies of pleiotropy and epistasis in *Escherichia coli*. II. Compensation for maldaptive effects associated with resistance to virus T4. Evolution 42:433–440

Lewis Z, Price TAR, Wedell N (2008) Review—sperm competition, immunity, selfish genes and cancer. Cell Mol Life Sci 65:3241–3254

Lewis Z, Champion de Crespigny FE, Sait SM, Tregenza T, Wedell N (2011a) *Wolbachia* infection lowers fertile sperm transfer in a moth. Biol Lett 7:187–189

Lewis Z, Wedell N, Hunt J (2011b) Evidence for strong intralocus sexual conflict in the Indian meal moth, *Plodia interpunctella*. Evolution 65:2085–2097

Long TAF, Rice WR (2007) Adult locomotor activity mediates intralocus sexual conflict in a laboratory-adapted population of *Drosophila melanogaster*. Proc R Soc Lond Ser B 274:3105–3109

Longdon B, Fabian DK, Hurst GDD, Jiggins FM (2012) Male-killing *Wolbachia* do not protect *Drosophila bifasciata* against viral infection. BMC Microbiol 12:S8 (doi:10.1186/1471-2180-12-S1-S8)

Louchart A, Tourment N, Carrier J (2011) The earliest known pelican reveals 30 million years of evolutionary stasis in beak morphology. J Ornithol 152:15–20

Mackay TFC, Stone EA, Ayroles JF (2009) The genetics of quantitative traits: challenges and prospects. Nat Rev Genet 10:565–577

Mainguy J, Cote SD, Festa-Bianchet M, Coltman DW (2009) Father-offspring phenotypic correlations suggest intralocus sexual conflict for a fitness-linked trait in a wild sexually dimorphic mammal. Proc R Soc Lond Ser B 276(1675):4067–4075

Markov AV, Lazebny OE, Goryacheva II, Antipin ML, Kulikov AM (2009) Symbiotic bacteria affect mating choice in *Drosophila melanogaster*. Anim Behav 77:1011–1017

McCart C, Buckling A, Ffrench-Constant RH (2005) DDT resistance in flies carries no cost. Curr Biol 15:R587–589

Merilä J, Sheldon BC, Ellegren H (1997) Antagonistic natural selection revealed by molecular sex identification of nestling collard flycatchers. Mol Ecol 6:1167–1175

Miller WJ, Ehrman L, Schneider D (2010) Infectious speciation revisited: impact of symbiont-depletion on female fitness and mating behavior of *Drosophila paulistorum*. PLoS Pathog 6:12

Miller W, Hayes VM, Ratana A, Petersenb DC, Wittekindta NE, J. Miller B et al (2011) Genetic diversity and population structure of the endangered marsupial *Sarcophilus harrisii* (Tasmanian devil). Proc Natl Acad Sci U S A 108:12348–12353

Murgia C, Pritchard JK, Yeon Kim S, Fassati A, Weiss RA (2006) Clonal origin and evolution of a transmissible cancer. Cell 126:477–487

Nichols RA, Butlin RK (1989) Does runaway sexual selection work in finite populations. J Evol Biol 2:299–313

Normark BB (2004) Haplodiploidy as an outcome of coevolution between male killing cytoplasmic elements and their hosts. Evolution 58:790–798

Novinski MA (1876) Zur Frage über die Impfung der krebsigen Geschwülste. Zentralblatt Medizin Wissenschaft 14:790–791

Parker GA (1979) Sexual selection and sexual conflict. In: Blum M, Blum N (eds) Sexual selection and reproductive competition in insects. Academic, New York

Parker GA, Partridge L (1998) Sexual conflict and speciation. Philos Trans R Soc Lond B Biol Sci 353:261–274

Pischedda A, Chippindale AK (2006) Intralocus sexual conflict diminishes the benefits of sexual selection. PLoS Biol 4(11):e356

Policansky D, Dempsey B (1978) Modifiers and "sex-ratio" in *Drosophila pseudoobscura*. Evolution 32:922–924

Policansky D, Ellison J (1970) "Sex ratio" in *Drosophila pseudoobscura*: spermiogenic failure. Science 169:888–889

Pomiankowski A, Hurst LD (1999) Driving sexual preference. Trends Ecol Evol 14:425–426

Powell J (1997) Progress and prospects in evolutionary biology: the *Drosophila* model. Oxford University Press, Oxford

Presgraves DC, Severance E, Wilkinson GS (1997) Sex chromosome meiotic drive in stalk-eyed flies. Genetics 147:1169–1180

Price TAR, Wedell N (2008) Selfish genetic elements and sexual selection: their impact on male fertility. Genetica 132:295–307

Price TAR, Bretman AJ, Avent TD, Snook RR, Hurst GDD, Wedell N (2008a) Sex ratio distorter reduces sperm competitive ability in an insect. Evolution 62:1644–1652

Price TAR, Hodgson DJ, Lewis Z, Hurst GDD, Wedell N (2008b) Selfish genetic elements promote polyandry in a fly. Science 322:1241–1243

Price TAR, Hurst GDD, Wedell N (2010a) Polyandry prevents extinction. Curr Biol 20:471–475

Price TAR, Lewis Z, Smith DT, Hurst GDD, Wedell N (2010b) Sex ratio drive promotes sexual conflict and sexual coevolution in the fly *Drosophila pseudoobscura*. Evolution 64:1504–1509

Price TAR, Lewis Z, Smith DT, Hurst GDD, Wedell N (2012) No evidence of mate discrimination against males carrying a sex ratio distorter in *Drosophila pseudoobscura*. Behav Ecol Sociobiol 66:561–568

Ravi Ram K, Wolfner MF (2007) Seminal influences: *Drosophila* Acps and the molecular interplay between males and females during reproduction. Integr Comp Biol 47:427–445

Rebbeck CA, Thomas R, Breen M, Leroi AM, Burt A (2009) Origins and evolution of a transmissible cancer. Evolution 63:2340–2349

Rogers DW, Denniff M, Chapman T, Fowler K, Pomiankowski A (2008) Male sexual ornament size is positively associated with reproductive morphology and enhanced fertility in the stalk-eyed fly *Teleopsis dalmanni*. BMC Evol Biol 8:236

SanMiguel P, Gaut BS, Tikhonov A, Nakajima Y, Bennetzen JL (1998) The paleontology of intergene retrotransposons of maize. Nat Genet 20:43–45

Siddle HV, Kriess A, Eldridge MDB, Noonan E, Clarke CJ, Pyecroft S, Woods GM, B. K (2007) Transmission of a fatal clonal tumour by biting occurs due to depleted MHC diversity in a threatened carnivorous marsupial. Proc Natl Acad Sci U S A 104:16221–16226

Simmons LW (2001) Sperm competition and its evolutionary consequences in the insects. Princeton University Press, Princeton

Smith DT, Hosken DJ, Rostant WG, Yeo M, Griffin RM, Bretman A, Price TAR, Ffrench-Constant RH, Wedell N (2011) DDT resistance, epistasis and male fitness in flies. J Evol Biol 24:1351–1362

Snook RR, Cleland SY, Wolfner MF, Karr TL (2000) Offsetting effects of *Wolbachia* infection and heat shock on sperm production in *Drosophila simulans*: analyses of fecundity, fertility and accessory gland proteins. Genetics 155:167–178

Stalker HD (1961) The genetic systems modifying meiotic drive in *Drosophila paramelanica.* Genetics 46:177–202

Sturtevant AH, Dobzhansky T (1936) Geographical distribution and cytology of "sex ratio" in *Drosophila pseudoobscura* and related species. Genetics 21:473–490

Swanson WJ, Vacquier VD (2002) Reproductive protein evolution. Annu Rev Ecol Syst 33:161–179

Tao Y, Masly J, Araripe L, Ke Y, Hartl D (2007) A *sex-ratio* meiotic drive system in *Drosophila simulans*. I. An autosomal suppressor. PLoS Biol 5:2560–2575

Teixeira L, Ferreira A, Ashburner M (2008) The bacterial symbiont Wolbachia induces resistance to RNA viral infections in *Drosophila melanogaster*. PLoS Biol 6:2753–2763

Trivers RL (1972) Parental investment and sexual selection. In Campbell B (ed) Sexual selection and the descent of man, 1871–1971. Aldine, Chicago, pp 136–179

Vagin VV, Sigova A, Li C, Seitz H, Gvozdev V, Zamore PD (2006) A distinct small RNA pathway silences selfish genetic elements in the germline. Science 313:320–324

Vala F, Egas M, Breeuwer JAJ, Sabelis MW (2004) *Wolbachia* affects oviposition and mating behaviour of its spider mite host. J Evol Biol 17:692–700

Van Vugt JJFA, de Jong H, Stouthamer R (2009) The origin of a selfish B chromosome triggering paternal sex ratio in the parasitoid wasp *Trichogramma kaykai*. Proc R Soc Lond B 276:4149–4154

Wedell N (2013) The dynamic relationship between polyandry and selfish genetic elements. Philos Trans R Soc Lond B Biol Sci 368:20120049

Wedell N, Kvarnemo C, Lessells CM, Tregenza T (2006) Sexual conflict and life histories. Anim Behav 5:999–1011

Wedell N, Wiklund C, Bergström J (2009) Coevolution of nonfertile sperm and female receptivity in a butterfly. Biol Lett 5:678–681

Werren JH (2011) Selfish genetic elements, genetic conflict, and evolutionary innovation. Proc Natl Acad Sci U S A 108 (Suppl. 2):10863–10870

Wilkinson GS, Presgraves DC, Crymes L (1998) Male eye span in stalk-eyed flies indicates genetic quality by meiotic drive suppression. Nature 391:276–278

Williams J, Lenington S (1993) Environmental and genetic factors affecting preferences of female house mice (*Mus domesticus*) for males that differ in *t*-complex genotype. Behav Genet 23:51–58

Wu C-I (1983) Virility deficiency and the sex-ratio trait in *Drososphila pseudoobscura*. I. Sperm displacement and sexual selection. Genetics 105:651–662

Yasui Y (1998) The 'genetic benefits' of female multiple mating reconsidered. Trends Ecol Evol 13:246–250

Zeh JA, Zeh DW (1996) The evolution of polyandry I: intra-genomic conflict and genetic incompatibility. Philos Trans R Soc Lond B Biol Sci 263:1711–1717

Chapter 9
Preference, Rationality and Interindividual Variation: The Persisting Debate About Female Choice

Frank Cézilly

Abstract Contemporary research on sexual selection remains deeply influenced by the controversy between Charles Darwin and Alfred Russel Wallace about the true nature and importance of female choice. After briefly reviewing the main points of disagreement between the two famous evolutionists, I discuss some methodological issues relevant to the contemporary study of female choice. I first use some recent controversy about sexual selection in the Indian Peafowl, *Pavo cristatus*, to illustrate several recurring problems and shortcomings in the empirical study of female preference for male characters. I then address the empirical evidence for rationality in female choice, and discuss how the recently emerged concept of animal personality may help to understand how inter-individual variation in female choice can be maintained in natural populations. Finally, I examine the possibility to develop a more integrated framework for the empirical study of female choice.

Keywords Intersexual selection · Mate choice · Personality · Preference · Rationality

9.1 Introduction

The study of female mate choice has received a very large attention over the last 30 years, both on theoretical and empirical grounds, to the point where it has become today one of the major topic addressed in the literature on animal behaviour (Gross 1994; Cézilly 2008). Most of this attention, though, has centred on the evolutionary causes and consequences of female choice, more than on female choice *per se*. For example, whereas some researchers have remained faithful to Darwin's original ideas in proposing that female preference was largely arbitrary (Lande 1981; Kirkpatrick 1982; Prum 2012), others have favoured the idea that females were choosy

F. Cézilly (✉)
UMR CNRS 6282 Biogéosciences, Université de Bourgogne, Dijon, France
e-mail: frank.cezilly@u-bourgogne.fr

T. Hoquet (ed.), *Current Perspectives on Sexual Selection,* History, Philosophy and Theory of the Life Sciences 9, DOI 10.1007/978-94-017-9585-2_9

because of direct and/or indirect benefits accruing from mating with the best possible males (Zahavi 1975; Hoelzer 1989; Kotiaho and Puurtinen 2007). A distinct line of research is concerned with whether female preference for a male trait could historically precede the evolution the trait: according to the sensory exploitation hypothesis (Ryan 1990; Egger et al. 2011), male traits might be positively selected if they match some pre-existing sensory bias in females. Whatever the working hypothesis about the origin and the functional consequences of female preferences for male traits, however, all theoretical models and empirical works incorporate the assumption that females indeed express a preference for male traits when favouring a mating option over another.

Yet, despite this rich and diverse literature, encompassing all kinds of biological models from invertebrates to human beings, some very basic questions about female preference for male traits and its evolutionary importance in shaping sexual dimorphism remain largely unanswered. The purpose of the present chapter is precisely to discuss the limitations of current approaches to female choice, to point out some gaps in our understanding of intersexual selection, and, finally, to emphasize the need for new research aimed at fillings those gaps. Contemporary research on sexual selection remains deeply rooted in past debates and I will thus start by providing a brief historical perspective on the importance of female choice in the theory of sexual selection, from which I will extract the major points of disagreement between such leading figures as Charles Darwin and Alfred Russel Wallace (see also Hoquet and Levandowsky, this volume). In the second part of the present chapter, I will discuss some methodological issues inherent to the empirical study of active female choice in relation to male characters and behaviour. The third part of the chapter will be devoted to the question of rational female choice, whereas the fourth one will evaluate the potential role of variation in personality between females on the study of female choice. In the conclusion, I will come back to some aspects of the controversy between Darwin and Wallace about sexual selection and female choice, and examine the possibility to develop a more integrated framework for the empirical study of female choice.

9.2 Intersexual Selection: Female Choice for Beauty?

The introduction of the process of sexual selection by Darwin in *The Origin of Species* in 1859 and its later development in *The Descent of Man* in 1871 might have been received with less scepticism at the time, if it was not for the emphasis that the famous evolutionist deliberately placed on female choice. Indeed, Darwin (1859, 1874) introduced two different processes that could account for the evolution of secondary sexual characters. First, he considered that sexual dimorphism can arise from the benefits males gain from possessing certain characters when competing between themselves for access to females, whether it corresponds to offensive or defensive weapons used during fights between males for access to females (i.e. interference competition), or sensory and locomotor organs allowing males to locate females before their rivals (i.e. exploitation competition), a process known today as *intrasexual selection*. Second, he regarded female preference for some male characters as a driving force leading to the amplification of such characters during the course of evolution, a process now defined as *intersexual selection* (Andersson 1994; Danchin and Cézilly 2008; Cézilly and Allainé 2010). As male characters preferred by females in many species looked to Darwin as ornaments with no direct utility for survival (or being even detrimental to fitness), he concluded to the existence of a "sense of beauty" in some female animals, that he compared to "that in lowest savages that admire and deck themselves with any brilliant, glittering or curious object" (Darwin 1874, p. 211).

At a time when the existence of a sense of aesthetics in women appeared doubtful to many an educated man, Darwin's proposition was somehow an iconoclast one (Miller 2000). For if the true nature of womanhood was seen as a sensitive one and devoted to beauty, women were simultaneously denied of any real artistic creativity or judgment (Losano 2008).

Elevating a supposed sense of beauty in female animals to the rank of an evolutionary force thus left many of Darwin's colleagues unconvinced, if not doubtful. Among them were standing St George Jackson Mivart and Alfred Russel Wallace for whom female tastes were far too versatile and inconsistent to be an evolutionary force in par with natural selection (Cronin 1991). Wallace, in particular, developed, in two consecutive books *Tropical Nature and Other Essays* (Wallace 1878) and *Darwinism* (Wallace 1889), a refutation of the role of female choice in the evolution of male ornaments. He first considered that many sexually dimorphic characters, such as for instance the "musical organs" of some insects, had for unique function to "enable the sexes to discover and recognise each other". Considering butterflies, he concluded that most cases of sexual dichromatism could be explained by the fact that the females, being more exposed to predation than males during egg laying, have adopted protective colorations, either though mimicking the conspicuous warning colorations of sympatric unpalatable butterfly species, or through resorting to camouflage. The latter phenomenon would also apply to birds, in which females, that often assume alone incubation and brooding duties, show duller and more cryptic colorations than males. In support of his assertion, Wallace quoted the case of hole-nesting birds in which females "are either coloured exactly like the males, or, when differently coloured, are equally conspicuous", although he admitted some exceptions to that rule. Contrary to what is sometimes claimed (e.g. Miller 2000), however, Wallace was not definitely "hostile" to female choice. He clearly conceded that the female "does exert some choice between very different males", but he found "no evidence that slight variations in the colour or plumes, in the way of increased intensity or complexity, are what determines the choice" (Wallace 1889). He rather thought that female choice was more influenced by the amount of activity of males, as this was a direct cue to their vigour. And because male vigour and stamina would influence not only their sexual attractiveness during the breeding season, but, more generally, their ability to escape predators and acquire resources at any time, he considered that natural selection was sufficient to explain male characters and female preference for them. Looking for a "*vera causa* for the origin of ornamental appendages of birds and other animals", Wallace, assuming that males have a higher metabolism than females, concluded that they were simply due to "a surplus of vital energy, leading to abnormal growths in those parts of the integument where muscular and nervous actions are greatest" (Wallace 1889). Ultimately, Wallace's reluctance to confer any aesthetic sense to animals was a logical consequence of his belief that the artistic faculties of human beings were part of his "spiritual nature" on which natural selection has no influence.

If Wallace's objections to intersexual selection were serious, they were not strong enough to convince all evolutionists to abandon the concept of intersexual selection, as shown by the positive account of Darwinian sexual selection provided by August Weissmann in *The Evolution Theory* (Weismann 1904). Actually, Wallace's views were also criticized, in particular by George and Elizabeth Peckham, two ardent American Darwinists, on the basis of their own studies of jumping spiders (Salticidae) and some comparative evidence in birds (Peckham and Peckham 1890). They essentially reproached to Wallace to draw firm conclusions from little empirical evidence, and pointed out several examples that ran counter to his views. They also criticized Wallace for being inconsistent in his line of reasoning, particularly for what concerns the relationship between male characters and energy, asking why, if the surplus of energy in males compared to females was a general phenomenon, sexual dimorphism was not more widespread. Still, several major objections raised by Wallace could not be refuted at the time, essentially because no experimental demonstration of female choice was available. Deciding whether females make a direct choice or simply consent to mate with the most persistent and enduring males was more a matter of opinion than of empirical evidence. Similarly, the ability of female animals to perform rational comparison and deliberation was merely a suggestion made by Darwin, not a scientific fact established on firm ground. In any case, it was not clear at the time whether females would necessarily agree in their preference for male characters. Finally, the proposition made by Wallace that sexual dimorphism was the result of natural selection favouring cryptic females, rather than sexual selection favouring gaudy males, awaited confirmation based on a quantitative examination of the distribution of sexual dimorphism in the Animal kingdom.

More than 150 years later, one may think that the revival of the interest in sexual selection that characterized behavioural ecology in the last 2 decades of the twentieth century (Gross 1994; Milam 2010), coupled with technological and methodological advances, has allowed researchers to clarify those points and go beyond them, such that today the disagreement between Darwin and Wallace should be more a matter of history than a contemporary issue. The following sections of the present chapter will show that it is not so.

9.3 The Peacock Tale: A Brief Critical Appraisal of the Empirical Study of Female Choice

In 1860 Darwin famously wrote (in a letter to the American botanist Asa Gray dated April, 3rd) that the sight of one feather of the peacock tail was enough to make him sick. He would probably feel even sicker today when considering the controversial evidence about the role of peahen behaviour in the evolution of the peacock tail. The Indian Peafowl, *Pavo cristatus*, is *a priori* an ideal model to test ideas about female choice. The male has a predominantly blue plumage with a fan-like crest of spatula-tipped wire-like feathers and a long train made up of elongated upper-tail covert feathers with colourful eyespots, whereas the female lacks the train and has a a duller, green and brownish plumage. In this lek-breeding species, the male provides no parental care, suggesting that females may obtain indirect benefits from selecting males on the basis of their phenotype. Indeed, the male train is raised into a fan and quivered in a display during courtship, and extensive variation in mating success between males has been observed in both captive (Loyau et al. 2005a), feral (Petrie and Halliday 1994; Takahashi et al. 2008) and natural populations (Yasmin and Yahya 1996).

First attempts to establish a relationship between male mating success and the degree of extravagance of the peacock tail were quite successful. Indeed one team in England (Petrie et al. 1991; Petrie and Halliday 1994) and another one in France (Loyau et al. 2005a) found, based on both direct observation of mating patterns and experiments manipulating the number of eye-spots, a positive correlation between the number or density of eye-spots a male had on his tail and his mating success. Furthermore, the same two research teams independently provided some evidence that tail ornamentation reflects the health and vigour of peacocks (Petrie 1992; Møller and Petrie 2002; Loyau et al. 2005b). This seemed sufficient to conclude that the ornamentation of the peacock tail is an honest signal of male quality and that females do not express a purely aesthetic sense, but actually benefit indirectly from choosing more ornamented males as mating partners. However, the evidence in favour of a role of peahen choice in the evolution of the peacock tail was revisited by Takahashi et al. (2008) who came with quite different conclusions. In a seven-year study of a feral population in Japan, they found absolutely no evidence that females expressed a preference for more ornamented males (Takahashi et al. 2008). In their discussion, Takahashi et al. (2008) picked up several inconsistencies between previous studies about the effect of the number of eyespots and train lengths on male mating success, and called for additional studies to further document the extent of preference of females for males with more elaborate tails. Adopting a relatively conservative attitude, Loyau et al. (2008) downplayed the importance of the results obtained by Takahasi et al. (2008), essentially through arguing that a single study was insufficient to refute the convergent findings of three previous ones.

However, a recent study (Dakin and Montgomerie 2011) provides support to the objections raised by Takahashi et al. (2008). Dakin and Montgomerie (2011) studied the geometric arrangement of tail feathers and natural variation in the number of eyespots displayed during courtship in three different feral populations in North America. In addition, they duplicated previous experiments consisting in removing a large number (>20) of eyespots from the outermost feathers of the peacock tail. Males with experimentally altered trains showed reduced mating success, but there was no effect of the natural variation of both train length and number of eyespots on male mating success. Such results suggest that removing eyespots alters drastically the appearance of the train, compared to the natural range of variation in peacock trains, making such manipulation of little value for the understanding of female choice. More worrying, Dakin and Montgomerie (2011), concluded, based on their own measurements and data from other studies, that peacocks typically grow a maximum of 169 feathers, and that this number is relatively invariant between males, thus contradicting previous reports of males with 171–177 eyespots in feral populations (Petrie et al. 1996). The reasons for such a large discrepancy remain obscure. The moral of the story is that we still do not know today whether peahens do prefer a particular trait in males (a rather ironic situation for such an iconic model), and a pessimistic Darwinian may argue that, in the end, little progress has been made since Darwin in our understanding of how the peacock tail did evolve.

Obviously, the "peacock tale" is not representative of all the research on female choice. But is it is exemplary of several recurring problems and shortcomings in the empirical study of female preference for male characters. First of all, the "peacock tale" suggests that is important to study the same phenomenon in different populations of the same species. The accumulation of positive results from the same research team working repeatedly with the same population might not necessarily be suspect, but might clearly incorporate some kind of autocorrelation. Different populations from the same species may face contrasted ecological constraints, thus making difference between populations of great interest in trying to understand how female choice can be modulated by external factors. The peacock tale also emphasizes the importance of publishing negative results if global evidence for the effect of mate choice has to be evaluated with some confidence using, for instance, meta-analyses. Recent evidence (Fanelli 2011) indicates an overall trend for a 20% increase in the proportion of positive results in the scientific literature between 1990 and 2007, suggestive of a lack of objectivity. Although the magnitude of the effect varies between disciplines and countries (Fanelli 2011), a similar study focused on the literature on female choice and sexual selection might be worth undertaking.

More importantly, the peacock tale reminds us that the practical advantages of reductionism should be evaluated against its drawbacks. Both Darwin and Wallace insisted that in most species males do not passively exhibit their ornaments, but instead engage in vigorous displays. There is certainly more in the peacock strut that the mere number of eyespots that is displayed. Indeed, display rate and total amount of time spent displaying, as well as the amount of vocalizations (Yasmin and Yahya 1996) or the size of other phenotypic characters such as crest feathers (Dakin 2011) may all combine in providing a complex sensation to females. Multidimensionality in male displays has attracted some interest from a theoretical point of view (Møller and Pomiankowski 1993; Candollin 2003; Van Doorn and Weissing 2004; Bro-Jørgensen 2010). Recent evidence indeed suggests that in various organisms display activity and vigour might be at least as important as ornaments in influencing female mating decisions (Byers et al. 2010; Cornuau et al. 2012). But to what extent morphological characters and display vigour jointly contribute to influence female decision to mate remains by large an open issue. Future studies may then benefit from developing a more comprehensive analysis of displays rather than atomizing them in different morphological and behavioural components studied in isolation of each other. In that respect, the use of video-recordings and computer-generated animations (Stamp Dawkins and Woodington 1997; Baldauf et al. 2009, Woo and Rieucau 2011) might prove particularly useful. For instance, O'Loghlen and Rothstein (2012) recently used video recording to study the preference of female brown-headed cowbirds, *Molothrus ater*, in response to males displays of varying intensity. In this species, males perform in front of females wingspread song displays that are generally less intense than versions of the same display that are directed at other males. During the experiment, females were presented with audiovisual recordings showing the same males performing both high intensity and low intensity, and their copulation solicitation display responses were recorded. Each pair of high and low intensity playback videos had the same audio track, such that differences in the responses of females were supposed to reflect differences in reaction to the motor component of the male display shown in the videos. All females responded with a higher intensity to the low intensity displays, suggesting that females may not necessarily prefer males with the most intense courtship displays (O'Loghlen and Rothstein 2012).

9.4 Is Female Choice Rational?

By definition, the word "choice" refers in behavioural ecology to the fact that an animal is engaging in one option when several are simultaneously available at a given time (Danchin et al. 2008). Assuming that females make a choice before mating does not imply any conscious decision on their part. However, whenever choice is not random, some information must have been used by females to sort out the different alternative mating options. In a large range of species, it has been indeed shown that females do not mate randomly (Andersson 1994). The theory of sexual selection by female choice assumes that females are acting *as if* they were balancing costs against benefits to select mating options that maximizes their own fitness (Andersson 1994; Iwasa and Pomiankowski 1994; Kirkpatrick 1996). Under the assumption that their preference functions are true adaptations, females should be able to rank potential mates according to one or few dimensions directly related to fitness, such that their probability of choosing one potential mate over another one should be a monotonic function of their respective values. In short, females are expected to act as rational decision makers when selecting between potential mates. It has been however suggested, from a theoretical point of view, that when several phenotypic dimensions of alternative mating options are simultaneously evaluated, rationality in female choice might be limited by their ability to combine several criteria in a single common currency for fitness (Jennions and Petrie 1997; van Doorn and Weissing 2004; Hamilton and Sullivan 2005; see also Tversky 1969).

Three dimensions of rationality in choice can be addressed, namely repeatability, independence from irrelevant alternatives, and transitivity (Bateson and Healy 2005; Kacelnick 2006; Houston et al. 2007). Repeatability is a measure of the consistency of behaviour within individuals, and, from a statistical point of view, is equal to the proportion of variance in a quantitative measure of behaviour that is explained by inter-individual differences (see Bell et al. 2009). Overall, the repeatability of behaviour tends to be higher both when behaviour is measured in the field rather than under laboratory conditions and when the time interval between measurements is short (Bell et al. 2009; David et al. 2012). Interestingly, in a recent meta-analysis Bell et al. (2009) found that although mate preference behaviour was one of the best studied behaviour, it was also the least repeatable. Indeed, experimental studies in insects (Reinhold et al. 2002; Greenfield et al. 2004), fish (Kodric-Brown and Nicoletto 1997; Cummings and Mollaghan 2006; Gabor and Aspbury 2008; Lehtonen and Lindstrom 2008; Gabor et al. 2011), amphibians (Kime et al. 1998), and birds (Johnsen and Zuk 1996; Forstmeier and Birkhead 2004; Holveck and Riebel 2007) have found only weak repeatability of female choice, whereas studies conducted in the wild have provided inconclusive results (see for instance Banbura 1992; Møller 1994).

Independence from irrelevant alternatives implies that the female choice between two males should be independent of the presence of an additional less attractive male (Bateson and Healy 2005). For example, if a female is preferring male A to male B in a binary choice, then the introduction of a third male C which is less attractive than both male A and male B should not affect the magnitude of the female's preference for A over B. When female choice is based on a single dimension, this is an obvious prediction. However, when females use more than one male trait to make a decision about who to mate with, the situation is somehow more complex. For instance, if the third male has a lower value than both male A and male B one dimension, but is inferior only to male A on the second dimension, females are predicted to increase their preference for male A if female choice is context dependent. Such a change in female preference is taken as evidence for the existence of comparative evaluation mechanisms, i.e. that females do not use absolute but relative values to discriminate between males. The independence from irrelevant alternatives has been seldom tested in animals in the context of mate choice. Royle et al. (2008) addressed the question in the green swordtail, *Xiphophorus helleri*, a sexually dimorphic fish species where adult males are characterized by a long ornamental tail streamer, the "sword." Female green swordtails tend to prefer larger bodied males, and after controlling for body size, show a strong preference for males with longer swords (Basolo 1990). In their experiment, Royle et al. (2008) compared preferences expressed by females when given a binary choice between two males of same total length, but of different body size and sword length, to those expressed in the presence of a third male who had either a larger body size than the long-sworded male or a longer sword than the large-bodied male. Females that preferred the larger sworded male in the binary trials reduced their preference for sword length in the presence of a third male, also with a larger sword. Similarly, females that preferred the larger bodied male in binary choice reduced their preference for body length in the presence of a third male with a large body size. Thus, contrary to expectations, the addition of a third male shifted preference of females away from the male phenotype they preferred in the binary comparison. However, in this experiment females showed no significant overall preference for either long-sworded or large-bodied males when given a binary choice between them, suggesting that their decision in a binary choice might have been influenced by other male characters.

Reaney (2009) addressed the same question in the fiddler crab *Uca mjoebergi*, using robotic crabs that closely mimicked real ones. In this species, females often encounter potential mates simultaneously, suggesting that they could make comparative evaluation of mating options. In order to attract females, males typically wave their enlarged, major claw. Previous studies have shown that the size of their major claws and the rate at which it is waved affect male attractiveness. The experimental design used by Reaney (2009) was similar to the one used by Royle et al. (2008), allowing this time control for both claw size and wave rate. The addition of an alternative option, potentially changing the relative attractiveness of two males previously encountered during a binary choice, had a significant effect on absolute preference, but not on relative ones, thus providing no support for comparative evaluation mechanisms. Thus, the available experimental evidence does not indicate deviations from rationality in female choice in relation to the presence of irrelevant alternatives. However, the very limited number of studies calls for additional experiments using a larger range of species. In particular, it would be interesting to compare the influence of irrelevant alternatives between species where females typically encounter males sequentially and species where males are encountered simultaneously.

Only a few studies have addressed the issue of transitivity in animal mate choice (Kirkpatrick et al. 2006; Regenwetter et al. 2011). Yet, transitivity is arguably the most fundamental axiom of rational choice (Bar-Hillel and Margalit 1988; Regenwetter et al. 2011), and analysis of transitivity has previously proved useful in the study of foraging decisions (e.g. Waite 2001; Schuck-Paim and Kacelnik 2002) or collective choice (Franks et al 2003; Pratt 2005) made by animals. In its simplest form, transitive female choice occurs if when a female prefers option A to B and B to C, then she prefers A to C (Navarick and Fantino 1972). However, transitivity in female choice can be expressed at different levels, depending on whether mating options can be ranked on an ordinal or an interval scale (Sumpter et al. 1999). In the case of three alternative mating options A, B, and C, weak stochastic transitivity (WST) will be observed if when $P(A, B) \geq 0.5$ and $P(B, C) \geq 0.5$, then $P(A, C) \geq 0.5$, with $P(x, y)$ being the probability of choosing option x from the choice set $\{x, y\}$. Evidence for WST validates only the assumption that the options can be ordered on a common scale, but not necessarily be quantified. By contrast, strong stochastic transitivity occurs when $P(A, C) \geq \max[P(A, B), P(B, C)]$ (Tversky and Russo 1969; Grace 1993; Houston 1997; Schuck-Paim and Kacelnik 2002), and is indicative that not only ordinal preference holds, but that quantitative measures of value can be assigned to mating options on a common scale (Sumpter et al. 1999). So far only one study, to our knowledge, has tested for weak and strong transitivity in female choice. Dechaume-Moncharmont et al. (2013) examined female preference for male size in the convict cichlid. In this species, adult males are about 30% larger than adult females (Noonan 1983), and positive size-assortative mating has been observed in the wild (Wisenden 1995; Alonzo et al. 2001). Male preference for larger, more fecund females has been shown (Nuttall and Keenleyside 1993), whereas female preference for male size varies between studies (Beeching et al. 2004; Gagliardi-Seeley et al. 2009; Dechaume-Moncharmont et al. 2011). Forty females were proposed a series of binary choice between three males of increasing size, and their preference measured from the relative amount of time spent in front of each male on each trial (see Dechaume-Moncharmont et al. 2011). Ninety percent of females showed transitivity in choice, a proportion well above what was expected by chance. In addition, the mean preference index was significantly higher when female convict cichlids had to choose between males of distant ranks (1 vs. 3) compared to when choosing between males of adjacent ranks (1 vs. 2 or 2 vs. 3), thus providing evidence for strong stochastic transitivity in mate choice.

Overall, the available empirical evidence suggests that females behave like rational decision makers when choosing about which male to mate with. Still, most often it is reasonable to assume that more than one single dimension in males will influence female choice. One interesting question for the future, then, is to understand how females are actually combining several male traits when choosing between mating alternatives, and to what extent cognitive abilities limit their ability to perform rational choice when preference is multidimensional.

9.5 Variation in Female Choice: A Role for Personality?

The intensity of inter-sexual selection depends to a large extent on whether or not female choice for male characters is unanimous. If all females in a population share the same directional preference for a particular male trait, the trait may evolve rapidly under directional selection. Conversely, if a certain proportion of females mate randomly whereas others are choosy, or if different females have different preference functions, then the evolution of male traits is more difficult to predict. Female choice might not be unanimous if, for instance, females use *relative* rather than *absolute* preference criteria, i.e. if the suitability of a male as a mating partner varies depending on female characteristics. This is likely to happen, for instance, if selection is favouring genetic and/or phenotypic compatibility between mates (Mays and Hill 2004; Neff and Pitcher 2005; Sinn et al. 2006; Puurtinen et al. 2005, 2009). Note, however, that relative choice does not prevent the use of a common single rule by all females when choosing mates. For instance, Dechaume-Moncharmont et al. (2013) showed that a self-referent matching rule, by which all females prefer males which are about 30% larger than themselves, was performing particularly well in predicting non-unanimous choices made by female convict cichlids when comparing between males of different size.

One source of variation in female choice for male traits may arise from differences in personality. Animals commonly differ in their reaction towards the same environmental stimuli, and such differences tend to remain stable across different contexts and situations. Different terms, such as "behavioural syndromes" (Sih et al. 2004), "temperament" (Réale et al. 2000) or "personality" have been used almost interchangeably to refer to consistent individual differences in a set of correlated traits (see for instance David et al. 2011). Empirical evidence indicates that variation in personality has both a genetic basis (van Oers et al. 2005) and fitness consequences (Smith and Blumstein 2008), such that natural selection may influence personality traits and behavioural syndromes. Personality may be relevant to the study of female choice for at least two reasons. From a theoretical point of view, behavioural compatibility between mates might be particularly important first to achieve mating, and, second, in the case of monogamous species, to coordinate parental behaviour and other activities, such as for instance territorial defence, between pair members. From a methodological point of view, variation in personality between females may affect the results obtained both in the field and in the laboratory.

Growing evidence suggests that the interaction between male and female personalities can significantly affect mating and, at least in birds, reproductive success. For instance, Sinn et al. (2006) observed that successful mating between male and female dumpling squids, *Euprymna tasmanica*, was determined by positive assortative mating for personality measured on a shy–bold axis. In the great tit, *Parus major*, fledglings from pairs consisting of two slow-explorers or two fast-explorers have been found to be in best condition (Both et al. 2005). A similar pattern has been observed in the Steller's jay, *Cyanocitta stelleri* (Gabriel and Black 2012a, b), with pairs more similar in explorative tendencies and in willingness to take risks being more likely to fledge offspring than dissimilar pairs. However, in this study, the benefits of assortative mating for personality did vary between breeding seasons, being particularly significant in a year with late breeding onset following a severe winter. A cross-fostering breeding experiment on captive zebra finches, *Taeniopygia guttata*, has shown that both parental personality traits and the combination of personalities within pairs had positive effects on the body mass and condition of foster offsprings (Schuett et al. 2011a), thus demonstrating that similarity in personality of can have important, non-genetic effects on reproductive success in socially monogamous, biparental species.

Personality could influence female choice in different ways. First, females with different personalities may vary in their sexual proceptivity and thus have different sexual arousal thresholds in reaction to male courtship. Gelez et al. (2003) showed, for instance, that domestic ewes, *Ovis aries*, that had been genetically selected for 'calm' temperament, were more active in both establishing a contact with and sexually soliciting a ram than females that had been genetically selected for 'nervous' temperament. Thus, between-female variation in promiscuity and selectivity towards prospective mates could arise as a consequence of more fundamental differences in personality, and eventually be maintained because of linkage between personality traits (see Patrick et al. 2012). Second, females should benefit from paying attention to the personality of prospective mates in relation to their own if behavioural compatibility is an important component of reproductive success. Limited evidence exists for this prediction so far. Schuett et al. (2011b) observed that female zebra finches with intermediate to high-exploratory tendencies prefered exploratory males over unexploratory ones, whereas male personality had no effect on the choice made by females with low-exploratory tendencies. In the field cricket, *Gryllus integer*, females tended to prefer 'bold' males over 'shy' ones, but no evidence was found for an association between male and female personalities (Kortet et al. 2012). One important subject of investigation for the future, then, will be to assess to what extent females may show preference for male personality, and whether such preferences are absolute or relative to females' own personalities. Third, females with different personalities may use different rules when choosing a mate if variation in cognitive abilities is related to variation in personality, as it has been recently suggested (Sih and Del Giudice 2012). According to this suggestion, females with different personalities may rely on different tactics to sample resources in their environment, including potential mates. For instance, a speed-accuracy trade-off may result in 'bold' females being quicker than 'shy' ones making a decision about which male to mate with, but being less accurate in their assessment of male quality. Male availability, environmental constraints, and time horizon available to complete mating and reproduction may then all contribute to favour different personality types in females, and hence modulate preference levels at the population level. Alternatively, differences in mate sampling tactics and choosiness might result from intrinsic differences in personality selected at a more global level and maintained as a consequence of spatially and temporally varying selection on correlated traits.

The existence of variation in personality between females (and males) may also have some practical consequences for the study of female choice. A large majority of experimental studies of female choice have been conducted in captivity. Although personality measured in captivity may reliably reflect personality in the wild (Herborn et al. 2010), one problem may lie in the representativity of the samples of individuals used in such experiments. Experimental studies using wild-caught animals might not be representative of the natural range of female behaviours simply because personality can affect capturability in the wild (Garamszegi et al. 2009; Carter et al. 2012). For instance, bolder individuals may differ from shy ones in their latency time to enter a baited trap, resulting in over-representation of bold individuals in a sample of trapped individuals. Although some practical solutions exist to reduce sampling bias induced by inter-individual variation in personality (Biro and Dingemanse 2008), its consequences for the study of female choice have not been fully assessed yet. Similarly, experimental studies drawing inference from animals bred in captivity might be biased by the influence of developmental conditions on personality. It has been shown in a large range of species that developmental conditions can affect personality at a later age (Krause and Naguib 2011; Mishra et al. 2011; Sih 2011). Therefore, the range of personalities available from a sample of captive-bred individuals may vary depending on the developmental conditions they experienced. To what extent this phenomenon can explain discrepancies that are sometimes observed between experimental studies of female choice based on the same species model, such a for instance the zebra finch (Burley and Coppersmith 1987; Collins et al. 1994; Forstmeier and Birkhead 2004; Simons and Verhulst 2011) remains to be evaluated. Interestingly, David and Cézilly (2011) showed that a female's personality in the zebra finch has a significant influence on its behaviour in a four-chamber choice-apparatus, classically used in experimental studies of female choice. Variation in exploratory tendencies of females explained variation in selectivity, preference strength and consistency, with highly exploratory females showing lower selectivity and lower, but more consistent, preference scores, compared to females with low-exploratory tendencies. Therefore, experimental studies of female choice based upon different ranges of female personalities may well reach different conclusions about female preference. The magnitude of such an effect remains to be evaluated.

9.6 Conclusion

Methods for experimentally testing animal preferences were not available to Victorian biologists such as Darwin and Wallace, such that evidence for female choice at that time was only indirect. Indeed, August Weissmann acknowledged that "direct observation of choosing is difficult, and that as yet there is little than can be said on this point" (Weismann 1904). Since then, a very large number of theoretical, observational, and experimental studies have assessed whether females show preference for particular traits in males and to what extent female choice can be the causative agent of the extravagance of such traits. Still, the issue is not quite settled, and, as seen above, recent developments in animal behaviour may shed new light on the co-evolution between male characters and female choice. In particular, the fact that choice made by females for male characters might be constrained by cognitive abilities selected in a global ecological context deserves further consideration. Besides, there is an urgent need for developing studies of the underlying genetic architecture of male complex displays, as we know virtually nothing about how genes that influence male motor performance and genes that influence male morphological characters co-evolve together, and how this co-evolution could be influenced by female preference. What is clearly needed too is a deeper investigation of the female sensory equipment and "emotional" reaction to male appearance. In that respect, the development of a neuroecological approach (Zimmer and Derby 2011) to female choice and sexual selection seems particularly desirable. This may however critically depends on the availability of refined techniques such as real-time imaging of the brain (see Henderson et al. 2012) to researchers in behavioural ecology.

Still, a crucial question remains: what is the overall importance of female choice in the evolution of sexual dimorphism in ornamentation? Has female choice been a major historical agent in the evolution of elaborate ornamentation in males, or has female choice been actually playing only a marginal role in the evolution of sexual dimorphism for ornamentation? Female choice was indeed advanced by Darwin as a mechanism to explain the evolution of sexual dimorphism. Under this scenario, males in a given species progressively acquire extravagant ornaments because they are preferred by a majority of females. Thus, this scenario posits that males were selected to become more colourful and ornamented through evolutionary time, rather than females to acquire duller appearance, as suggested by Wallace. Although this idea has been widely accepted (Andersson 1994), phylogenetic evidence suggests a more complex situation. Indeed, Wiens (2001) pointed out that the evolutionary loss of male traits is a widespread phenomenon that deserves further consideration in relation to the role of female choice in the evolution of sexual dimorphism. For instance, in a comparative study of the evolution of sexual colour dimorphism in passerine birds, Price and Birch (1996) concluded that the rate of transition from dimorphism to monomorphism was actually higher than the reverse. Similarly, comparative evidence indicates that changes in female plumage occurred more frequently than changes in male plumage during the course of evolution both in Thraupidae (Burns 1998; but see Burns and Shultz 2012) and in Icteridae (Hoffmann et al. 2008). Finally, Soler and Moreno (2012) found that a change in nesting habits (from open to cavity nesters) in European passerines influence the likelihood of changes in both dichromastism and plumage conspicuousness in males, but not in females. They concluded that, as predicted by Darwin, variation in the conspicuousness of male plumage in open-nesting species can be accounted for by sexual selection, whereas, as predicted by Wallace, females of monochromatic species show more conspicuous plumage than those of dichromatic species, particularly in cavity-nesting ones. Recent studies of sexual dichromatism in Lepidopterans also provide a more balanced appreciation about the relative importance of Darwin's and Wallace's arguments. Kunte (2008), for instance, provided evidence that sexual dimorphism in *Papilio* butterflies is better explained by female-limited Batesian mimicry (with females being mimetic and males non-mimetic) and with the deviation of female wing colour patterns from ancestral patterns conserved in males. On the other hand, Oliver and Monteiro (2011) found that derived monomorphism in ornamentation in the butterfly genera *Bicyclus* and *Junonia* could correspond either to a loss of the ornament in one sex or to the gain of the same ornament in the other one. Overall, recent evidence provides a more nuanced appreciation of the role of 'ecological' and 'sexual' selective mechanisms in the evolution of sexual dimorphism, thus marking a *renaissance* of Wallace's ideas (Punzalan and Hosken 2010). Still, assessing the reality and actual importance of female choice in the evolution of sexual dimorphism remains a challenge for evolutionary biologists. Hopefully, continuous development of new concepts and new techniques in the field of animal behaviour will offer several opportunities to shed new light on an old (but still hot) debate in the not too distant future.

References

Alonzo JJ, McKaye KR, van den Berghe E (2001) Parental defense of young by the convict cichlid, *Archocentrus nigrofasciatus* in lake Xiloá, Nicaragua. J Aquaric Aquat Sci 9:209–228

Andersson M (1994) Sexual selection. Cambridge University Press, Cambridge

Baldauf SA, Kullmann H, Thunken T, Winter S, Bakker TCM (2009) Computer animation as a tool to study preferences in the cichlid *Pelvicachromis taeniatus*. J Fish Biol 75:738–746
Banbura J (1992) Mate choice by females of the swallow *Hirundo rustica*: is it repeatable? J Ornithol 133:125–132
Bar-Hillel M, Margalit A (1988) How vicious are cycles of intransitivity preference? Theor Decis 24:119–145
Basolo AL (1990) Female preference for male sword length in the green swordtail, *Xiphophorus helleri* (Pisces, Poeciliidae). Anim Behav 40:332–338
Bateson M, Healy SD (2005) Comparative evaluation and its implications for mate choice. Trends Ecol Evol 20:659–664 (http://www.staff.ncl.ac.uk/melissa.bateson/Bateson_Healy_2005.pdf)
Beeching SC, Wack CL, Ruffner GL (2004) Female convict cichlids (*Archocentrus nigrofasciatus*) prefer to consort with same-sized males. Ethol Ecol Evol 16:209–216
Bell AM, Hankinson SJ, Laskowski KL (2009) The repeatability of behaviour: a meta-analysis. Anim Behav 77:771–783
Biro PA, Dingemanse NJ (2008) Sampling bias resulting from animal personality. Trends Ecol Evol 24:66–67
Both C, Dingemanse NJ, Drent PJ, Tinbergen JM (2005) Pairs of extreme avian personalities have highest reproductive success. J Anim Ecol 74:667–674
Bro-Jørgensen J (2010) Dynamics of multiple signalling systems: animal communication in a world of flux. Trends Ecol Evol 25:292–300
Burley N, Coppersmith CB (1987) Bill color preferences of zebra finches. Ethology 76:133–151
Burns KJ (1998) A phylogenetic perspective on the evolution of sexual dichromatism in tanagers (*Thraupidae*): the role of female versus male plumage. Evolution 52:1219–1224
Burns KJ, Shultz AJ (2012) Widespread cryptic dichromatism and ultraviolet reflectance in the largest radiations of neotropical songbirds: implications of accounting for avian vision in the study of plumage evolution. The Auk 129:211–221
Byers J, Hebets E, Podos J (2010) Female mate choice based upon male motor performance. Anim Behav 79:771–778
Candollin U (2003) The use of multiple cues in mate choice. Biol Rev 78:575–595
Carter AJ, Heinsohn R, Goldizen AW, Biro PA (2012) Boldness, trappability and sampling bias in wild lizards. Anim Behav 83:1051–1058
Cézilly F (2008) A history of behavioural ecology. In: Danchin E, Giraldeau L-A, Cézilly F (eds) Behavioural ecology. Oxford University Press, Oxford, pp 3–27
Cézilly F, Allainé D (2010) La sélection sexuelle, In: Thomas F, Lefèvre T, Raymond M (eds) Biologie évolutive. De Boeck, Bruxelles, pp 387–422
Collins SA, Hubbard C, Houtman AM (1994) Female mate choice in the zebra finch –the effect of male beak colour and song rate. Behav Ecol Sociobiol 35:21–25
Cornuau JH, Rat M, Schmeller DS, Loyau A (2012) Multiple signals in the palmate newt: ornaments help when courting. Behav Ecol Sociobiol 66:1045–1055
Cronin H (1991) The ant and the peacock: altruism and sexual selection from Darwin to Today. Cambridge University Press, Cambridge
Cummings M, Mollaghan D (2006) Repeatability and consistency of female preference behaviours in a northern swordtail, *Xiphophorus nigrensis*. Anim Behav 72:217–224
Dakin R (2011) The crest of the peafowl: a sexually dimorphic plumage ornament signals condition in both males and females. J Avian Biol 42:405–414
Dakin R, Montgomerie R (2011) Peahens prefer peacocks displaying more eyespots, but rarely. Anim Behav 82:21–28
Danchin E, Cézilly F (2008) Sexual selection, another evolutionary process. In: Danchin E, Giraldeau L-A, Cézilly F (eds) Behavioural ecology. Oxford University Press, Oxford, pp 363–426
Danchin E, Giraldeau L-A, Cézilly F (2008) Behavioural Ecology. Oxford University Press, Oxford
Darwin C (1859) On the origin of species by means of natural selection, or the preservation of favoured races in the struggle for life. John Murray, London

Darwin C (1874) The descent of man, and selection in relation to sex, 2nd edn. John Murray, London
David M, Cézilly F (2011) Personality may confound common measures of mate-choice. PLoS ONE 6:e24778
David M, Auclair Y, Cézilly F (2011) Personality predicts social dominance in female zebra finches, *Taeniopygia guttata*, in a feeding context. Anim Behav 81:219–224
David M, Auclair Y, Cézilly F (2012) Assessing short- and long-term repeatability and stability of personality in captive zebra finches using longitudinal data. Ethology 118:1–11
Dechaume-Moncharmont F-X, Cornuau JH, Keddar I, Ihle M, Motreuil S, Cézilly F (2011) Rapid assessment of female preference for male size predicts subsequent choice of spawning partner in a socially monogamous cichlid fish. Comptes Rendus Biologie 334:906–910
Dechaume-Moncharmont F-X, Freychet M, Motreuil S, Cézilly F (2013) Female mate choice in convict cichlids is transitive and consistent with a self-referent directional preference. Front Zool 10:69
Egger B, Klaefiger Y, Theis A, Salzburger W (2011) A sensory bias has triggered the evolution of egg-spots in cichlid fishes. PLoS ONE 6:e25601
Fanelli D (2011) Negative results are disappearing from most disciplines and countries. Scientometrics 90:891–904
Forstmeier W, Birkhead TR (2004) Repeatability of mate choice in the zebra finch: consitency within and between females. Anim Behav 68:1017–1028
Franks NR, Mallon EB, Bray HE, Hamilton MJ, Mischler TC (2003) Strategies for choosing between alternatives with different attributes: exemplified by house-hunting ants. Anim Behav 65:215–223
Gabor CR, Aspbury AS (2008) Non-repeatable mate choice by male sailfin mollies, *Poecilia latipinna*, in a unisexual-bisexual mating complex. Behav Ecol 19:871–878
Gabor CR, Parmley MH, Aspbury AS (2011) Repeatability of female preference in a unisexual-bisexual mating system. Evol Ecol Res 13:145–157
Gabriel PO, Black JM (2012a) Behavioural syndromes, partner compatibility and reproductive performance in Steller's Jays. Ethology 118:76–86
Gabriel PO, Black JM (2012b) Reproduction in Steller's jays (*Cyanocitta stelleri*): individual characteristics and behavioral strategies. The Auk 129:377–386
Gagliardi-Seeley J, Leese J, Santangelo N, Itzkowitz M (2009) Mate choice in female convict cichlids (*Amatitlania nigrofasciata*) and the relationship between male size and dominance. J Ethol 27:249–254
Garamszegi LZ, Eens M, Török J (2009) Behavioural syndromes and trappability in free-living collared flycatchers, *Ficedula albicollis*. Anim Behav 77:803–812
Gelez H, Lindsay DR, Blache D, Martin GB, Fabre-Nys C (2003) Temperament and sexual experience affect female sexual behaviour in sheep. Appl Anim Behav Sci 84:81–87
Grace RC (1993) Violations of transitivity: implications for a theory of contextual choice. J Exp Anal Behav 60:185–201
Greenfield MD, Siegfreid E, Snedden WA (2004) Variation and repeatability of female choice in a chorusing Katydid, Ephippiger ephippiger: an experimental exploration of the precedence effect. Ethology 110:287–299
Gross MR (1994) The evolution of behavioral ecology. Trends Ecol Evol 9:358–360
Hamilton PS, Sullivan BK (2005) Female mate attraction in ornate tree lizards, *Urosaurus ornatus*: a multivariate analysis. Anim Behav 69:219–224
Henderson LA, Stathis A, James C, Brown R, McDonald S, Macefield VG (2012) Real-time imaging of cortical areas involved in the generation of increases in skin sympathetic nerve activity when viewing emotionally charged images. Neuroimage 62:30–40
Herborn KA, Macleaod R, Miles WTS, Schofield ANB, Alexander L, Arnold KE (2010) Personality in captivity reflects personality in the wild. Anim Behav 79:835–843
Hoelzer GA (1989) The good parent process of sexual selection. Anim Behav 38:1067–1078
Hoffmann CM, Cronin TW, Omland KE (2008) Evolution of sexual dichromatism. 1. Convergent losses of elaborate female coloration in new world orioles (*Icterus* spp.). The Auk 125:778–789

Holveck M-J, Riebel K (2007) Preferred songs predict preferred males: consistency and repeatability of zebra finch females across three test contexts. Anim Behav 74:297–309
Houston AI (1997) Natural selection and context-dependent values. Proc Biol Sci 264:1539–1541
Houston AI, McNamara JM, Steer MD (2007) Do we expect natural selection to produce rational behaviour? Philos Trans R Soc Lond B Biol Sci 362:1531–1543
Iwasa Y, Pomiankowski A (1994) Evolution of multiple sexual preferences by the handicap process. Evolution 47:853–867
Jennions MD, Petrie M (1997) Variation in mate choice and mating preferences: a review of causes and consequences. Biol Rev 72:283–327
Johnsen TS, Zuk M (1996) Repeatability of mate choice in female red jungle fowl. Behav Ecol 7:243–246
Kacelnick A (2006) Meanings of rationality. In: Hurley S, Nudds M (eds) Rational animals? Oxford University Press, Oxford, pp 87–106
Kime NM, Rand AS, Kapfer M, Ryan MJ (1998) Consistency of female choice in the túngara frog: a permissive preference for complex characters. Anim Behav 55:641–649
Kirkpatrick M (1982) Sexual selection and the evolution of female choice. Evolution 82:1–12
Kirkpatrick M (1996) Good genes and direct selection in the evolution of mating preferences. Evolution 50:2125–2140
Kirkpatrick M, Rand A, Ryan MJ (2006) Mate choice rules in animals. Anim Behav 71:1215–1225
Kodric-Brown A, Nicoletto PF (1997) Repeatability of female choice in the guppy: response to live and videotaped males. Anim Behav 54:369–376
Kortet R, Niemelä PT, Vainikka A, Laakso J 2012. Females prefer bold males; an analysis of boldness, mate choice and bacterial resistance in the field cricket *Gryllus integer*. Ecol Parasitol Immunol 1:ID 235580
Kotiaho JS, Puurtinen M (2007) Mate choice for indirect genetic benefits: scrutiny of the current paradigm. Funct Ecol 21:638–644
Krause ET, Naguib M (2011) Compensatory growth affects exploratory behaviour in zebra finches, *Taeniopygia guttata*. Anim Behav 81:1295–1300
Kunte K (2008) Mimetic butterflies support Wallace's model of sexual dimorphism. Proc Biol Sci 275:1617–1624
Lande R (1981) Models of speciation by sexual selection on polygenic traits. Proc Natl Acad Sci U S A 78:3721–3725
Lehtonen TK, Lindstrom K (2008) Repeatability of mating preference in the sand goby. Anim Behav 75:55–61
Losano A (2008) The woman painter in victorian literature. The Ohio State University Press, Columbus
Loyau A, Saint Jalme M, Sorci G (2005a) Intra- and intersexual selection for multiple traits in the peacock (*Pavo cristatus*). Ethology 111:810–820
Loyau A, Saint Jalme M, Cagniant C, Sorci G (2005b) Multiple sexual advertisement honsetly reflect health status in peacocks (*Pavo cristatus*). Behav Ecol Sociobiol 58:552–557
Loyau A, Petrie M, Saint Jalme M, Sorci G (2008) Do peahens not prefer peacocks with more elaborate trains? Anim Behav 76:e5–9
Mays HL Jr, Hill GE (2004) Choosing mates: good genes versus genes that are a good fit. Trends Ecol Evol 19:554–559
Milam EL (2010) Looking for a few good males: female choice in evolutionary biology. Johns Hopkins University Press, Baltimore
Miller G (2000) The mating mind: how sexual choice shaped the evolution of human nature. Heineman, London
Mishra S, Logue DM, Abiola IO, Cade WH (2011) Developmental environment affects risk-acceptance in the hissing cockroach, *Gomphadorhina portentosa*. J Comp Psychol 125:40–47
Møller AP (1994) Repeatability of female choice in a monogamous swallow. Anim Behav 47:643–648
Møller AP, Petrie M (2002) Condition dependence, multiple sexual signals, and immunocompetence in peacocks. Behav Ecol 13:248–253

Møller AP, Pomiankowski A (1993) Why have birds got multiple sexual ornaments. Behav Ecol Sociobiol 32:167–176

Navarick DJ, Fantino E (1972) Transitivity as a property of choice. J Exp Anal Behav 18:389–401

Neff BD, Pitcher TE (2005) Genetic quality and sexual selection: an integrated framework for good genes and compatible genes. Mol Ecol 14:19–38

Noonan KC (1983) Female mate choice in the cichlid fish *Cichlasoma nigrofasciatum*. Anim Behav 31:1005–1010

Nuttall DB, Keenleyside MHA (1993) Mate choice by the male convict cichlid (*Cichlasoma nigrofasciatum*; pisces, cichlidae). Ethology 95:247–256

Oliver JC, Monteiro A (2011) On the origins of sexual dimorphism in butterflies. Proc Biol Sci 278:1981–1988

O'Loghlen A, Rothstein SI (2012) When less is best: female brown-headed cowbirds prefer less intense male displays. PLoS ONE 7:e36130

Patrick SC, Chapman JR, Dugdale HL, Quinn JL, Sheldon BC (2012) Promiscuity, paternity and personality in the great tit. Proc Biol Sci 279:1724–1730

Peckham GW, Pekham EG (1890) Additional observations on sexual selection in spiders of the family *Attidae*, with some remarks on Mr. Wallace's theory of sexual ornamentation. Occas P Nat Hist Soc Wis 1:117–151

Petrie M (1992) Peacocks with low mating success are more likely to suffer predation. Anim Behav 44:585–586

Petrie M, Halliday T (1994) Experimental and natural changes in the peacock's (*Pavo cristatus*) train can affect mating success. Behav Ecol Sociobiol 35:213–217

Petrie M, Halliday T, Sanders C (1991) Peahens prefer peacocks with more elaborate trains. Anim Behav 41:323–331

Petrie M, Cotgreave P, Stewart I (1996) Variation in the train morphology of peacocks (*Pavo cristatus*). J Zool 238:365–371

Pratt S (2005) Behavioral mechanisms of collective nest-site choice by the ant *Temnothorax curvispinosus*. Insectes Soc 52:383–392

Price T, Birch GL (1996) Repeated evolution of sexual color dimorphism in passerine birds. The Auk 113:842–848

Prum RO (2012) Aesthetic evolution by mate choice: Darwin's *really* dangerous idea. Philos Trans R Soc Lond B Biol Sci 367:2253–2265

Punzalan D, Hosken DJ (2010) Sexual dimorphism: why the sexes are (and are not) different. Curr Biol 20:R972–R973

Puurtinen M, Ketola T, Kotiaho JS (2005) Genetic compatibility and sexual selection. Trends Ecol Evol 20:157–158

Puurtinen M, Ketola T, Kotiaho JS (2009) The good-genes and compatible-genes benefits of mate choice. Am Nat 174:741–752

Réale D, Gallant BY, Leblanc M, Festa-Bianchet M (2000) Consistency of temperament in bighorn ewes and correlates with behaviour and life-history. Anim Behav 60:589–597

Reaney LT (2009) Female preference for male phenotypic traits in a fiddler crab: do females use absolute or comparative evaluations? Anim Behav 77:139–143

Regenwetter M, Dana J, Davis-Stober CP (2011) Transitivity of preferences. Psychol Rev 118: 42–56

Reinhold K, Reinhold K, Jacoby KJ (2002) Dissecting the repeatability of female choice in the grasshopper *Chorthippus biguttulus*. Anim Behav 64:245–250

Royle NJ, Lindström J, Metcalfe NB (2008) Context-dependent mate choice in relation to social composition in green swordtails *Xiphophorus helleri*. Behav Ecol 19:998–1005

Ryan MJ (1990) Sexual selection, sensory systems and sensory exploitation. Oxf Surv Evol Biol 7:157–195

Schuck-Paim C, Kacelnik A (2002) Rationality in risk-sensitive foraging choices by starlings. Anim Behav 64:869–879

Schuett W, Dall SRX, Royle NJ (2011a) Pairs of zebra finches with similar 'personalities' make better parents. Anim Behav 81:609–618

Schuett W, Godin J-GJ, Dall SRX (2011b) Do female zebra finches, *Taeniopygia guttata*, choose their mates based on their 'personality'? Ethology 117:908–917
Sih A (2011) Effects of early stress on behavioral syndromes: an integrated adaptive perspective. Neurosci Biobehav Rev 35:1452–1465
Sih A, Del Giudice M (2012) Linking behavioural syndromes and cognition: a behavioural ecology perspective. Philos Trans R Soc B Biol Sci 367:2762–2772
Sih A, Bell A, Johnson JC (2004) Behavioural syndromes: an ecological and evolutionary overview. Trends Ecol Evol 19:372–378
Simons MJP, Verhulst S (2011) Zebra finch females prefer males with redder bills independent of song rate—a meta-analysis. Behav Ecol 22:755–762
Sinn DL, Apiolaza LA, Moltschaniwskyj NA (2006) Heritability and fitness related consequences of squid personality traits. J Evol Biol 19:1437–1447
Smith BR, Blumstein DT (2008) Fitness consequences of personality: a meta-analysis. Behav Ecol 19:448–455
Soler JJ, Moreno J (2012) Evolution of sexual dichromatism in relation to nesting habits in European passerines: a test of Wallace's hypotheses. J Evol Biol 25:1614–1622
Stamp Dawkins M, Woodington A (1997) Distance and the presentation of visual stimuli to birds. Anim Behav 54:1019–1025
Sumpter CE, Temple W, Foster TM (1999) The transitivity of choice between different response requirements. J Exp Anal Behav 72:235–249
Takahashi M, Arita H, Hiraiwa-Hasegawa M, Hasegawa T (2008) Peahens do not prefer peacocks with more elaborate trains. Anim Behav 75:1209–1219
Tversky A (1969) Intransitivity of preferences. Psychol Rev 76:31–48
Tversky A, Russo EJ (1969) Substitutability and similarity in binary choices. J Math Psychol 6:1–12
Van Doorn GS, Weissing FJ (2004) The evolution of female preference for multiple indicators of quality. Am Nat 164:173–186
van Oers K, de Jong G, van Noordwijk AJ, Kempanaers B, Drent PJ (2005) Contribution of genetics to the study of animal personalities: a review of case studies. Behaviour 142:1191–1212
Waite T (2001) Intransitive preferences in hoarding gray jays (*Perisoreus canadensis*). Behav Ecol Sociobiol 50:116–121
Wallace AR (1878) Tropical nature and other essays. McMillan, London
Wallace AR (1889) Darwinism: an exposition of the theory of natural selection with some of its applications. Macmillan & Co (2nd edn 1905), London
Weismann A (1904) The evolution theory. [Translated by JA Thomson, MR Thomson] (2 vol) E. Arnold, London
Wiens JJ (2001) Widespread loss of sexually selected traits: how the peacock lost its spots. Trends Ecol Evol 16:517–523
Wisenden BD (1995) Reproductive behaviour of free-ranging convict cichlids, *Cichlasoma nigrofasciatum*. Environ Biol Fishes 43:121–134
Woo KL, Rieucau G (2011) From dummies to animations: a review of computer-animated stimuli used in animal behavior studies. Behav Ecol Sociobiol 65:1671–1685
Yasmin S, Yahya HSA (1996) Correlates of mating success in Indian peafowl. The Auk 113:490–492
Zahavi A (1975) Mate selection. A selection for a handicap. J Theor Biol 53:205–214
Zimmer RZ, Derby CD (2011) Neuroecology and the need for broader synthesis. Integr Comp Biol 51:751–755

Chapter 10
Reaction Norms of Sex and Adaptive Individual Flexibility in Reproductive Decisions

Malin Ah-King and Patricia Adair Gowaty

Abstract Biology at large is in the midst of a revolution in our understanding of the determination of phenotypes. Epigenetics has shifted our focus from genetic determinism to ecological origins of gene expression. We argue that this shift should be incorporated into sexual selection, changing the conceptualization of sex from a discrete trait to a developmental reaction norm. "Sex is a reaction norm" implies that the variation within and between the sexes is a result of genetic, epigenetic and environmental influences on developmental plasticity of phenotypes. "Choosy females" and "indiscriminate males" constitute one of the best examples of assumed strict sex differences that are in fact phenotypically plastic in response to environmental, social and internal factors. Here we summarize the empirical evidence, which empiricists have explained with trade-off hypotheses: individuals trade-off energy of reproductive decision-making with diverse, usually unitary factors: predation risk or density or OSR, etc. Gowaty and Hubbell's (2009) Switch-Point Theorem simplifies and unifies these trade-offs into a single hypothesis and works as an integrative framework, both for reinterpreting earlier findings and as a pointer to new directions for sexual selection research. We conclude that it is time to pay more attention to morphological, physiological, and behavioural phenotypes as developmentally plastic and/or individually flexible.

Keywords Phenotypic plasticity · Mate choice · Sexual selection · Switch-point theorem · Stochastic demography · Sex differences

---

M. Ah-King (✉)
Centre for Gender Research, Uppsala University, Box 527, 751 20, Uppsala, Sweden
e-mail: Malin.Ah-King@gender.uu.se

P. A. Gowaty
Department of Ecology and Evolutionary Biology and Institute of Environment and Sustainability, 621 Charles E. Young Dr. S., Los Angeles, CA 90095, USA

P. A. Gowaty
Smithsonian Tropical Research Institute, Unit 9100, BOX 0948, Panama 34002-9998, USA

P. A. Gowaty
Polistes Foundation, Inc. 133 Washington Street, Belmont, MA 02478, USA

T. Hoquet (ed.), *Current Perspectives on Sexual Selection,* History, Philosophy and Theory of the Life Sciences 9, DOI 10.1007/978-94-017-9585-2_10

10.1 Introduction

Biology as a whole is in the midst of a paradigm shift, reflecting the rise in our deeper understanding of the "determination" of phenotypes. Driven largely by ecological-evolutionary-developmental biologists—the "eco-evo-devos"—and the molecular geneticists who study "gene regulation", today's exciting conversations are about "differential gene expression", not differences in genes, but differences in gene *expression* that regulate the production of phenotypes (Gilbert 2001). Regulators of gene expression and thus phenotypic variation include intrinsic regulators, but there is now an impressive list of environmental factors—glandular secretions, epigenetic marks or behavioural variation that are transmitted from the parental phenotype or environment to offspring—that alter, *i.e.*, regulate patterns of gene expression, sometimes over generations without any changes in gene frequencies (Bonduriansky 2012). With attention to the environmental regulation of gene expression that produces fitness enhancing, adaptive phenotypes has come attention to varieties of phenotypic variants that arise from differential regulation of the same gene or genotypes.

Phenotypic plasticity and *developmental plasticity* are terms that capture the idea that "organisms are active processes, moving targets from fertilization to death" (p. 235, Fausto-Sterling 2000). *Polyphenism* is another general term indicating an either/or switch in morphological phenotypes within a single population that results from environmental effects or cues (e.g., temperature, day length, seasonality) that regulate the expression of a common genotype. *Genetic imprinting* is a more specific example of an autosome expressing differently depending on some "ecological" or "social" factor: in the case of *imprinting* an autosome expresses differently in an offspring depending on whether it was inherited from its mother or its father, a bit of nature that demonstrates that genes are environments too. To add to the excitement are the new and extensive mechanistic discussions of how individuals make behavioural "decisions", which are the alternative paths to fitness enhancement that individuals may take in real time, moment-to-moment as their ecological, demographic and social circumstances change. We use the term "*adaptive flexibility of individuals*" to re-emphasize that behaviour is the ultimate in the continua of examples of developmental phenotypic plasticity. We use the platform of "phenotypic variation via ecological regulation of gene expression" to emphasize that as far as the vast majority of genes (autosomes) are concerned, there is nothing so like a male as a female and vice-versa, a truism that sex differences researchers, including those using the classical paradigms of sexual selection, must grapple. Could it be that ecological and social forces, rather than differences in fixed gene differences, make the sex differences that so fascinate us?

Here we (1) discuss how the very modern view from developmental plasticity characterizing sex as a reaction norm puts the lie to typological views of "essential" sex differences, (2) We briefly review the "typological history of sexual selection", (3) recount theoretical and empirical challenges to the usual view of sexual selection that dominated the twentieth Century, and (4) briefly describe how the Switch Point Theorem (SPT) can simplify and unify accounts of reproductive decision-making, particularly the behaviour of accepting or rejecting potential mates. (5) We then turn around and show how an altered view of the arrows of causation can facilitate the development of new ways to ask questions about the ecological and evolutionary origins and adaptive significance of behavioural variants. We thereby suggest that even the most prosaic of within-sex variation in traits may function to facilitate variation in number of mates and within-sex variance in fitness.

10.2 Sex is a Reaction Norm

As biologists and by convention, we define an individual's sex in terms of the size of the gamete it produces. But in practice biologists and others often-time assign to females and males sets of associated traits that we think of as "sexed" (even if they are not). Mothers and hens are, after all, almost always females; cocks and stallions are, after all, almost always males. But, there are exceptions even to these statistically significant rules, for example some males and some females have versions of genitalia or bodies in-between male-female anatomies (Dreger 2007). And, the variations may often not be from mutation, but from differential gene regulation, some of which may enhance fitness. Much of the variation within and the overlap between sexes in what we call "sexed" traits come from environmental signals that cue developmental cascades (including the mechanisms of epigenetics that work throughout life to moderate gene expression). These myriad developmental cascades often depend on very small (part per trillion) nudges from environmental or social forces to instantiate the effects of organizational hormones in eggs, fetuses, and neonates and the regulatory actions of hormones mediating adult behaviour, physiology, morphology and life-history[1]. In fact, when one looks closely some of these environmental nudges to developmental cascades flip on and off switches that make it abundantly clear that there is nothing so similar to a female as a male or another female. An example is the hormone-like effects of a atrazine (a industrial pesticide) run-off in streams that renders 90 % of male frogs infertile and 10 % fertile females (Hayes et al. 2010). We should not be surprised that some of the most profound sex differences in within-sex and between-sex variations come from environmental and social signals: after all, females and males share in common the vast majority of their genes and often share very similar or identical ecological circumstances: When genes and environments interact it is no surprise that there is nothing so like a male as a female and vice-versa, and no surprise that many of the sex differences we count as so important are environmentally induced. Sex is a reaction norm (Ah-King and Nylin 2010).

Phenotypic plasticity implies a reaction norm. A reaction norm is the range of phenotypic expressions that one genotype can give rise to, in response to different environmental conditions (e.g. Gotthard and Nylin 1995; Stearns and Hoekstra

[1] For example, West-Eberhard's observation the two distinct female phenotypes in social insects, reproductively active queens and sterile workers, may have originated as facultative response to environmental or social conditions (West-Eberhard 1987).

2005). Mechanisms of phenotypic plasticity produce switches that are sensitive to both/either genetic and environmental input on further development of an already existing phenotype (West-Eberhard 2003). Consider that crocodile eggs are unsexed, so that the sex of a crocodile is determined by an environmental cue and is a function of the temperature eggs experience during incubation. In crocodiles sex is a reaction norm of eggs to temperature differentials. Conceptualizing sex as a reaction norm need not imply that sex is completely plastic. For example, some species will sometimes fail to change sex, and may be insensitive to known environmental effects while others are very sensitive to frequent changes in ambient conditions. There is a thus a continuum of species between extremes of strict genetic sex determination and environmental sex determination (Ah-King and Nylin 2010) with many species falling in-between. Of course, those interested in the ubiquity of epigenetically organized gene-environment interactions will wonder if there is ever "strict genetic determination".

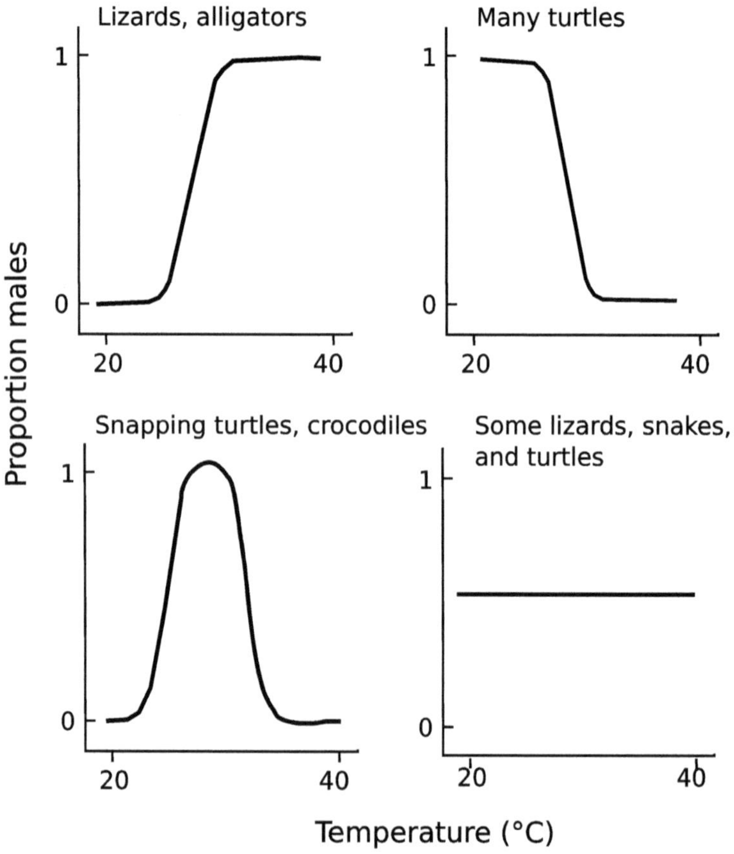

Fig. 10.1 Temperature effects on sex determination in different species. In lizards, alligators, turtles, snapping turtles and crocodiles, incubation temperature affects the sex ratio. After (Bull 1983)

It is useful to consider the plasticity of traits under the reaction norm perspective in terms of five attributes. A trait subject to change due to environmental influences on development (1) can be *reversible* or irreversible (e.g. sex change). (2) Traits may differ in the *amount* of plasticity having large or small responses to environmental change. (3) Traits may differ in the *rapidity* of their response, (4) the *shape* of the response (e.g. environmental sex determination response curves to temperature) and (5) in *competence*, that is, "the ability of the developmental system to respond to environmental stimuli only during particular time "windows" in the ontogenetic trajectory" (Schlichting and Pigliucci 1998). Hence, a reaction norm may be flexible in many different axes, for example, sex determination may be environmentally influenced during a specific time-window during egg development, the shape of the response curves to temperature may differ between species or even individuals, and in many species or individuals the effect may be irreversible, often depending on whether the original epigenetic effect is organizational or regulatory (Crews and McLachlan 2006). The reaction norms perspective can be rephrased as the idea that all phenotypes emerge out of environmental interactions with pre-existing phenotypes.

It is important to remember too that natural selection may cause changes in mechanisms affecting the relative importance of endogenous and exogenous environments on the expression of a trait (*i.e.* genetic accommodation). An example is in the multiple origins during evolution of both environmental sex determination and genetic sex determination (West-Eberhard 2003), and the observation that switches between these systems in some clades are common (e.g. lizards and turtles: West-Eberhard 2003; fishes: Mank et al. 2006). This evolutionary flexibility demonstrates that sex is a reaction norm not only in the proximate sense that environment and genes interact to affect phenotypes, but also in the sense that over evolutionary time and, even sometimes, ecological time, sex and sex-related traits vary along continua. Plumage characteristics in birds provide a ready example: within-species plumage variation ranges from monomorphic to spectacular dimorphism and dichromatism of peacocks (*Pavo cristatus*). In many species the plumages are not sex-specific, as in monomorphic species. Furthermore, in many hummingbirds, in which both females and males may have "male-like" or "female-like" plumages (Bleiweiss 2001). Nor should we forget that further with-in individual switches in plumage, condition, and morphology and behaviour, such as mating competition, courtship, incubation, and parental care are induced by pathogens, parasites (Beckage 1997) and environmental toxins (Hayes et al. 2010), each of which can have profound effects on fitness variances. "Sex as a reaction norm" is a dynamic and non-typological way of incorporating current knowledge about within-and between-sex, individual, and species-level variation. "Sex as a reaction norm" is a dynamic and non-typological way of incorporating current knowledge about within-and between-sex, individual, and species-level variation.

From an evolutionary perspective, the production of eggs (female function) and sperm (male function) are the only traits universally applying to all males and all females, except when individuals do both, either as a general pattern of the species or in response to toxic endocrine disrupters. Selection can act on any of components of the "sexed" traits so that characteristics can evolve just like any other characters. Therefore, the expectation that sex differences fall into discrete classes is unjustified by the ubiquity of variation. It is a paradox that we biologists are knowledgeable about an astounding diversity of sex and sexuality among animals but continue nonetheless perpetuate unfounded stereotypes in our explanations of sex differences (Ah-King 2011).

10.3 A Brief History of Sexual Selection

10.3.1 Darwin was not an Essentialist

Darwin's biggest books (1859, 1871) are full of descriptions of variability but also provided some followers with the typology of typical "sex roles": females as coy and choosy, and males as eager and competitive with each other (Darwin 1871). In Darwin's sexual selection book (Darwin 1871), which was his defense against the critics of natural selection (Ghiselin 2003), Darwin, nonetheless, also discussed species in which the "typical sex role pattern" is reversed, for example, as it is in phalaropes (*Phalaropus tricolor*) with colorful females and dull colored incubating males. During his lifetime as his theory developed, Darwin modified his definitions of sexual selection applying his idea of sexual selection more broadly as he and his experience matured (Gowaty 2011). In *the Origin of Species* (Darwin 1859) he said that sexual selection "depends, not on a struggle for existence, but on a struggle between the males for possession of the females; the result is not death to the unsuccessful competitor, but few or no offspring." Later, in *The Descent of Man and Selection in Relation to Sex*, he said "…sexual selection… depends on the advantage which certain individuals have over other individuals of the same sex and species, in exclusive relation to reproduction" (Darwin 1871). Obviously, the narrowest (oldest) definition refers to mechanisms of sexual selection acting only on males and leading to variance in reproductive success among males, which we now call narrow-sense sexual selection, while the latter definition, Darwin's most mature, is a gender-neutral idea that acknowledges that sexual selection may occur in one or both sexes at different times or simultaneously. Thus, it is ironic that some of Darwin's followers in the twentieth Century were more sex-typological than he was, as the Darwin-Bateman paradigm (Dewsbury 2005) illustrates.

10.3.2 Linkages of Causality and Extensions to the Darwin-Bateman Paradigm

The Darwin-Bateman paradigm (Dewsbury 2005) is a set of ideas linked together almost a century after Darwin's first suggestion of sexual selection. The Darwin-Bateman world-view includes three main notions (1) males are fixed for eagerness to mate and indiscriminate in mate choice and females are fixed for choosy mating and discreet, most-often unenthusiastic acquiesce to mating, (2) males gain more by increasing mate number than females do, and (3) male reproductive success is more variable than female reproductive success. Linkages after Bateman (Parker et al. 1972; Trivers 1972) were meant to explain the then-thought-to-be rule that males are eager and females less so, *etc*. Thus, this world-view might be renamed the Darwin-Bateman-Parker-Trivers (D-B-P-T) "linkages of causality" and the notions rephrased as (1) the sex with the larger gametes and/or higher PI is choosy and coy while the sex with the smaller gametes and/or lower PI is indiscriminate and choosy, (2) given that gamete size to often larger in females, males are generally eager to mate and indiscriminate in patterns of acceptance and rejection (A&R) of potential mates[2], while females reject potential mates more often and seem far less eager, (3) males gain more by increasing mate number than females do, and (4) male reproductive success is more variable than female reproductive success. The order of these ideas captures the theorized direction of the arrows of causation from selection via sex-differentiated PI to sex-differences in behaviour to sex differences in within-sex variances in number of mates to within-sex variances in fitness. The linkage of causes results, it is said, in typical fitness variances for females and males.

10.3.3 Theoretical Challenges to the D-B-P-T Casual Links

Sutherland (Sutherland 1985) was the first to challenge the linkage between choosy females/competitive males to within-sex variances in either sex. Sutherland's statistical model of binomial probabilities for number of mates variances showed that fixed sex differences plus chance variation in encounters with potential mates can lead to sex differences in fitness variances even in the absence of any mechanisms of sexual selection. Latencies, which were fixed within-sex, could not contribute to within sex variances in fitness, but stochastic within-sex variation in encounters with potential mates could. Thus, even though Sutherland assumed intrinsic sex differences, his model demonstrated that chance effects on encounters and random mating could explain Bateman's results, which were about within-sex fitness variances. Given that Bateman did not watch behaviour (Bateman 1948; Dewsbury 2005), Sutherland's elegant claim that chance can explain Bateman's results remains a strong alternative explanation for Bateman's claims[3]. Chance produces fitness variances even in the absence of sexual selection.

[2] We operationalize "mate choice" throughout this manuscript in order to characterize what we mean and what investigators usually observe: namely, acceptances or rejections of potential mates or patterns of acceptance or rejection of potential mates. In contrast to other authors, we consider assessment of mates a *cognitive* process that is different from the *motor pattern* of accepting or rejecting a potential mate.

[3] More recently others have challenged Bateman's results and the uses to which his conclusions have been put on other grounds (See Dewsbury 2005; Gowaty et al. 2012; Snyder and Gowaty 2007; Tang-Martinez and Ryder 2005).

Hubbell and Johnson (1987) mating theorem then challenged the direction of the key linkage in the D-B-P-T predicted causal flow that goes from choosy/indiscriminate behaviour to within-sex fitness variances. Their two-part paper did a number of things. First, their analytical mating theorem showed that stochastic variation in survival and encounters could produce lifetime fitness variances under random mating. Second, they showed that an ESS for acceptance and rejection of potential mates could have evolved before the evolution of anisogamy: that pre-evolved, intrinsic sex differences are unnecessary for the evolution of sex-differentiated behaviour[4] and consequent within-sex fitness variances. Their results emphasized that stochastic effects on variances in mating were a demographic certainty even in the absence of sexual selection. In other words (cf Gowaty and Hubbell 2005), Hubbell and Johnson (1987) demonstrated that stochastic fitness variances would favour individuals—regardless of their sex—when they adaptively and flexibly expressed choosy and indiscriminate behaviour given the demographic environments they experienced. Simple assumptions of stochastic variation in mate encounter and individual survival probabilities each affecting individuals' time available for mating can determine lifetime variation in individual mating success and lifetime within-sex variance in mating success and fitness. In turn background probabilities of fitness variances can influence whether individuals express choosy or indiscriminate behaviour. Once again the conclusion was: chance in the absence of sexual selection can produce fitness variances. Perhaps more important was the lesson that unless we account for the stochastic effects on *e.g.* survival and encounters we cannot distinguish the relative importance of sexual selection on fitness variances (Gowaty and Hubbell 2005; Gowaty and Hubbell 2009; Sutherland 1985).

After both Sutherland and Hubbell and Johnson, Clutton-Brock and Parker's (1992) algebraic models about how the operational sex ratio (OSR) along with the potential reproductive rate (PRR) might shape accessment of potential mates and competition shared similarity of assumptions with both Sutherland (1985) and Hubbell and Johnson (1987). With Sutherland, Clutton-Brock and Parker (1992) shared assumptions about pre-existing sex differences indicated by their PRR term as well as a sometimes-fluctuating demography term, OSR. Hubbell and Johnson's math had no assumptions of sex of individuals, but did assume stochastic variation in demography (by chance death and chance movements individuals enter and leave populations), extending the reach of these models to calculate *lifetime* means and variances in fitness.

Unlike Clutton-Brock and Parker (1992), Sutherland (1985) and Hubbell and Johnson (1987) were more or less ignored until around 2001, perhaps because Sutherland (1985) was a full-frontal challenge to Bateman's principles and hard for many to swallow; while Hubbell and Johnson (1987) used a then-unfamiliar modelling platform and perhaps said too much in one paper to be digestible. Nonetheless,

[4] Clutton-Brock (2007) and Kokko and Jennions (2008) each also say that causality may be reversed so that sexual selection (mate competition and choice) may influence investment; they buttress their insights with data showing that sex differences in egg/sperm size do not determine the parental care system; instead, these systems coevolve and may have complex feedbacks.

Clutton-Brock and Parker's ideas stimulated a new era in empirical testing that has demonstrated conclusively that in most tested species within-sex variation in choosy and indiscriminate behaviour is developmentally plastic, part of the reaction norm of sex (Ah-King and Gowaty ms).

10.3.4 Empirical Challenges to the Casual Links in the D-B-P-T Predictions and Some Current Biases

Self-conscious empirical challenges to the D-B-P-T expectations seldom appear in print, which could be because investigators are shy of pointing out that their results challenge dominant ideas or perhaps because reviewers and editors, uncomfortable with results outside of "normal science" reject them. Historically, the belief in Bateman's "principles" retarded for a long time the exploration of multiple mating in females, sources of variation in female reproductive success, the cost of sperm production, male mate choice and sexual selection in females at large (*e.g.* Gowaty 1997; Gowaty 2003; Hrdy 1981; Stutchbury and Neudorf 1998; Tang-Martinez and Ryder 2005). But, today it is difficult to sustain those basic claims. Empirical evidence now shows that females are often enthusiastic about mating (e.g. Hrdy 1986; Small 1993), often mate multiply (Arnqvist and Nilsson 2000; Griffith et al. 2002), sex roles are not fixed (Forsgren et al. 2004; Gwynne and Simmons 1990), and males—even in species with no male PI—also gain fitness benefits from discriminating between potential mates (Altmann 1997; Bonduriansky 2001; Johnson 1982; Johnson 1983; Johnson and Hubbell 1984). Since the 1990s the initial focus on male ornaments and male-male competition has broadened to include questions about the force of variation in females in the evolutionary dynamics of sex (Gowaty 1996; Gowaty 1997; Gowaty and Buschhaus 1998; Gowaty et al. 2003; Rhainds 2010; Rice 1996) and to female variance in reproductive success (Gowaty 1996; Gowaty 1997; Gowaty et al. 2003). Females are said to control male reproductive success even under or especially during sperm competition (Birkhead and Møller 1998; Eberhard 1996; Parker 1970; Pizzari, Chap. 20 this volume) and they are in evolutionary equipoise with males under sexually-antagonistic allelic evolution (Arnqvist and Rowe 2005). Thus, the predictions from PI and anisogamy theories about fixed sex differences in behaviour and fitness (Gowaty and Hubbell 2013b) often fail to be met by the facts of nature, yet it is curious how few investigators actually say that.

To complicate further our interpretations about the direction of the arrows of causation investigators have imagined many more mechanisms of sexual selection, apart from female-female competition and male choice that Darwin also mentioned. For example, male's attempts to coerce females and female resistance (Gowaty 1997) or for that matter females' attempts to manipulate male reproductive decisions (Gowaty and Hubbell 2010) produce simultaneous opportunities for the production of high fitness variances both among-males and among-females.

Despite many successful theoretical challenges, expectations from D-B-P-T linkages still largely dog empirical research. For example, Schärer et al. (2012)

express their belief in anisogamy as the ultimate cause of sex differences in "sex-roles". Obviously, without females and males, there will be no sex-specific selection, however, anisogamy, which we use to categorize individuals as either male or female, cannot explain the enormous variation in sexual strategies among males and females (Gowaty and Hubbell 2005; Ah-King 2012). A correlation between investment in gametes and sexual strategies is simply not evidence of causation (Gowaty and Hubbell 2005; Ah-King 2012). Thus it is necessary to test the relative importance of intrinsic differences (including anisogamy) versus environmental determinism and stochastic effects before concluding that anisogamy is the source for sex differences in behaviour (Gowaty and Hubbell 2005; Gowaty and Hubbell 2009).

In current sexual selection studies, stochastic effects on variance in reproductive success are often overlooked. One example is Rodríguez-Muñoz et al.'s (2010) ground-breaking study, in which they constantly video-monitored behaviour of a wild population of field crickets, *Gryllus campestris,* and conducted parentage analyses on the offspring surviving to the subsequent year to estimate lifetime reproductive success. Rodríguez-Muñoz et al. found no statistically significant sex differences in variance in number of mates, but males had slightly, but significantly, higher variance in reproductive success than females. Rodríguez-Muñoz et al. (2010) provided two possible explanations for observed sex differences in reproductive success: post-mating sexual selection on males or differences in offspring viability. Rodríguez-Muñoz et al. (2010) overlooked the alternative explanation of stochastic demography (Gowaty and Hubbell 2005, 2009) as a possible explanation of sex differences in fitness variances. Thus, they did not solve the puzzle of relative apportionment of stochastic or sexually selected effects on fitness variances. Had they done so, they may have discovered, after accounting for the inevitability of stochastic effects, residual highly significant effects due to sexual selection.

10.4 Within-Sex Phenotypic Plasticity in Acceptance and Rejection (A&R) of Potential Mates

A sea-change in our view of fixed differences in the sexes has resulted from the many elegant empirical studies showing that within-choice phenotypic plasticity is common varying with environmental, social and internal circumstances (some examples are in Table 10.1). Most of the earliest studies of changes from choosy to random mating tested how population level independent variables such as the PRR or the OSR affected between-sex differences in behaviour (Jiggins et al. 2000). Later, investigators paid attention to variation in predation risk, disease risk, chooser's parasite load, chooser's age and experience, etc, which all resulted in phenotypic plasticity in A&Rs.

Table 10.1 summarizes empirical studies that we considered in our review paper (Ah-King and Gowaty ms). We did the review to evaluate (1) the frequency of phenotypic plasticity in acceptance and rejection of potential mates and what

Table 10.1 A summary table of species showing phenotypic plasticity in reproductive decision-making in response to environmental, social, demographic and internal factors (for references see Ah-King & Gowaty ms). Empiricists have explained this variation with trade-off hypotheses: individuals trade-off energy of reproductive decision-making with diverse, usually unitary factors: predation risk or density or OSR or encounter rate or chooser condition

Potential cause	Species, *scientific name*
Predation risk	Crickets, *Gryllus integer* Water striders, *Aquarius remigis* Amphipods *Gammarus duebeni* Sand gobies, *Pomatoschistus minutus* Black goby *Gobius niger* Guppies, *Poecilia reticulata* Panamanian bishop, *Brachyrhaphis episcopi* Pipefish, *Syngnathus typhle* Tungara frogs, *Physalaemus pustulosus* Fiddler crab, *Uca mjoebergi* Fiddler crab, *Uca beebei*
Body condition of chooser	Katydids bushcricket, *Requena verticalis:* wolf-spiders *Schizocosa ocreata* and *S. rovneri* swordtail fish, *Xiphophorus birchmanni* zebra finches, *Taeniopygia guttata*
Parasite load of chooser	damselfly *Calopteryx haemorrhoidalis* upland bullies, *Gobiomorphus breviceps* spadefoot toads, *Scaphiopus couchii* guppies, *Poecilia reticulata* pipefish, *Syngnathus typhle* wild turkeys, *Meleagris gallopavo*
Age of chooser	House crickets, *Acheta domesticus* Tanzanian cockroaches, *Nauphoeta cinerea* Real's wood white, *Leptidea reali* Guppies, *Poecilia reticulata*
Habitat quality	Cockroach, *Nauphoeta cinerea* marine iguanas, *Amblyrhynchus cristatus*
Population density	Fruitflies, *Drosophila melanogaster* Butterflies, *Acraca encedon* Pill bugs, *Armadillidium vulgare* Katydids Bushcrickets *Xederra charactus* Speckled wood butterfly, *Pararge aegeria* Ladybeetles, *Coleomegilla maculata* Guppies, *Poeciliareticulata* Pipefish, *Syngnathus typhle* Fiddler crab, *Uca uruguayensis* Kestrels, *Falco thmuncuhts*
Relative attractiveness or availability of resources	Beaugregory damselfish, *Stegastes leucostictus* Threespine sticklebacks, *Gasterosteus aculeatus* Common goby, *Pomatoschistus microps*

Table 10.1 (continued)

Potential cause	Species, *scientific name*
Experience	*Drosophila paulistorum* Field crickets, *Teleogryllus oceanicus* Bark beetles, *Ips pini* Moths, *Helicoverpa armigera* Damselflies, *Enallagma* Wolf spider, *Schizocosa* Threespine sticklebacks, *Gasterosteus aculeatus* Guppies, *Poecilia reticulata* Lincoln's sparrows, *Melospiza lincolnii* Zebra finches, *Taeniopygia guttata* Red-sided garter snakes *Thamnophis sirtalis parietalis*
OSR	Katydids Butterfly, *Acraea encedon* Spider, *Zygiella x-notata* Japanese medaka, *Oryzias latipesm* Two spotted gobies, *Gobiusculus flavescens* Flagfish, *Jordanella floridae* Pipefish, *Syngnathus typhle* Guppies, *Poecilia reticulata* Common goby, *Pomatoschistus microps*

Table 10.2 The environmental correlates of phenotypic plasticity may simply be more-complex proxies for the hypothesized inducing variables in the SPT. Examples show how complex proxies may be affecting parameters of the SPT

Environmental correlate	SPT parameter
Predation risk, parasite load, condition, age, food deprivation	Survival probability
OSR, ASR, population density, predation risk, guarding/territoriality, density of opposite sex, attractiveness of resources/chooser, predation risk, parasite load	Encounter probability
OSR, ASR	Number of potential mates
Experience, population density	*w-distribution*

environmental forces have been associated with phenotypic plasticity in mating and (2) if there exists current empirical scope suggesting that within-sex phenotypic plasticity was better characterized as sex-neutral individual flexibility. Our review covered almost 200 studies of a variety of taxa: birds, fish, spiders, frogs, lizards, crustaceans and insects (Ah-King & Gowaty, ms). The reviewed papers described both laboratory and field studies showing that within-sex mating decisions often change from choosy to random. Table 10.1 is a brief summary of some of the species that exhibit within-sex phenotypic plasticity and the variables associated with it (Ah-King and Gowaty ms). The environmental, social and internal factors can be simplified or factored into the five unifying variables of the SPT. For example, predation risk can be factored into effects on survival probability or decreased encounters with potential mates since many animals decrease their movements in the vicinity of predators (Table 10.2).

Despite this crescendo of evidence for phenotypic plasticity and given the D-B-P-T assumptions, investigators still predict that the choosing ones are usually females who prefer the most extreme expression of a sexually selected trait in males (also criticized by Cézilly in Chap. 20). One example of how current researchers handle the variability in mating decisions was an Association for the Study of Animal Behaviour (ASAB) (December 1–2, 2011) conference in London entitled "Why do animals mate with the "wrong" partner?" The organizers questioned whether the "mistakes" are interesting or not? They asked if the mistakes were perhaps not mistakes but constraints mediating mechanisms of mate choice (which is what the SPT predicts)? They wondered if "mistakes" were hidden adaptations? Or were these "mistakes" maybe ignored for a good reason? The talks described the "wrong" partner as unattractive partners, flexible acceptance or rejection of potential mates altering because of age, parasite exposure, rearing experience, hunger or adult sex ratio, and phenomena such as mate choice copying, hybridization, and same-sex sexuality. Here we claim that what was considered acceptance of the "wrong" partner choice may actually be an optimal decision that enhances fitness (increased RS or higher survival) under current demographic and ecological circumstances. Ecological and demographic factors, such as predation risk, OSR, age, parasite load and resource influence the mating behaviour of many animals changing acceptance/rejection of potential partners (Forsgren et al. 2004; Gwynne and Simmons 1990; Poulin 1994). Below we emphasize a theoretical framework for understanding individually adaptive flexibility in acceptance and rejection of potential mates that enhances fitness.

The review shows that within-sex phenotypic plasticity is common in many taxa, in response to environmental, social and internal factors (Ah-King & Gowaty, ms). Both females and males show phenotypically plastic choosy behaviour, sometimes mating at random and sometimes rejecting potential mates. Individuals can be both competitive and choosy at the same time.

Generalities that emerged from Ah-King and Gowaty (ms) included:

- "The "choosy sex" is phenotypically plastic in tested species sometimes mating at random, other times rejecting potential mates.
- Males show phenotypically plastic choosy behaviour, as do females.
- Females show phenotypically plastic competitive behaviour, just as males do.
- Population density effects may be due to either number of potential mates *n* or to encounter rate with potential mates *e*.
- OSR may not adequately capture the inducing variable of individual behaviour *e.g.,* even if there are more sexually available females than males, males may vary in their *encounters* with sexually available, receptive females, or the *w-distribution* may affect individuals' decisions.
- Predator risk is complex, because the number of potential mates *n* may decline in the presence of a predator, an individual's survival probability *s* decreases, and *e* may be reduced.
- Anisogamy and parental investment theories failed to predict within-sex phenotypic plasticity.
- Almost all studies fit one or more of the SPT's simplifying parameters of *s, e, n, l*, and *w-distribution.*"

10.5 The Switch-Point Theorem Simplifies and Unifies Studies of Within-Sex Phenotypic Plasticity

In the recent debate about the state of sexual selection research Cornwallis and Uller (2010) urged for a more integrated direction of sexual selection research, including influences of heterogeneous environments and phenotypic plasticity in both theory and empirical research. Others (Safran et al. 2010) argued that there is already a rich empirical literature on these matters and suggest that what is really needed is a strong integrative framework to handle the dynamic effects of environmental heterogeneity and plasticity into sexual selection theory. We agree that empirical studies have apparently preceded theoretical development, however, a strong integrative framework does already exist, namely Gowaty and Hubbell's (2005, 2009, 2010, 2013a; Gowaty 2012) models of individually adaptive flexibly in acceptance and/or rejection of potential mates.

The Switch-Point Theorem (SPT) is an alternative hypothesis to anisogamy and parental investment hypotheses for the origins of choosy and indiscriminate mating and within-sex fitness differences (Gowaty and Hubbell 2009). The SPT is gender-neutral: the model works the same way for any individual, *i.e.*, it makes no necessary *a priori* assumptions about sex differences.[5] The rules governing outcomes will be the same for individuals experiencing the same ecological and social constraints, just as they will be different for individuals experiencing different ecological and social constraints. What it proves theoretically is that (1) demographic environments that vary stochastically, as they all do, inevitably select against individuals fixed for choosy or fixed for indiscriminate behaviour, (2) selection will favour flexible individuals able to adaptively modify their acceptance/rejection behaviour in ecological time as their circumstances change. Using the SPT allows investigators to predict what individuals do given their encounters with potential mates and their likelihood of survival against a background of fitness that would be conferred for any given mating. Thus, one way that the SPT simplifies things is that it focuses on a simpler unit of biological organization, individuals, not sexes.

The SPT unifies things in that it factors more complex proxy variables into their components having to do with time available for mating (Table 10.2). The SPT assumes that time is finite for all individuals and it assumes a simple life story in which individuals at the beginning of their reproductive careers enter a series of stages with some probabilities: receptivity to mating, encountering a potential mate and mating or not, after which an individual may return to receptivity or in some cases experience a latency period after mating before they re-enter latency. An individual's survival probability determines whether it survives while

[5] A misconception about gender-neutral models is the claim that such models ignore sex differences (Schärer et al. 2012). However gender-neutral assumptions can predict sex differences whenever individuals of different sexes experience different ecological or social constraints. The gender-neutral models also predict what happens when ecological constraints on indivdiuals are relaxed, something that rigid assumptions of sex differences do not do. Gender-neutral models do not ignore sex differences, but rather predict variation among individuals due to sex blind, stochastic demography. (For a reply on Schärer et al.'s questioning of gender-neutral models see Ah-King 2012.)

in a particular state. When the individual dies it enters an absorbing state from which it cannot exit. So, the life history and reproductive careers of individuals are a simple set of required stages for iteroparous organisms. Previously-hypothesized environmental, social, demographic and internal factors that influence mating decisions (Table 10.1) can be simplified into effects on the SPTs parameters (Table 10.2): encounter probability, survival probability, latencies after mating to receptivity to mating again (which can be required or optional), number of potential mates in the population, and the distribution within the population of fitness that would be conferred from random mating among all individuals in the population. Each of these parameters reflect individuals' ecological and social constraints on time available for mating. According to the SPT, an individual is expected to assess fitness differences of mating with potential mates and be sensitive to time it has left to reproduce, so that any given "decision" to accept or reject is made in terms of a trade-off between fitness that will be conferred and the time available for mating. Thus, the SPT unifies a set of trade-off hypotheses that investigators have offered as explanations for their observations (Table 10.1).

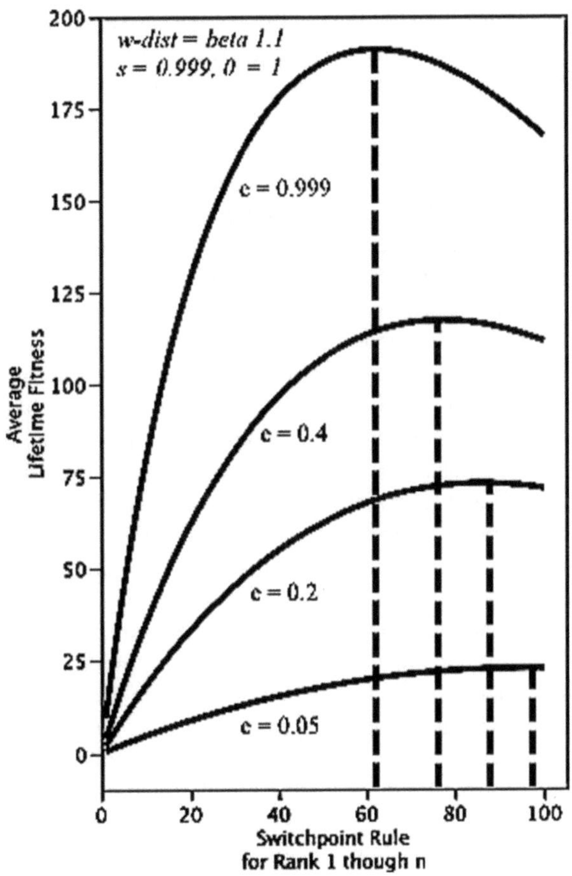

Fig. 10.2 Switch points under varying encounter probabilities

Thus, according to the SPT, all individuals access potential mates before an individual should, no matter its sex and if all else is equal, reject more potential mates when it experiences (1) higher survival probability (*e.g.* decreased predation risk), or (2) increased encounters with potential mates (*e.g.* by increased population density), or (3) increased latency (*e.g.* lower reproductive rate), (4) with increased number of potential mates in the population, or (5) if the distribution of fitnesses conferred is flat or somewhat left skewed (so that a larger proportion of potential mates in the population result in low fitness). In fact, all five of these variables determine an individual's switch point for accepting or rejecting a potential mate.

The assumptions of the SPT (Gowaty and Hubbell 2009) are: (1) time available for mating is finite for all individuals, and is characterized in terms of constraints in time available for mating. The parameters of earlier theories (variation in ecological, social and internal factors) are simplified and unified into the effects on time available for mating on five parameters (survival probability s, encounter probability e, previously experienced post-mating time outs o, number of potential mates n, and the distribution of fitness that would be conferred under random mating, the w-distribution). (2) Individuals assess the likely fitness that would be conferred by alternative potential mates (w-distribution) before accepting or rejecting a potential mate. (3) The SPT assumes individuals rank potential mates in accordance with their unique fitness assessment and (4) individuals encounter potential mates at random with respect to their rank. The analytical solution of the simultaneous effects of these five parameters on time available for mating results in an optimal switch point rule—the fraction of potential mates that a focal individual finds acceptable to mate maximizes lifetime fitness under current circumstances. If these assumptions are met, then individuals are making adaptively flexible mating decisions, changing sometimes moment-to-moment given current ecological and social circumstances.

The SPT is not a model of morphological trait evolution; however it can be used to inform ideas about morphological trait evolution (Gowaty and Hubbell 2009). Stochastic variation in its parameters give rise to variance in lifetime number of mates and in lifetime reproductive success. If stochastic demographic effects are strong, selection will be random in regard to traits. So, as Hubbell and Johnson (1987) emphasized, before we are able to distinguish a correlation between a factor/trait and fitness, we need to account for the "uninteresting" effect of stochastic demography (see Snyder and Gowaty 2007) or we will overestimate the opportunity for sexual selection (as Bateman did). Reasoning from the SPT however, suggests ways that selection might act: selection may favor individuals with traits that increase their likelihood of encounter with potential mates, such as easily seen plumage and loud calls. In contrast to classic predictions, the fitness payout for these secondary sexual characters may be not in acquiring *more* mates but in increasing an individual's opportunity to mate with *better quality* mates (Gowaty and Hubbell 2009). And, if this is so, a novel prediction follows: fancy males may say "no", *i.e.* be "choosier" and reject more females than less fancy males.

Hence, the SPT informs us about what many empirical studies have already concluded: focusing only on sexual traits fails to tell the whole story. Fitness consequences of fancy traits clearly depend on social and ecological circumstances. In addition, the SPT may explain previous failures to associate some observed sexual traits with reproductive success (Cornwallis and Uller 2009). Experiments may have included uncontrolled variation in the SPT's parameters that could explain unexpected or "wrong" results. The SPT also predicts that individuals will appear not to discriminate between potential mates if mating with either has equivalent fitness consequences. This detail emphasizes that assessment is different from the behaviour of accepting or rejecting potential mates.

Running the SPT allows us to predict what a given individual will do given the environmental and social constraints and opportunities they experience at any given time, but, what is of equal interest is what flexible individuals do in the context of a dynamic population of interacting individuals. To study the behavioural outcomes within and between individuals all interacting in the same dynamic population, Gowaty and Hubbell (2005, 2010) built DYNAMATE$^{@}$ and used DYNAMATE$_{@}$ in a series of numerical experiments. DYNAMATE$^{@}$ is an agent-based model of individuals dynamically interacting in a population. DYNAMATE$^{@}$ can simulate the outcomes of behaviour and fitness within and between individuals for an arbitrary amount of time or until all individuals in a population die. As individuals mate, enter latencies or die, they become unavailable as potential mates for some amount of time or for infinity, when they die. DYNAMATE$^{@}$ can compete the SPT's rules of induced behaviour and consequent fitness under deterministic or stochastic demography with the static rules of the D-B-P-T causal links. It is of some interest here to point out that consistent with the sensitivity analysis of the SPT, in DYNAMATE$^{@}$ assigned sex differences in l seldom have notable effects on sex-differences in fitness variances: again s and e exert the most notable changes in sex-differentiated variances in fitness. However, it is equally notable that within-sex variation in l can have very large effects on *within-sex* fitness variances. The SPT thus predicts that within-sex selection may act strongly on variation in l.

10.6 New Directions and Conclusions

The papers in Table 10.1 and in Ah-King and Gowaty (ms) suggest that an individual's ecological, developmental, social and endogenous circumstances may matter more for their mating decisions than the fancy traits of their potential mates. Is there is a different function to a bearer's fancy traits than beguiling females? Might an important function be simply increasing the bearers' encounters with potential mates, increasing the range of *their* acceptances and rejections. The SPT predicts that some fancy traits may evolve to increase the bearer's encounters with mates with potential mates with whom they will have higher quality offspring, thereby enhancing their individual fitness via increases in the quality of their offspring or via the lottery ticket mechanism increasing the likelihood of healthy offspring (Gowaty and Hubbell 2009; Gowaty et al. 2010).

Adaptive responses of individuals to their likelihood of survival, their likelihood of encounters with potential mates are also sources of information to potential mates. What do potential mates learn about another who hides or outwits a predator? Mate preferences tuned to environmental and social conditions can explain the maintenance of heritable variation in sexually selected traits, such as, genetic diversity in cricket mating calls (Hedrick and Dill 1993), body size variation (Basolo 2004), and genetic variation of colour patterns in time cycles such as in the side-blotched lizard, *Uta stansburiana* (Alonzo and Sinervo 2001).

When individuals with different phenotypes are successful breeders during different years, preferences may change accordingly. For example, male traits correlated with female acceptance of potential mates in lark buntings (*Calamosiza melanocorys*) shift dramatically between years (Chaine and Lyon 2008). Furthermore, the traits that females prefer correlate with high nesting success, suggesting that fitness indicators switch between years (Chaine and Lyon 2008). In lark buntings what do changes in probabilities of high nesting success do to males' acceptances or rejections? Adaptive variation, such as shifting between preferred traits, may reduce or even eliminate male trait evolution.

We are beginning to understand acceptances and rejections of potential mates and other reproductive decisions as integral parts of life-history. For example, female guppies (*Poecilia reticulata*) under predation pressure either accept more mates or become altogether sexually unreceptive (Godin and Briggs 1996). These are two extreme responses, and suggest that there is underlying variation among females that may influence these two extremes. Thus, animals adjust their life histories in response to ecological factors to maximize lifetime reproductive success.

Furthermore, acceptances and rejections may differ between different social, ecological or genetic contexts (Fig. 10.3). Both males and females often mate multiply and the criteria for acceptance of a social partner and extra-pair partners may differ. For example, female pied flycatchers accept as primary mates

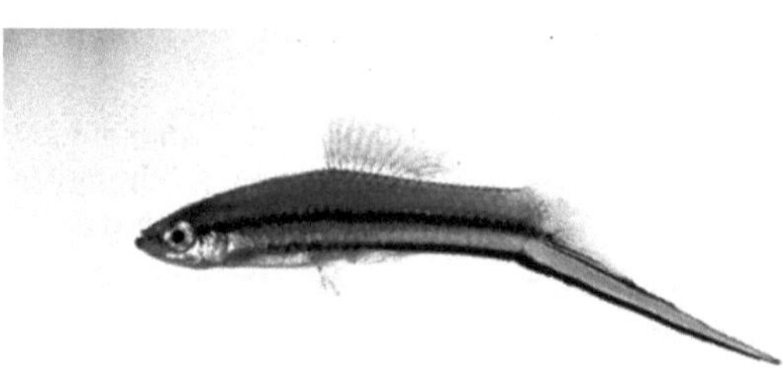

Fig. 10.3 In an experiment with the green swordtail *Xiphophorus helleri*, females were shown videorecordings of males with and without swords. Females first preferred long-sworded males, but after having seen a long-sworded male being eaten by a predator, their preferences switched to males without swords (Johnson and Basolo 2003). Photo by the Xiphophorus Genetic Stock Center, Texas State University–San Marcos, San Marcos, Texas

males with good territories or an ornament signaling good parental abilities, but when females solicit extra pair copulations their mating decision may be based on genetic qualities alone (Qvarnstrom et al. 2000). Sometimes, it might even be advantageous to have a social partner of the same sex. In black swans, same-sexed male pairs have higher breeding success, are more aggressive, have larger territories and share incubation time more evenly than opposite sex pairs (Bagemihl 1999).

Multiple mating also makes cryptic female acceptances and rejections possible (Eberhard 1996). Very little is known about the extent to which females determine the outcome of sperm competition, which possibly also varies with environmental, social and intrinsic factors. For example, female soldier flies (*Merosargus cingulatus*) regulate the timing of oviposition and thereby determine paternity (Barbosa 2009). Since there is last male precedence in *M. cingulatus*, ovipositioning directly after mating gives a high number of offspring for the last mated male.

Predictions of flexible acceptance or rejection of potential mates (Gowaty and Hubbell 2009) are also corroborated in post-copulatory studies of A&R, in what may be called cryptic male A&R of potential mates. An example is that dominant male fowls *(Gallus gallus)* transfer more quality sperm to ornamented females, while subdominants do not adjust ejaculates as a function of female ornaments (Cornwallis 2007).

Since the expression of sexually selected traits are context-dependent, benefits from choosing the most ornamented partner might differ between environments. Genes that are good in one environment might have a negative effect in another. One potentially fruitful way to proceed would be to investigate effects of genes and environment on acceptance and rejection decisions. Is there a reaction norm for acceptance and rejection of mates? Or as the SPT quantitatively and theoretically predicts, it is possible that individuals' assessment of the quantitative rank of potential mates does not change on an axis of their best to their worst mate, but their switch along that axis may change as the demographic circumstances (their encounters, their survival likelihood) change.

Much of recent research on sexual selection has focused on females mating with males relative to their fancy traits. But the evidence for sexual selection characters correlating with fitness benefits is contradictory (Cornwallis and Uller 2009). With a different approach, in a series of experiments on mallards, fruit flies and mice, Gowaty and co-workers showed that mate preferences in both males and females predict offspring survival (Gowaty et al. 2007). The investigators randomly picked "choosers" from sets of breeding age virgins and evaluated their assessment behaviour when experimentally exposed to two potential mates also picked at random with respect to phenotypes. Thus, the investigators were blinded to the traits of potential mates that mediated the choosers' assessments. The point of the experiments was to inform the fitness effect for choosers independent of knowledge of the mediating traits. After testing for choosers' assessments, the investigators placed choosers randomly with either the potential mate they preferred or did not preferred and compared the resulting fitness of breeders. Choosers with their preferred partners had higher offspring viability.

In the future investigations of sexual selection would draw subjects at random with respect to their sex, simultaneously testing behaviour of females and males, controlling environmental conditions that may affect individual variation in reproductive decision making. One could then simultaneously address sex differences and sex similarities to evaluate if adaptations of reproductive decision-making are sex-limited adaptations or flexible adaptive responses that can be induced in individuals independent of their sex. In addition, the role of stochastic demography in reproductive decision-making should be more frequently included in studies of sexual selection. Could it be that ecological constraints on individual's time for mating, and not fixed sex differences *per se*, result in sex differences in mate choice? It is therefore important to investigate whether sex differences in ecological constraints, *e.g*. lower survival of female butterfly larvae (Darwin 1871), are related to systematic differences in reproductive decisions. So, if individuals of different sexes have different exposure to pathogens or predators or have different feeding niches, these may be associated with differences in survival, encounter rate, time outs before remating, number of potential mates or *w-distribution* leading to usual differences in reproductive decisions that are nevertheless induced by variation in the same cues with different values? Hence, the perceived fixed-typical sex differences in acceptance and rejection patterns observed in nature may have more to do with consistent ecological and social constraints on time available for mating than with the assumptions of fixed sex differences that are key to understanding the D-B-P-T linkages. The view from the "reaction norms of sex" gives priority not simply to genes, but to the interactions of genes, epigenetic effects, and the environmental conditions that induce flexible behaviour, perhaps by the rules of the SPT. The SPT is consistent with the sex as a reaction norm perspective particularly with adaptively flexible behaviour, and it provides novel, quantitative (not just qualitative), testable predictions about the nature of acceptance and rejection behaviour and interactions between the sexes. It provides a tool to distinguish between causal factors for sexual strategies in females and males, enabling crucial tests of contrasting predictions between traditional sexual selection predictions and alternative theories (Gowaty 2012).

So what is left of sexual selection? We suspect that there remains much to investigate, and that the basic structure of Darwin's most mature definition of sexual selection (broad sense) still seems a useful guide for discerning if broad sense selection occurs. Recall that the basic description of the broad-sense selection process—variation in within-sex traits and selection pressures leading to differential within-sex reproduction—remains intact, no matter what we call specific within-sex hypotheses (Gowaty 2011; Gowaty this volume). What seems missing from many tests of sexual selection is thorough testing of all the assumptions of given sexual selection hypotheses, as most tests are partial leaving questions about either (1) within-sex trait variation, (2) the complete set of selective mechanisms, and (3) the components of fitness that selection acts through. In addition new within-sex hypotheses predict additional mechanisms of selection including sex-neutral mechanisms. Classical formulations of sexual selection can now be contrasted to alternative theories, enabling tests that may resolve questions about sex differences that have dogged researchers since Darwin (Gowaty 2012).

Phenotypic plasticity in mate choice is a relatively new field of study. New investigations are currently accumulating on context- and condition-dependent acceptance and rejection of the sex with the higher parental investment or the bigger gametes (Ah-King & Gowaty ms). The new studies show that in contrast to expectations from PI and anisogamy theories within-sex phenotypic plasticity and individual flexibility occur. Yet, an open question is related to the distribution and abundance of individual adaptive flexibility.

Beginning with sex differences to explain further sex differences is sometimes a circular argument. By using gender-neutral assumptions we may distinguish effects on fitness variances depending on fixed differences, stochastic demographic effects and effects due to selection by whatever name (Gowaty and Hubbell 2009; Gowaty 2012).

Sex, like many other phenotypic traits, such as predator effects on the morphology of crucian carp (*Carassius carassius*) (Brönmark and Miner 1992), is a reaction norm. Environmental and social circumstances affect sex determination, traits associated with one sex and sexual strategies. As we have emphasized here another reaction norm is individually adaptive acceptance and rejection of potential mates.

We conclude with a call for more attention to morphology, physiology, and behaviour as developmentally plastic and/or individually flexible in the context of (1) within-sex variation, (2) mechanisms of within-sex reproductive competition, and (3) their associated components of fitness, that is, in the context of sexual selection.

Acknowledgements Great thanks to Thierry Hoquet and the organizers, presenters and participants of the conference on "What's left of sexual selection?" in Paris (May 2011). We also thank Stephen P. Hubbell for many conversations and for his inspiring original models. MA was supported by the GenNa-program at the Centre for Gender Research, Uppsala University during this work, by a grant from the Royal Swedish Academy of Sciences and the Swedish Research Council. PAG acknowledges previous career awards from NIH and NSF Grants IBN-9631801, IBN- 0911606, IOS-1121797 each of which supported empirical work testing ideas related to the models, which provided partial support during the years of developing and exploring the assumptions of her ideas *Attributions:* MA collected, curated, reviewed, and summarized citations on the substantial experimental literature on phenotypic plasticity in acceptance and rejection of potential mates. MA made tables of results. MA prepared Fig. 10.1; PAG prepared Fig. 10.2. MA and PAG wrote the paper.

References

Ah-King M (2011) Biologins paradox: föränderliga kön och rigida normer [The paradox of biology: flexible sexes and rigid norms.] Lambda Nordica 4:26–52

Ah-King M (2013). On anisogamy and the evolution of "sex-roles". Trends in Ecology and Evolution 28(1):1–2

Ah-King M, Gowaty PA ms Random and choosy mating: Within-sex developmental plasticity or individual flexibility?

Ah-King M, Nylin S (2010) Sex in an evolutionary perspective: just another reaction norm. Evol Biol 37:234–246

Alonzo SH, Sinervo B (2001) Mate choice games, context-dependent good genes, and genetic cycles in the side-blotched lizard, *Uta stansburiana*. Behav Ecol Sociobiol 49:176–186
Altmann J (1997) Mate choice and intrasexual reproductive competition: contributions to reproduction that go beyond acquiring more mates. In: PA Gowaty (Ed) Feminism and evolutionary biology, boundaries, intersections and frontiers. Chapman & Hall, New York, pp 320–333
Arnqvist G, Nilsson T (2000) The evolution of polyandry: multiple mating and female fitness in insects. Animal Behav 60:145–164
Arnqvist G, Rowe L (2005). Sexual conflict. Princeton University Press
Bagemihl B (1999) Biological exuberance: animal homosexuality and natural diversity. Profile Books, London
Barbosa F (2009) Cryptic female choice by female control of oviposition timing in a soldier fly. Behav Ecol 20:957–960
Basolo AL (2004) Variation between and within the sexes in body size preferences. Animal Behav 68:75–82
Bateman AJ (1948) Intra-sexual selection in drosophila. Heredity 2:349–368
Beckage NE (ed) (1997) Parasites and pathogens: effects on host hormones and behaviour, volume 1. Chapman & Hall, New York
Birkhead TR, AP Møller (1998) Sperm competition and sexual selection. Academic Press, San Diego
Bleiweiss R (2001) Asymmetrical expression of transsexual phenotypes in hummingbirds. Proc Royal Soc Lond B 268:639–646
Bonduriansky R (2001) The evolution of male mate choice in insects: a synthesis of ideas and evidence. Biol Rev 76:305–339
Bonduriansky R (2012) Rethinking heredity, again. Trends Ecol Evol 27:330–336
Brönmark C, Miner JG (1992) Predator-induced phenotypical change in body morphology in crucian carp. Science 258:1348–1350
Bull JJ (1983) Evolution of sex determining mechanisms. Reading, Mass. The Benjamin/Cummings Publishing Company Inc, San Francisco
Chaine AS, Lyon BE (2008) Adaptive plasticity in female mate choice dampens sexual selection on male ornaments in the lark bunting. Science 319:459–462
Clutton-Brock TH (2007) Sexual selection in males and females. Science 318:1882–1885
Clutton-Brock TH, Parker GA (1992) Potential reproductive rates and the operation of sexual selection. Q Rev Biol 67:437–456
Cornwallis CK, Uller T (2010) Towards an evolutionary ecology of sexual traits. Trends Ecol Evol 25:145–152
Crews D, McLachlan JA (2006) Epigenetics, evolution, endocrine disruption, health, and disease. Endocrinology 147(6)(Supplement):S4–S10
Darwin C (1859) The origin of species. Murray, London
Darwin C (1871) The descent of man and selection in relation to sex. Murray, London
Dewsbury DA (2005) The darwin-bateman paradigm in historical context. Integr Comp Biol 45:831–837
Dreger A (2007) Sex beyond the karyotype. In: DL Kleinman, KA Cloud-Hansen, C Matta, J Handelsman (eds) Controversies in science and technology, volume 2: from chromosomes to the cosmos. Mary Ann Liebert, Inc
Eberhard WG (1996) Female control: sexual selection by cryptic female choice. Princeton University Press, Princeton and Chichester
Fausto-Sterling A (2000) Sexing the body: gender politics and the construction of sexuality. Basic Books, New York
Forsgren E, Amundsen T, Borg ÅA, Bjelvenmark J (2004) Unusually dynamic sex roles in a fish. Nature 429:551–554
Ghiselin MT (2003) The triumph of the darwinian method. Dover Publications Inc, Mineola
Gilbert SF (2001) Ecological developmental biology: developmental biology meets the real world. Develop Biol 233:1–12

Godin JJ, Briggs SE (1996) Female mate choice under predation risk in the guppy. Animal Behav 51:117–130

Gotthard K, Nylin S (1995) Adaptive plasticity and plasticity as an adaptation: a selective review of plasticity in animal morphology and life history. Oikos 74:3–17

Gowaty PA (1996) Multiple mating by females selects for males that stay: another hypothesis for social monogamy in passerine birds. Animal Behav 51:482–484

Gowaty PA (1997) Sexual dialectics, sexual selection and variation in reproductice behaviour. In: PA Gowaty (Ed) Feminism and evolutionary biology, boundaries, intersections and frontiers. Chapman & Hall, New York, pp 351–384

Gowaty PA (2003) Sexual natures: how feminism changed evolutionary biology. Signs 28:901–921

Gowaty PA (2011) What is sexual selection and the short herstory of female trait variation. Behav Ecol 22:1146–1147

Gowaty PA (2012) In press. A new theoretical framework for sex-role science. In: MLF Fisher, JR Garcia, RS Chang (Eds) Evolution's empress: darwinian perspectives on the nature of women. Oxford University Press, New York

Gowaty PA, Buschhaus N (1998) Ultimate causation of aggressive and forced copulation in birds: female resistance, the code hypothesis, and social monogamy. Am Zool 38:207–225

Gowaty PA, Hubbell SP (2005) Chance, time allocation, and the evolution of adaptively flexible sex role behaviour. Integr comp biol 45:931–944

Gowaty PA, Hubbell SP (2009) Reproductive decisions under ecological constraints: it's about time, Proc Natl Acad Sci USA 106(Suppl 1):10017–10024

Gowaty PA, Hubbell SP (2010) Killing time: a mechanism of sexual selection and sexual conflict. In: A Cordoba-Aguilar, J Leonard (Eds) The evolution of primary sexual characters in animals. Oxford University Press, New York, pp 79–96

Gowaty PA, Hubbell SP (2013a) Bayesian animals sense ecological constraints to predict fitness and organize individually flexible reproductive decisions. Behav Brain Sci 36(3):215–6

Gowaty PA, Hubbell SP (2013b) The evolutionary origins of mating failures and multiple mating. Entomologia Experimentalis et Applicata 146(1):11–25

Gowaty PA, R Steinichen, WW Anderson (2003) Indiscriminate females and choosy males: within- and between-species variation in drosophila. Evolution 57:2037–2045

Gowaty PA, Anderson WW, Bluhm CK, Drickamer LC, Kim Y-K, Moore AJ (2007) The hypothesis of reproductive compensation and its assumptions about mate preferences and offspring viability. Proc Natl Acad Sci USA 104:15023–15027

Gowaty PA, Kim Y-K, Rawlings J, Anderson WW (2010) Polyandry increases offspring viability and mother productivity but does not decrease mother survival in Drosophila pseudoobscura. Proc Natl Acad of Sci USA 107:13771–13776

Gowaty PA., Kim Y-K, Anderson WW (2012) No evidence of sexual selection in a repetition of Bateman's classic study of Drosophila melanogaster. Proceedings of the National Academy of Sciences USA, In press

Griffith SC, Owens IPF, Thuman, KA (2002) Extra pair paternity in birds: a review of interspecific variation and adaptive function. Mol Ecol 11:2195–2212

Gwynne DT, Simmons LW (1990) Experimental reversal of courtship roles in an insect. Nature 346:172–174

Hayes TB, Khoury V, Narayan A, Nazir M, Park A (2010) Atrazine induces complete feminization and chemical castration in male african clawed frogs (Xenopus laevis). Proc Natl Acad Sci USA 107:4612–4617

Hedrick AV, Dil LMl (1993) Mate choice by female crickets is influenced by predation risk. Animal Behav 46:193–196

Hrdy SB (1981) The woman that never evolved. Harvard University Press, Cambridge

Hrdy SB (1986) Empathy, polyandry and the myth of the coy female. In: R Bleier (Ed) Feminist approaches to science. Pergamon Press, New York, 119–146

Hubbell SP, Johnson LK (1987) Environmental variance in lifetime mating success, mate choice, and sexual selection. Am Nat 130:91–112

Jiggins FM, Hurst GDD, Majerus MEN (2000) Sex-ratio-distorting wolbachia causes sex-role reversal in its butterfly host. Proc Royal Soc B-BiolSci 267:69–73.

Johnson LK (1982) Sexual selection in a brentid weevil. Evolution 36(2):251–262

Johnson LK (1983) Reproductive behaviour of *Claeoderes bivittata* (Coleoptera: Brentidae). Psyche 90(1–2):135–149

Johnson LK, Hubbell SP (1984) Male choice—experimental demonstration in a brentid weevil. Behav Ecol Sociobiol 15:183–188

Johnson JB, Basolo AL (2003) Predator exposure alters female mate choice in the green swordtail. Behav Ecol 14:619–625

Kokko H, Jennions MD (2008) Parental investment, sexual selection and sex ratios. J Evol Biol 21:919–948

Mank JE, Promislow DE, Avise JC (2006) Evolution of alternative sex-determining mechanisms in teleost fishes. Biol J Linnean Soc 87:83–93

Parker GA (1970) Sperm competition and its evolutionary consequences in the insects. Biol Rev 45:525–567

Parker GA, Baker RR, Smith VGF (1972) The origin and evolution of gamete dimorphism and the male-female phenomenon. J Theoritical Biol 36:529–553

Poulin R (1994) Mate choice decisions by parasitized female upland bullies, *Gobiomorphus breviceps*. Proc Royal Soc Lond Ser B-Biol Sci 256:183–187

Qvarnström A, Tomas P, Sheldon BC (2000) Adaptive plasticity in mate preference linked to differences in reproductive effort. Nature 405:344–347

Rhainds M (2010) Female mating failures in insects. Entomologia Experimentalis et Applicata 136(3):211–226

Rice WR (1996) Sexually antagonistic male adaptation triggered by experimental arrest of female evolution. Nature 381:232–234

Rodríguez-Muñoz R, Amanda B, Tregenza T (2010) Natural and sexual selection in a wild insect population. Science 328:1269–1272

Safran RJ, Vitousek MN, Hauber ME, Ghalambor CK (2010) Sexual selection: a dynamic state of affairs response to the comments of cornwallis and uller in the article: towards an evolutionary ecology of sexual traits. Trend Ecol Evol 25:429–430

Schärer L, Rowe L, Arnqvist G (2012) Anisogamy, chance and the evolution of sex roles. Trend Ecol Evol 27, 260–264

Schlichting CD, Pigliucci M (1998) Phenotypic evolution, a reaction norm perspective. Sinauer Associates, Sutherland

Small MF (1993) Female choices; sexual behaviour of female primates. Cornell University Press, Ithaca

Snyder BF, Gowaty PA (2007) A reappraisal of bateman's classic study of intrasexual selection. Evolution 61:2457–2468

Stearns SC, Hoekstra RF (2005) Evolution: an introduction. Oxford University Press, Oxford

Stutchbury BJM, Neudorf DL (1998) Female control, breeding synchrony, and the evolution of extra-pair mating systems. In: PG Parker, NT Burley (eds) Avian reproductive tactics: female and male perspectives. Washington DC, American Ornithologists' Union, pp 103–121

Sutherland WJ (1985) Chance can produce a sex difference in variance in mating success and account for bateman's data. Animal Behav 33:1349–1352

Tang-Martinez Z, Ryder TB (2005) The problem with paradigms: Bateman's worldview as a case study. Integr Comparative Biol 45:821–830

Trivers RL (1972) Parental investment and sexual selection. In: B Campbell (ed) Sexual selection and the descent of man. Aldine, Chicago, pp 136–179

West-Eberhard MJ (1987) The epigenetical origins of insect sociality. In: E Rembold (ed) Chemistry and biology of social insects. Verlag J. Peperny, München

West-Eberhard MJ (2003) Developmental plasticity and evolution. Oxford University Press, New York

Part III
Prospects: Animal Aesthetics?

Chapter 11
The Role of Sexual Autonomy in Evolution by Mate Choice

Richard O. Prum

Abstract The field of sexual selection is dominated by research in which natural selection on mating preferences is assumed. As a result, there has been no need to recognize any of the consequences of mate choice that can be independent of natural selection. One such consequence is the evolution of sexual autonomy- defined here as the capacity of an individual organism to pursue its mating preferences independent of sexual coercion from the opposite sex. Here, I propose that the concept of sexual autonomy was implicit in Darwin's work on mate choice, as evidenced by early criticisms of sexual selection by Darwin's' contemporaries. Subsequently, Fisher, Lande, and Kirkpatrick provided models of the origin of sexual autonomy through mate choice. Here, I propose that sexual autonomy evolves via the indirect costs of sexual coercion, and can involve evolution of either resistance to coercion or new aesthetic preferences. I review two distinct examples of this phenomenon—antagonistic genital coevolution in waterfowl (*Anatidae*) and the evolution of bowerbird (*Ptilonorhynchidae*) architecture—both of which protect females from sexual coercion. Waterfowl vaginal complexity evolves as a mechanism of resistance to protect females from unwanted forced fertilization, and reinforces female capacity for autonomous mate choice. Bowerbird bowers evolve by female preference, and function to physically protect females from male sexual attack. Female preferences for bower architecture provide an aesthetic mechanism for reinforcing female sexual autonomy. Recognizing the evolution of female sexual autonomy provides a new avenue for investigating the evolution by the indirect benefits of mate choice.

Keywords Sexual conflict · Sexual coercion · Forced copulation · Aesthetic evolution · Mating preferences · Antagonistic coevolution · Waterfowl · Bowerbirds

R. O. Prum (✉)
Department of Ecology and Evolutionary Biology, and Peabody Museum of Natural History, Yale University, New Haven, Connecticut, USA
e-mail: richard.prum@yale.edu

T. Hoquet (ed.), *Current Perspectives on Sexual Selection,* History, Philosophy and Theory of the Life Sciences 9, DOI 10.1007/978-94-017-9585-2_11

11.1 Introduction

In the past 30 years, the science of mate choice has developed from an obscure topic to a major field of evolutionary biology. There have been extensive theoretical and empirical debates, and substantial progress in understanding the mechanisms of evolution of intersexual traits and preferences. Yet, I think the field of sexual selection is in a substantial intellectual crisis (Prum 2010, 2012). Through the wholehearted embrace of adaptive mechanisms of mate choice–i.e. good genes and direct benefits–and the *a priori* rejection of arbitrary Fisherian and sensory/cognitive bias mechanisms, the majority of sexual selection researchers have abandoned testability in favor of the confirmation of adaptative mate choice paradigm (Prum 2010, 2012). Display traits are assumed to have evolved adaptive signal information or adaptive design. Failure to confirm these assumptions is interpreted merely as failure to have yet demonstrated what the field "knows" to be true. The result is a weak, confirmationist science, and a distorted view of the importance of adaptation in nature (Prum 2010, 2012). I have proposed the formal adoption of the Lande-Kirkpatrick (LK) mechanism as the null model of the evolution of trait and preferences. The LK null model assumes the existence of genetic variation in display traits and mating preferences and the absence of natural selection on mating preferences (Prum 2010). Use of a null hypothesis in sexual selection will insure that adaptive mate choice is actually tested rather than accepted as a matter of faith.

In parallel, I have advocated that the full incorporation of the LK null into sexual selection research will also reinstate Darwin's original aesthetic concept of sexual selection by mate choice in to modern evolutionary biology (Prum 2012). By entertaining the possibility, indeed the likelihood, of the evolution of arbitrary sexually selected traits, we reestablish Darwin's view of aesthetic evolution as including the evolution of traits that are merely attractive and provide no adaptive benefit to individuals who prefer them (Prum 2012). However, aesthetic evolution explicitly includes the mechanisms of honest signaling through encoding information about good genes and direct benefits as well as the LK null process.

I follow Darwin (1871) in conceptualizing sexual selection as a distinct evolutionary mechanism from natural selection. Thus, the indirect, Fisherian benefit of sexually attractive offspring is not due to natural selection, and is therefore not adaptive. Adaptation is a term that should be exclusively used to refer to evolution by natural selection. The key distinction between adaptive mate choice and an arbitrary aesthetic evolution by sexual selection is whether or not preferences come under natural selection (Prum 2010, 2012).

Recently, Patricia Brennan and I have been analyzing the relationship between sexual selection by mate choice, sexual selection by intrasexual competition, and sexual conflict over fertilization. Based on our analyses of sexually antagonistic coevolution in waterfowl genitalia which occurs through sexual conflict over fertilization (Brennan et al. 2007, 2010), we have proposed a vital role for the indirect costs of sexual coercion in species with simultaneous mate choice and male-male competition through coercion (Brennan and Prum 2012). If sexual coercion leads to forced fertilization in females with evolved mating preferences, then females will have male offspring that lack preferred male display traits. These offspring will be less successful in subsequent competition to attract female mates creating an indirect cost to females of sexual coercion. Recognition of the indirect costs of sexual coercion as a mechanism for the evolution of female resistance to coercion has been previously obscured by inappropriately and unproductively narrow conception of sexual conflict (Brennan and Prum 2012). Thus, genital coevolution in waterfowl is extremely different from classic examples of sexually antagonistic genital coevolution through sexual conflict, such as bedbugs (Cimicidae). Some male bedbugs have evolved a body piercing intromittent organ; females in these lineages have co-evolved reinforcing body armor and other anatomical adaptations to resist traumatic insemination injury (Carayon 1966, Stutt and Siva-Jothy 2001, Arnqvist and Rowe 2005). Whereas resistance to sexual coercion in bedbugs functions to prevent direct fitness losses and perhaps as a barrier to be overcome by the 'best,' most coercive males, the derived genital morphologies of female ducks evolve in order to reassert the opportunity for their individual mate choices for specifically preferred males. Based largely on invertebrate models which lack clear female mating preferences, narrow-sense sensory bias have synonymized female preference and resistance, and male display and coercion. (For a notable exception, see Alonzo 2008). The framework fails to explain many obvious examples of sexual conflict (Brennan and Prum 2012), including primates and humans (see numerous data sets in Muller and Wrangham 2009).

In this chapter, I further explore additional concepts at the interface between mate choice, intrasexual competition, sexual coercion, and sexual conflict. Again, this exploration is necessitated by the adoption of a null hypothesis framework in which mate choice is not assumed to be adaptive, and therefore distinct evolutionary process that are not accounted for by natural selection can arise. Specifically, I will propose that the concept of *sexual autonomy* should play an important role in our understanding of evolution by sexual selection. I argue that the concept of sexual autonomy was nascent in Darwin's work. Furthermore, recognizing the concept of sexual autonomy is required to constructively understand the relationship between mate choice, intrasexual competition, and sexual conflict over mating and fertilization.

Here, I propose a scientific concept of sexual autonomy, and I review the historical origins of the concept within Darwin's work and in the explicit attacks on sexual autonomy by Darwin's early critiques. I then present verbal description of the role of the indirect costs of sexual coercion on the evolution of sexual autonomy. I illustrate the concept with discussions of two distinct examples of the evolution of sexual autonomy through the indirect costs of coercion from two very different avian breeding systems–waterfowl (*Anatidae*) and bowerbirds (*Ptilonorhynchidae*). My goal is to establish that the recognition of sexual autonomy in evolutionary biology does productive intellectual work, and clarifies the complex relationship between both mechanisms of sexual selection and sexual conflict over reproduction.

11.2 Sexual Autonomy and Sexual Agency

An important development in the revived study of sexual selection in the late twentieth century has been the explicit advocacy for adopting female perspectives (e.g. Hrdy 1981, Smuts 1985, Gowaty 1997a). However, this process has not been completed. Gowaty (1997b) has criticized the general failure to fully incorporate "proactive female agency" into sexual selection theory and analysis. But there has not been, to my knowledge, an attempt to define the concepts of sexual agency and sexual autonomy as evolving properties of mating systems, and appropriate theoretical concepts in sexual selection.

I define *sexual autonomy* as the capacity of an organism to pursue its individual mating preferences independent of sexual coercion from the opposite sex. Sexual autonomy does not constitute complete sexual control, or release from sexual competition within a sex. In contrast, sexual preferences may not be realized because of competition within one sex for access to preferred mates of the other sex. Thus, rejection by a preferred mate is not a limit, constraint, or threat to sexual autonomy. Rather, sexual autonomy can be constrained by sexual coercion in which other individuals attempt to physically control an individual's sexual behavior, force fertilization (including rape), or otherwise disrupt, intimidate, or interfere with the independent sexual preference of another individual.

What organisms can exhibit sexual autonomy? Inherent in the definition of sexual autonomy is the concept of individual sexual agency–the sensory capacity to detect variation among potential mates, the cognitive capacity to evaluate those sensory experiences and to have an individual sexual motivation, and the physical capacity to act on those motivations. It is possible that many organisms may lack the cognitive capacity for independent sexual agency or the ecological opportunity to exercise it. However, even cognitively complex organisms that have no evolved sexual preferences would also lack motivation necessary for sexual agency. Therefore, the absence of sexual agency and consequent sexual autonomy might arise because of the lack of necessary cognitive or neural complexity, or because of the absence of the evolved mating preferences at all. Lastly, sexual autonomy may not exist in a given species for a third reason–the complete dominance or control over fertilization by male-male competition or sexual coercion. In this case, there may be cognitive capacity, physical capacity, and even evolved mating preferences, but there still may be an absence of social/ecological opportunity to act upon them because of the success of sexual coercion.

11.3 The Darwinian Roots of Sexual Autonomy

The concept of sexual autonomy has been inherent in discussion of sexual selection since Darwin's (1871) Descent of Man, but it has failed to be recognized explicitly in sexual selection theory. Darwin's theory of evolution by mate choice relied explicitly on individual aesthetic responses to courtship display traits. Darwin hypothesized that courtship displays would "charm" and "delight" female sensibilities, resulting in differential sexual success among males that would lead to the evolution of secondary sexual ornamentation. Darwin lacked a complete theory of the evolution of preferences, but Fisher (1915, 1930) initiated one that was outlined in detail by Lande (1981) and Kirkpatrick (1982). Darwin (1871) explicitly recognized sexual selection as a distinct evolutionary mechanism from natural selection, in contradiction to nearly all current conceptions.

Darwin avoided rhetorical recognition of sexual autonomy through mate choice, perhaps out of personal anxiety over the radical social implications of the concept. However, his theory of sexual selection clearly proposed it. In essence, Darwin concluded that female sexual autonomy has been a major force in the evolution of biodiversity including humans. Furthermore, Darwin was clear that evolution by mate choice was a potentially independent force that *could not* be completely subsumed within, or synonymized with, natural selection (Prum 2012).

Even if Darwin avoided stating the more radical implications of his theory with an explicit concept of sexual autonomy, its implication was not lost on his early critics. Contemporary reviewers criticized both the proposal of female sexual autonomy and its potential independence from natural selection. For example, in his 1871 review of *The Descent of Man*, St. George Mivart (1871, p. 55, emphasis added) wrote of sexual selection in insects:

> Mr. Darwin gives a number of instances of sexual characters, such as horns, spines, etc., in beetles and other insects; but there is no fragment of evidence that such structures are in any way due to feminine caprice.

Of Darwin's report of the personal preferences of a captive peafowl for a specific pied male, Mivart (1871, p. 59, emphasis added) wrote:

> [S]uch is the instability of *vicious feminine caprice*, that no constancy of coloration could be produced by its selective actions.

Mivart's vocabulary requires further analysis, because the meanings of both vicious and caprice have changed in common English usage over the last 140 years. In our modern sense, the word vicious means deliberately violent, nasty, or dangerous, but its original meaning was immoral, depraved, or wicked (Soanes and Stevenson 2005). Vicious meant having the characteristics of vice. *Caprice* has come to refer to an entertaining fancy or light-hearted whim, but its original meaning was a prompt, arbitrary "turn of mind made without apparent or adequate motive" (Soanes and Stevenson 2005).

Thus, in response to Darwin's proposals on the existence and evolutionary consequences of female sexual autonomy, Mivart characterized the proposed female mating preferences in strongly culturally-conditioned language, as immoral, sinful, depraved, and lacking appropriate justification.[1] Because of his low evaluation of female cognitive abilities, including women, Mivart further concluded that female

[1] The last of Mivart's objections to Darwin's aesthetic theory of mate choice remains the main criticism of contemporary evolutionary biologists to the aesthetic LK mechanism as a null model. Adaptationist biologists regard the arbitrary, self-organized evolution of mating preferences in the

sexual preferences would be so malleable that they would never result in the evolution of the peacocks tail. Fickle females will want one thing 1 min and another the next, and could never constitute a coherent evolutionary force. In summary, Mivart concluded that instability and inconsistency of feminine caprice would prevent this mode of action from being evolutionarily effective.

Mivart's language confirms that his attack is aimed at Darwin's proposal of female sexual autonomy among "brute" animals:

> "even in Mr. Darwin's specially-selected instances, there is not a tittle of evidence tending, however slightly, to show that any brute possesses the representative reflective faculties." "...it cannot be denied that, looking broadly over the whole animal kingdom, there is no evidence of advance in mental power on the part of brutes." (1871, pp. 75–76)

Mivart asserts that animals lack the requisite cognitive complexity, or "free will" required for sexual agency, and thus autonomy. He concludes that animals cannot be active players, or selective agents, in their own sexual evolution because they lack the requisite cognitive power or free will to make sexual choices based on sensory evaluation.

Mivart also established another successful future intellectual trend by apparently being the first person to view Darwin as a traitor to his own legacy. By proposing his aesthetic theory sexual selection by mate choice in *The Descent of Man*, Darwin was betraying to the true adaptationist cause of Darwinism:

> The assignment of the law of 'natural selection' to a subordinate position is virtually an abandonment of the Darwinian theory; for the one distinguishing feature of that theory was the all-sufficiency of 'natural selection.

This comment provides clear evidence from a Victorian contemporary of Darwin that Darwin viewed sexual selection as an evolutionary mechanism distinct from natural selection, in direct contradiction to the view of modern "Darwinism" (Prum 2012). If, as Mivart claimed, Darwin assigned natural selection to a subordinate position, we should ask, "Subordinate to what?" The answer is that Mivart thought Darwin placed natural selection as subordinate to the immoral, arbitrary, and independent force of female sexual autonomy. If independent female sexual agency could affect the outcome of mating, and therefore mate choice constituted an evolutionary force independent of natural selection, then natural selection was subordinate to the "vicious feminine caprice." This was unacceptable.

St. George Mivart was English Roman Catholic biologists and anatomist, and an early vocal supporter of Darwin's theory of natural selection, but he broke with Darwin over the issue of sexual selection. From our modern perspective, Mivart's moralistic language and theistic perspective may seem scientifically irrelevant, but his use of these rhetorical tools documents well the socially radical content of Darwin's hypothesis of female mate choice and sexual autonomy as an evolutionary force. Regardless, Mivart's scientific criticisms were influential, and were largely echoed by Darwin's adversary over sexual selection–Alfred Russel Wallace.

---

absence of additional adaptive as an insufficient, illegitimate, or inappropriate explanation of the origin of diversity in preference.

A. R. Wallace had a complex response to Darwin's proposal of sexual selection that strengthened over the years into a strong critique (for an excellent historical review see Cronin 1991). For many reasons, Wallace argued that sexual selection by mate choice was unlikely to occur and should be rare. However, Wallace was unable to reject the evolutionary mechanism entirely (Cronin 1991). When mate choice did occur, Wallace (1895, pp. 378–379) declared that:

> The only way in which we can account for the observed facts is by supposing that colour and ornament are strictly correlated with health, vigor, and general fitness to survive.

Although Wallace is largely credited with destroying Darwin's theory of the evolution by mate choice, this quote demonstrates that Wallace is actually the first person to clearly articulate the modern adaptive mate choice mechanism of the evolution of mating preferences through traits that indicate good genes and direct benefits (Prum 2012). These seemingly contradictory facts can be reconciled by understanding that our modern adaptationist view of sexual selection is essentially Wallacean, and not Darwininan. The strictly adaptationist view of mate choice is as stridently anti-Darwinian today as it was when articulated by Wallace more than a century ago (Prum 2012). Interestingly, Wallace called the strikingly modern mechanism of the evolution of ornaments that are "strictly correlated with health, vigor, and general fitness to survive" natural selection not sexual selection (Prum 2012).

Like Mivart, Wallace saw Darwin's proposal of sexual selection by mate choice as an abandonment of Darwin's primary accomplishment–their shared accomplishment of a greater, all powerful theory of natural selection. In the introduction to his book Darwinism, Wallace (1889, pp. xi–xiii) wrote,

> Even in rejecting that phase of sexual selection depending on female choice, I insist on the greater efficacy of natural selection. This is pre-eminently the Darwinian doctrine, and I therefore claim for my book the position of being the advocate of pure Darwinism.

Here, Wallace claimed to be more purely Darwinian than Darwin. His insistence on the greater efficacy of natural selection is essentially the same intellectual position that characterizes contemporary adaptive mate choice research which involves the largely untested, universal assumption of differential cost to mate preferences, and the a priori rejection of a non-adaptive, aesthetic null model of sexual selection (Prum 2010, 2012).

Wallace ultimately objected to the same broad elements of Darwin's aesthetic view of sexual selection as Mivart–the existence of female sexual autonomy, the possibility of purely aesthetic traits, and the potential for evolutionary mechanisms independent of natural selection. Like Mivart, Wallace attacked the plausibility of female mating preferences by arguing that female animals were cognitively incapable of mate choice. Like Mivart, Wallace was a believer in the divine creation of human beings, but he had a different take on human specialness. While denying mate choice in other animals, Wallace reserved the capacity for mate choice to human females. In an odd version of proto-feminist eugenics, Wallace went on to argue that women were capable of mate choice, and he foresaw that female choice for indicators of quality in humans would contribute to rational and eugenic improvement of humans. He advocated that the advance of higher education, financial independence, and suffrage for women would result in eugenic improvement of the human species.

In summary, Wallace's maintained that there were no evolutionary consequences of mate choice that were not congruent with natural selection and, therefore, the concept of sexual selection by mate choice should be abandoned and subsumed *within* natural selection. His strongly expressed advocacy for a purer "Darwinism" in the absence of aesthetic mate choice and female sexual autonomy came to dominate evolutionary biology for nearly a century. Thus, from these contemporaneous critiques of Darwin's *Descent of Man*, we can see that fear of female sexual autonomy contributed to the placement of sexual selection by mate choice within natural selection, and therefore under the control of natural selection. Furthermore, this unitary conceptual framework led to the exclusion of Darwin's arbitrary, aesthetic conception of mate choice from evolutionary biology.

One of the few notable exceptions to the abandonment sexual selection by mate choice for the next century was in the work of R. A. Fisher (1915, 1930). Fisher really proposed two verbal models–one for the evolutionary origin of mating preference for traits that are an initial, accidental 'index' of superiority; and a second model for the evolutionary elaboration of female preferences and evolutionary decoupling of the trait from any initial quality information. Fisher asserted that, once mating preferences have evolved, the action of those mating preferences will erode the traits original value as an index of superiority, and therefore the eliminate the natural selection in favor of the preference itself. But trait and preference will still coevolve through the indirect benefit of sexually attractive offspring. Fisher hypothesizes that the mere existence of mating preferences will unhinge subsequent preference evolution from their original naturally selected advantages. Accordingly, mating preferences will evolve purely through their indirect sexual selection through their genetic correlation with the trait.

Fisher proposed that mating preferences do not have to evolve because the particular male that the female chooses is any better than any male. In fact, preferred males may be worse (Lande 1981; Kirkpatrick 1982). But if the female's male offspring inherit his attractive trait, then her preference will increase in the population through her offspring's sexual success. The advantage that drives the evolution of preference does not accrue directly to the female's own survival, her number of offspring, or their survival. Rather the advantage to preference accrues through her offspring's fitness, specifically through the fitness advantage of having sexually attractive male offspring.

Fisher's recognition of the evolution of mating preferences through indirect sexual selection is an early demonstration of the evolutionary origin of sexual agency–i.e. the capacity for mate choice–and its evolutionary consequences–i.e. indirect selection on preference leading to arbitrary elaboration and the loss of adaptive value of traits or preferences. Fisher's verbal theory and quantitative models of mate choice by Lande (1981) and Kirkpatrick (1982) document why a concept of sexual autonomy is required. If the action of mate choice can lead to the evolutionary elaboration of traits and mating preferences independent of natural selection on preferences, then the consequences of mate choice cannot be entirely described as aspects of natural selection. The active agents in this additional, potentially arbitrary and non-adaptive, evolutionary process are the individuals themselves making autonomous mating decisions.

Of course, organisms may make adaptive mate choices as a result of sexual autonomy. The concept of sexual autonomy does not exclude the possibilities of adaptive mate choice for traits that indicate good genes or direct benefits. However, the good genes and direct benefits models do exclude the possibility of no natural selection on mating preferences. In this regard, modern adaptive models of the evolution of mate choice follow Wallace's view that sexual selection is merely a form of natural selection; mate choice is another means to the same adaptive goal. Consequently, if mate choice is always adaptive, there is never the need to recognize sexual agency or autonomy of the individuals making mate choices. If mate choice behavior can be entirely understood and characterized by adaption through natural selection, then evolutionary theory can be viewed as complete. Thus, the lack of recognition of sexual autonomy in sexual selection research is a modern repetition of Wallace's rejection of Darwin's hypothesis of sexually autonomous mate choice selection as an independent evolutionary mechanism. Why bother conceptualizing female sexual autonomy when the phenomenon can already be clearly understood through natural selection alone? Of course, this is the view that effectively shut down progress in understanding sexual selection for over a century. I think that this is not a productive path for the field to take.

11.4 The Evolution of Sexual Autonomy

Sexual autonomy is a manifestation of mating preferences, but it is not synonymous with mating preferences. Rather, sexual autonomy deals with the additional aspects of the interaction between mating preferences and sexual coercion. So, when and how does sexual autonomy evolve? I maintain that sexual autonomy evolves through indirect sexual selection on either resistance or mating preferences.

Before elaborating, it is important to state that the evolution of resistance by natural selection to reduce the direct harm of sexual coercion can occur, but it does not enhance sexual autonomy. This is the traditional mechanism of the evolution of resistance in narrow-sense sexual conflict (e.g. Arnqvist and Rowe 2005). However, the ultimate control over fertilization remains male-male competition. Females merely act in passive response to reduce the direct fitness costs of being a victim of male sexual coercion. Regardless of the contributions that this form of resistance to female fitness, this mechanism does not enhance female sexual autonomy–i.e. the opportunity to exercise independent mate choice free from coercion.

Now, to think about the evolution of sexual autonomy, let's imagine a species with a system of arbitrary male display traits and female preferences, and a simultaneous independent strategy by unsuccessful males to obtain fertilizations by direct sexual coercion of females. The result will be trait and preference coevolution via an LK null process, but that process will be constrained not by natural selection on preferences but by male sexual coercion. Some proportion of offspring will be fathered by coercive copulations which violate sexual autonomy of individual females. Brennan and Prum (2012) hypothesize that females whose freedom to choose is violated by direct sexual coercion will have male offspring who fail to inherit genes for that female's preferred values of male display traits. Consequently, these male offspring will not be preferred by other females in the population, who have evolved similar mating preferences as the mother. Because the male offspring of sexually coerced fertilizations will not enjoy the benefits of sexually advantageous display traits, this creates an indirect fitness cost to the sexually coerced female. Since the display traits are arbitrary and confer no other advantages beyond the advantage of being sexually attractive, the indirect cost of sexual coercion can obviously be recognized as *sexual selection* on female phenotype to avoid or evade this indirect fitness cost of coercion. However, this conclusion is still true if the preferences themselves are under natural selection for good genes indicators.

How are we to define and distinguish mating preference from resistance to coercion in the context of sexual conflict over fertilization? In general, I define a mating preference as an affirmative, behavioral/cognitive mating response to a non-coercive sexual stimulus. Mate choice is a cognitive process (Ryan et al. 2009). That's why it is called choice. Any use of the term "mate choice" that does not actually involve a cognitive decision is an abuse of the concept of choice. Thus, this definition excludes various biochemical or anatomical mechanisms of post-copulatory intersexual selection which have been called 'cryptic mate choice' (see below). In contrast, resistance is a response to a physical, behavioral, or biochemical attempt to coerce or force fertilization by subverting the autonomous process of female choice. Resistance to sexual coercion functions to reduce the direct harm of coercion or to reinstate the opportunity for affirmative mate choice. Resistance does not constitute a form of choice. In this framework, cryptic female choice would be a form of resistance, or a form of non-cognitive mate selection that is not choice. Decades of sloppy application of the concept of mate choice to include non-cognitive mechanisms of selection do not make the practice correct.

From the other (frequently male) perspective, a display trait is any component of the phenotype which is evaluated by, and subject to, uncoerced mating preferences. Sexual coercion, on the other hand, has been defined as the use of force, or the threat of force, against another individual that increases the probability of fertilization (Smuts and Smuts 1993). Thus, coercion constitutes invasive control over autonomous mating preferences. I would expand this definition to include not just physical force (i.e. behavior) but also other biological mechanisms of control including anatomical, physiological, and biochemical mechanisms of coercion (Clutton-Brock and Parker 1995).

An intellectual impediment to the recognition of female sexual autonomy is the current "narrow sense" sexual conflict framework (described in Brennan and Prum 2012). Advocates of sexual conflict in the narrow sense would likely reject these definitions of trait, preference, coercion, and resistance as artificial, unproductive, and unnecessary. They advocate that the entire complex phenomenon of mating interactions should be viewed as a single, simplistic, intellectually "flat" landscape–the mating bias function–lacking in any true distinctions between preference and resistance, or display and coercion. Although it may be mathematically simpler or intellectually expedient to view all forms of mating bias as "two sides of the same coin" (e.g. Gavrilets et al. 2001; Kokko 2005), this biologically naive framework simply fails to account for the complexity of behavioral and anatomical complexity of organismal phenotype (Brennan and Prum 2012). Further, the narrow sense view of sexual conflict defines out of existence the indirect costs of sexual coercion by recognizing only direct harm in the definition of sexual conflict (e.g. Arnqvist and Rowe 2005).

11.5 Evolution of Resistance or Choice Can Further Sexual Autonomy

Sexual autonomy evolves indirectly through sexual selection itself. To understand how, let's return to our imaginary population described above in which arbitrary female mate choice and male sexual coercion are simultaneously occurring, and male coercion disrupts female choice and creates an indirect costs of sexual coercion over fertilization. The result will be selection on females to evolve mechanisms to evade these indirect costs by neutralizing male sexual coercion. I want to distinguish between two types of evolutionary responses to the indirect costs of sexual coercion, which can affect sexual autonomy. One should be easy to understand (but will still pose an intellectual struggle for those who define sexual conflict solely in terms of direct harm), whereas the other is somewhat more novel.

One mechanism for the evolution of sexual autonomy is the evolution of resistance. Behavioral, anatomical, physiological, or biochemical resistance can evolve in order to alter the equilibrium between male coercion and female mate choice in favor of female sexual autonomy. Resistance to sexual coercion evolves because females who can prevent sexually coerced fertilization via resistance will have the opportunity to mate with males with preferred traits. They will avoid the indirect costs of coerced fertilization. Their genes for resistance will therefore evolve through the indirect Fisherian advantages gained by the attractive male offspring that result from their autonomous choices. In this manner, female resistance can evolve to reduce the effectiveness of sexual coercion and to expand sexual autonomy. The evolved function of this form of resistance is not to reduce the direct harm of sexual coercion nor to replace mate choice, but to reinstate the conditions for mate choice to operate autonomously.

Another possible outcome of the indirect costs of sexual coercion is the evolution of aesthetic preferences that further sexual autonomy. This process is also indirect. Imagine a preference for an aesthetic male display trait that is either arbitrary or an indicator of good genes, but is not an indicator of any direct benefit; however, this male trait is also associated with reduced efficiency or capacity for male sexual coercion. If the population is in an active state of sexual conflict over fertilization (either through an ongoing, coevolutionary arms race, or in an evolutionary equilibrium between sexual coercion and mate choice), then any variation in mating preferences that increases female sexual autonomy will evolve because such preferences will create an additional advantage to female preference against male sexual coercion. Such preferences evolve not only because of the pure Fisherian benefit of having sexually attractive offspring, but because of the additional benefit of expanding the *opportunity for autonomous female choice* by evading the constraints established by male sexual coercion. Thus, mating preferences that have an ancillary benefit of advancing female interests in ongoing sexual conflict with males will evolve because they expand choice against male control. We can even imagine that sexually coercive males are more fit (i.e. have higher survival, etc.) or have better genes than the males that females prefer; however females will still evolve to preferences for arbitrary traits that reinforce their sexual autonomy because of the sexual advantages of attractive offspring, as shown by Lande and Kirkpatrick. Because this mechanism operates by the indirect cost of sexual coercion, this mechanism will also apply to female preferences for traits that indicate good genes.

By this indirect evolution mechanism, a female does not prefer the specific aesthetic trait to advance her own individual autonomy because she must already have sexual autonomy in order to enact her a mating preference. Further, by this mechanism, a female is not using her sexual autonomy to look for a subordinate social partner that she can socially dominate in order, for example, to use that social control to mate with other more genetically desirable males. Such a process would only lead to the evolution of mating preferences for the males that were actually preferred. Thus, the evolution of sexual autonomy is an indirect consequence of mate choice, and not a direct benefit to the female who actually makes a specific choice. During a coevolutionary arms race or equilibrium between male sexual coercion and female sexual autonomy, arbitrary mating preferences that incidentally expand freedom of mate choice will evolve because they tip the balance in favor of female determination resulting in an expansion in female sexual autonomy.

What determines whether sexual autonomy evolves by resistance, by aesthetic mate choice, or fails to evolve at all? The answer lies in the specific details of the biology of each species, and the dynamic interplay between the evolved responses by males and females to sexual conflict and sexual selection.

In summary, mate choice creates the opportunity for sexual autonomy through the independent sexual agency of individuals. Once established, sexual autonomy can evolve to reinforce itself through the evolution of mechanisms of resistance or aesthetic preferences for display traits that reduce male sexual coercion. In the following sections, I will review two contrasting examples of this process from the breeding systems of living birds–the coevolution of anatomical mechanisms of resistance and forced copulation in waterfowl (Anatidae), and the evolution of aesthetic bower architecture in polygynous bowerbirds (*Ptilonorhynchidae*).

11.6 Sexual Coercion and Genital Coevolution in Waterfowl

The breeding systems of waterfowl include both female mate choice based on male secondary sexual displays, and male coercion through violent, forced extra-pair copulations (FEPCs). Female preference and resistance behaviors are clearly differentiated phenotypically from one another, as are male display behavior and sexual coercion behavior. In other words, there is no difficulty distinguishing any of them from the others.

Many males and female waterfowl have elaborate genitalia that reveal a strong and dynamic pattern of antagonistic coevolution (Brennan et al. 2007). Unlike most birds, male waterfowl have retained the primitive archosaur penis. However, the waterfowl penis has numerous distinct features including explosively rapid lymphatic erection, counter-clockwise corkscrew shape, and a wide variety of keratinized surface features from ribs and ridges to fully developed spiky hooks (Brennan et al. 2010). In ducks, penile erection and vaginal intromission are the same event, and take about one third of a second to accomplish (Brennan et al. 2010). Brennan et al. (2007) demonstrated that penis size and surface elaboration has coevolved with previous unknown waterfowl vaginal complexity. Females of many ducks have one or more cul-du-sac out-pocketings just inside the vagina from the cloaca, and sometimes one or more clockwise twists in the vagina further from the cloaca. These vaginal elaborations have coevolved with longer and more elaborate penises, and higher rates of sexual coercion through forced copulations (Brennan et al. 2007). The phylogenetic pattern of genital coevolution in waterfowl is complicated and dynamic, with multiple independent instances of coevolutionary genital elaboration and diminution in different waterfowl lineages (Brennan et al. 2007). These results provide strong evidence of sexually antagonistic coevolution through sexual conflict over fertilization in waterfowl.

In subsequent experimental work on the erection mechanism of duck penis, Brennan et al. (2010) showed that the explosively rapid male intromission proceeded normally in straight or counter-clockwise spiraling glass tubes, but was greatly impeded by glass tubes of the same diameter but with a 135° bend or a clockwise spiral. This evidence documents that the shapes of the coevolved waterfowl vaginal morphologies pose functional challenges to male erection/intromission. Further, Brennan et al. (2010) documented that during solicited, voluntary copulations, female ducks dilate the muscles of the cloaca, presumably to facilitate intromission.

Waterfowl provide strong example of sexually antagonistic coevolution over mating. However, the data are strongly incongruent with the standard "narrow sense" view of sexual conflict as defined solely by the direct harm of sexual coercion to the female (Gavrilets et al. 2001; Arnqvist and Rowe 2005; Fuller et al. 2005; Chapman 2006). Female waterfowl strongly resist FEPCs, and are not infrequently killed in the process (McKinney et al. 1983). The genital coevolutionary responses of female waterfowl to male sexual coercion do not function in reducing direct harm to the female. Rather, behavioral and anatomical resistance in female waterfowl have evolved by sexual conflict in the broad sense–which includes both direct and indirect costs and benefits, and recognizes the independence of preferences and resistance, and traits and coercion.

This hypothesis is further supported by genetic data which show that even in species with very high rates of FEPCs, the fertilization success of FEPCs is extremely low. In the Ross's Goose (Chen rossii) and Lesser Snow Goose (*Chen caerulescens*) respectively, FEPCs accounted for 33 and 38 % of all successful copulations but they only result in 2–5 % of young (Dunn et al. 1999). In wild mallards, where up to 40 % of observed copulations can be forced copulations (Evarts and Williams 1987), only 3 % of offspring were produced by "extra-pair" copulations (Burns et al. 1980). In wild Gadwalls (*Anas strepera*), forced copulations are relatively common, but only 4 % of offspring were produced by extra-pair copulations (Peters et al. 2003). It is unlikely that within pair copulations are significantly underreported in these studies, because within pair copulation in waterfowl is associated with conspicuous pre- and post-copulatory display behavior (e.g. Johnson 2000).

Brennan and Prum (2012) proposed the critical role of selection on females to avoid the indirect costs of forced fertilizations in the evolution of complex vaginal morphologies. They further proposed that elaborations in penis length and surface structure evolve by male-male competition, or intrasexual selection, as males compete with each other for coercive fertilizations. Anatomical features that advance the efficiency of coercive fertilization create indirect fitness losses to females because they will have offspring from males that have not been preferred. Male offspring that do not inherit preferred display trait values will be less likely to be chosen by other females who can choose, resulting in indirect fitness losses to the female who is the victim of sexual coercion. Consequently, females will be selected to evolve anatomical and behavioral resistance.

But what is the nature of this selection? Females are suffering the loss of fitness due to the absence of the sexual advantages of attractive traits in their offspring. These traits could include good genes and direct benefit indicators, but they could also include arbitrary aesthetic traits which provide no additional naturally selected advantage. In the latter case, regaining the indirect fitness advantages of an arbitrary mating preference cannot be seen as a form of natural selection. Although mathematical models of such a evolutionary process have yet to be formulated, it is clear that the indirect fitness advantages of an autonomous preference for an arbitrary trait would still exist if the sexually coercing males were fitter than the preferred males. Therefore, the selection for the evolution of resistance acting through the indirect benefits of mating preferences is sexual selection, not natural selection. This makes this mechanism of the evolution of resistance even more distinct from natural selection for elimination of the direct harm of sexual coercion, which is featured in sexual conflict in the narrow sense (Arnqvist and Rowe 2005).

Brennan and Prum (2012) also discuss the hypothesis that female genital anatomy in waterfowl has not evolved in order to screen coercive males and maximize the indirect benefits of coercion to females (Adler 2009). This "resistance as choice" hypothesis essentially proposes that forced copulation and resistance is adaptive for females, because the female will have offspring from the most competitive and successful coercers, and her male offspring will inherit those genes for being successful as sexual coercion (Eberhard 1996, 2002; Eberhard and Cordero 2003). Yet again, the genetic data cited above documents the overwhelmingly strong selection against fertilization by forced copulation achieved by female genital morphologies. Furthermore, the idea of an indirect benefit to sexual coercion predicts that many females should have clutches composed largely or entirely of offspring from extra-pair copulations, yet this is never found. Thus, the paternity data demonstrate the great unlikelihood that there is any residual indirect benefit to females from forced copulations (Brennan and Prum 2012).

Interestingly, the indirect benefits of sexual autonomy only occur if females control can determine the specific identities of their mates. Thus, the antagonistically coevolved genital morphologies can only evolve if the males that females prefer are *different individuals* from those males that are successful at sexual coercion. If they were not different, there would be no indirect benefit to excluding them from fertilizing their eggs. This fact presents a conundrum to advocates of universal adaptive sexual selection under all circumstances. Frequently, it is simultaneously claimed that male-male competition will result in an adaptively "better" male mate, and that female mate choice will also result in an adaptively "better" male mate. But the situation in waterfowl demonstrates that these males cannot be the same individuals. The Panglossian belief in adaptive sexual selection leads to an evolutionary conundrum. If all outcomes are adaptive, then how do you explain sexual conflict over the specific outcome of mating?

Although the reality of all indirect benefits of mate choice have been seriously questioned due to the absence of robust evidence of good genes (e.g. Kotiaho and Puurtinen 2007), the evolution of genital morphologies that protect the loss of female sexual autonomy provide excellent evidence of indirect benefits of mate choice. Currently, the evolution of female preferences in waterfowl are entirely consistent with the conclusion that the males that they prefer are not extrinsically better in anyway, other than that they are preferred by other females *and* they are not the same individuals that are most successful at sexual coercion. This implies that the features that females prefer in males are *not identical* to the characteristics that contribute to success at sexual coercion–including increased aggression, strength, vigor, body size, or penis size. Such attributes are frequently cited as correlated with male genetic quality. Thus, since these commonly hypothesized attributes of quality are not preferred by female waterfowl, it is entirely plausible that waterfowl preferences and displays constitute an arbitrary, aesthetic radiation in traits and preferences through the LK null sexual selection mechanism.

11.7 Bower Evolution

The bowerbirds (*Ptilonorhynchidae*) include 20 species in eight genera which are endemic to Australia and New Guinea (Frith and Frith 2004). They vary in body mass from 75 g (size of a thrush) to 250 g (size of a small crow). Like other famously polygynous avian lineages, such as birds of paradise, manakins and cotingas, bowerbirds feed largely on forest fruits. In 17 or 20 species, all of the parental care is done by the female, and males display in solitary territories in which they build an enduring courtship structure called a bower (Frith and Frith 2004).

A bower is a secondary sexual construction–a secondary sexual component of the extended phenotype of the bowerbirds. Coined by Richard Dawkins (1999), the extended phenotype includes all of the consequences of the organisms genome interacting with its environment, including its impacts on its environment. Dawkins defined the extended phenotype to document the full range of adaptive consequences of genes. However, if the extended phenotype becomes involved in secondary sexual display, then it can become subject to aesthetic evolution by mate choice which includes both adaptive and arbitrary sexual selection mechanisms. Here, we will explore the evolution of bower diversity among bowerbird lineages and the role of bower evolution in enhancing female sexual autonomy.

The most basal, or earliest branch of the bowerbirds, are the three species of catbirds (*Aileurodeus*). Catbirds are monogamous, have biparental care, and enduring pair bonds. Catbird nest construction is carried out exclusively by the female (Frith and Frith 2001). Thus, prior to the evolution of advanced male architectural capabilities, ancestral male bowerbirds likely had no role in nest construction. The architectural capacities of male bowerbirds are entirely a result of intersexual selection by mate choice on the extended phenotype after the evolutionary loss of the ancestral, monogamous pair bond.

The rest of the bowerbirds–the sister group to the catbirds–are polygynous with female only parental care. One rather basal, polygynous species, the Tooth-billed Bowerbird *Scenopeetes dentirostris*, does not build a bower at all, but merely clears a court a couple meters wide on the forest floor which he decorates simply with large green leaves spaced out from one another. A second rather derived species, the Archbold's Bowerbird *Archboldia papuensis*, builds a simple court covered by a veil of hanging vines. Archbold's Bowerbird decorates its courtyard with the pennant feathers from the King Saxony Bird of Paradise, the elytra of brilliant beetles, butterflies, and land snail shells, but makes no stick construction.

Males of the remaining bowerbird species make stick display constructions that can be classified into two natural groups, which share no other architectural features in common. The first is the avenue bower. A simple avenue bowers consist of a pair of parallel stick walls with a narrow pathway, or avenue, between them. On one or both ends of the central avenue, the male displays an assortment of found materials that vary tremendously among species and even among populations. In some species, it is fruit and leaves, in others, it is bones, insects and feathers. Often the materials are laid on a bed of straw or pebbles. There are a several derived variations in avenue bower architecture. The double-avenue bower made by Lauterbach's Bowerbird Chlamydera lauterbachii has two parallel paths on a raised platform, and the found materials are displayed within the avenues themselves. The Spotted Bowerbird *Chlamydera maculata* build a wide 'boulevard' bower in which the central pathway is especially wide and the 'walls' are formed by a transparent screen of sticks rather than a solid mass.

The second class of bowers consists of the maypole bowers. A maypole bower is built around a central sapling or small tree with a conical pile of horizontal sticks placed around this central support. The stack of brown sticks is broadest at the base and narrows at the top to form a structure that is like a conical bottle brush. Around the base of the maypole, the male has a circular path, or runway, which is clear of materials and allows for rapidly running around the maypole. The court around the runaway is decorated with gathered materials that can include flowers, fruits, beetles and butterflies, or fungus. In some species, the twigs are also decorated with hanging materials, like regurgitated fruit pulp or caterpillar frass.

As in the avenue bowers, there are additional, derived variations on the maypole bower. The Golden Bowerbird (*Prionodura newtoniana*) builds a double maypole bower with two conical stacks of sticks built up like twin peaks around neighboring saplings. The Vogelkop Bowerbird (*Amblyornis inornatus*) and the Streaked Bowerbird (*Amblyornis subalaris*) build hut bowers in which the central pole and the circular runway are mostly covered by thatched roof with an oval doorway. In hut bowers, the gathered objects are displayed outside the doorway of the court, frequently in well organized piles, sorted by color, size and texture.

Phylogenetically, court ornamentation evolved in the common ancestor of all polygynous bowerbirds, but it appears that the maypole and avenue bowers were independently evolved in two separate lineages (Kusmierski et al. 1993, 1997). Maypole and avenue bowers each have a single phylogenetic origin, but hut bowers are convergently evolved, and bower construction has been lost once in *Archboldia* (Kusmierski et al. 1993, 1997). Furthermore, male courtship in bowerbirds is variable among species. Typically males perform a series of energetic displays which are elaborate, but many lack the stereotyped form of birds of paradise or manakins. Studies by Gerry Borgia and others on multiple species have established that the features of the bower and the bower decorations are under strong sexual selection (Borgia 1985; Borgia et al. 1985; Uy and Borgia 2000; Uy et al. 2001; Patricelli et al. 2003; Madden and Balmford 2004). Thus, in bowerbirds, the extended phenotype has been evolutionarily co-opted into male sexual advertisement.

But why have bowers evolved at all? Why have bowers continued to diversify among species, and even populations, of bowerbirds? The literature has moved forward substantially from the mid-twentieth century hypothesis of "physiological coordination" between the sexes (Marshall 1954; Gilliard 1969; Diamond 1982) Initially, Gerry Borgia and colleagues hypothesized that the bowers were an indicator of male status, quality, and good genes (Borgia 1985; Borgia et al. 1985; Pruett-Jones and Pruett-Jones 1994). Despite great progress on understanding bowerbird courtship and mate choice, I do not think there is yet any evidence that female preferences are under natural selection in bowerbirds, and would propose that the entire radiation is consistent with the arbitrary Lande-Kirkpatrick null sexual selection process. Borgia's arguments against such a view are that (1) female mate choice is costly, (2) the LK mechanism falls apart with costs of mate choice, and (3) younger and older female bowerbirds have different preferences, which should limit the development of the genetic correlations between preference and trait that drive the LK mechanism. However, as Prum (2010, 2012) has described, it is not the existence of generalized costs of mate choice that are critical, but differential costs of mate choice among preferences.[2] There is no evidence of differential costs due to the specific, genetic components of female mating preferences in bowerbirds. Indeed, the differential mating costs between young and older females are explicitly not genetic costs (Uy et al. 2000). Further, recent modeling by Bailey and Moore (2012) documents that social and environmental flexibility in mate choice will enhance, not constrain, a Fisher process and the arbitrary evolution of mate choice. So, there is currently no evidence inconsistent with the view of bowerbirds as an arbitrary, aesthetic radiation. However, this discussion will focus instead on another hypothesized novel function and mechanism for the origin and evolution of bowers proposed by Gerry Borgia in 1995.

Borgia (1995) observed that the intense, energetic, and violent displays of the bowerbirds often startle or frighten visiting females. Borgia hypothesized that females are exposing themselves to the threat of forced copulations when they perch on the court to observe the male and his decorations as close range. He argued that bowers evolve through female preferences for protection from the threat of violent physical display, sexual coercion, and forced copulation. Borgia cited lots of natural history evidence in favor of this 'threat reduction' hypothesis. For example, because of either maypole or avenue bower architecture, if a male attempts copulation with a female before she is receptive, then the female is protected. Depending on the bower design, the female can either fly out of the front of an avenue bower when the male tries to mount her from behind, or she can hop to the side to maintain the central maypole between her and the threatening male. So, many varieties of bower architecture protect females from sexual attack, and thus provide a safe manner for observing a male, his energetic displays and song, and his collected materials at very close range. Given that the two classes of bower architecture appear to be independently evolved within bowerbirds (Kusmierski et al. 1993, 1997), Borgia's threat reduction hypothesis implies that this female protection function evolve twice independently within the family, establishing two, distinct, alternative protective designs–the maypole and the avenue.

Borgia (1995) further noted that female visits to the elaborate avenue bowers of male Satin Bowerbirds often last for several minutes, whereas in Tooth-billed Bowerbird (*Scenopetes dentirostris*) which have a simple, open court without a bower, a female arrives on the court only to mate, and she is immediately and aggressively mounted by the male. The longest observed female visit to a male's court is 3.8 s. Thus, mate choice in Tooth-billed Bowerbird is based on observation of the male and his ornaments at a greater distance. Consequently, the ornaments of the male Tooth-billed Bowerbird are simpler, larger, and displayed on a coarser spatial scale.

---

[2] We are all going to die, but that doesn't mean that we are under natural selection (Prum 2012). All biological existence involves costs, but that is not a description of the mechanisms of adaptive evolution. Individuals with different mating preferences must have differential costs. Commonly, the existence of generalized costs of mate choice are inappropriately assumed to be synonymous with differential costs. To my knowledge, these differential costs are frequently assumed and very rarely tested.

Borgia and Presgraves (1998) supported the threat reduction hypothesis of bower function in an investigation of the unique 'boulevard' bower of the Spotted Bowerbird (*Chlamydera maculata*). In the Spotted Bowerbird bower, the avenue is broader so that the female can sit sideways in the avenue. In association, the sides of the bower are made not made of a solid wall of sticks but are a thinner see-through screen of lighter straw. During visits to a male's bower, females sit sideways within the bower and to watch the male display through the transparent, side wall-screen of straw. Coevolved with the changes in bower architecture, the displays of male Spotted Bowerbirds are more physically energetic and aggressive than other *Chlamydera* species, including a rapid running rush toward the bower, which sometimes results in a male colliding with the bower itself. When Borgia and Presgraves destroyed one random wall of each male's bower, females continued to observe display males and the male continued to display to female through the remaining wall of the bower, rather than through the open side. Thus, Borgia and Presgraves concluded that the coevolution of the uniquely, derived bower structure and the more physically threatening male display behavior in Spotted Bowerbird confirmed the threat reduction hypothesis.

Gail Patricelli, Gerry Borgia, and colleagues subsequently developed a research program with robotic female models to explore the threat reduction hypothesis, and to investigate the role of female-to-male communication during courtship display (Patricelli et al. 2002, 2003, 2004a). They observed that female Satin Bowerbirds (*Ptilonorhynchus violaceus)* appeared to communicate their level of comfort with male display by crouching in the bower. They used robotic stuffed female bowerbird models with remotely controlled motors to produce natural looking crouching, looking around, and wing fluffing movements. By placing the robots in the bower and regulating its posture and movements, Patricelli et al. (2002) demonstrated male behavioral responsiveness to female crouching, indicating that the female behavior functions as a signal to males. Furthermore, males that were more responsive to female behavior during the robotic experiments were also less likely to startle wild females during natural visits by females to their bower. Lastly, controlling for variation in male responsiveness, Patricelli et al. (2002) showed that males with greater display intensity in the robotic experiments were also more sexually successful in natural courtship.

Patricelli et al. (2003) then showed that females were more tolerant of intense courtship display with attractive males (i.e. males that were ultimately chosen more frequently by other females). Further, females become more tolerant of intense display as their mate choice process– narrowing down to fewer and fewer, more attractive males– proceeds (Patricelli et al. 2004a). Patricelli et al. (2002, 2003, 2004a, 2004b) analyzed the function of female communication in Satin Bowerbirds solely in terms of the direct benefits to females. For example, Patricelli et al. (2004a) supported the prediction that effective communication (measured in reduced startling responses) made female mate searching more efficient. But, as an evolutionary explanation, this observation assumes that female bowerbirds are actually under natural selection to reduce their mate searching time, for which there is no evidence. However, Patricelli et al. (2004b) did propose that the advantages of mate search efficiency explicitly includes the reduction of risk of sexual coercion and forced copulation. Although they did not emphasize it as I will, the conceptual framework of Patricelli et al. is consistent with the role of the indirect benefits of sexual autonomy on the evolution of bower architecture.

In conclusion, as Borgia (1995) proposed, the evolutionary origin and radiation of bowers appears to be related to the bower's function in protecting females from sexual coercion and forced copulation while allowing them to observe male display and bower decorations at very close proximity. Borgia's threat reduction hypothesis has great explanatory power, but his proposed evolutionary mechanism has not explicitly identified the role of the indirect benefit of female sexual autonomy. Borgia (1995) hypothesized that bowers are worthwhile compromise for males–who trade fitness through forced copulations for fitness through female visitations. According to Borgia's model, high quality males gave up the benefits of forced copulation because they could still gain sexual advantages through bower display, and low quality males are forced to follow suit. But this model, assumes that the males that are successful at forced copulations are somehow the same males that are successful through female choice. However, as in waterfowl, sexual coercion only has an indirect cost if females actually prefer different males from those that are successful at coercion.

Borgia's (1995) model obscures the fact that the bowers likely evolved through female choice, and that bowers function to facilitate autonomous female choice. Those components of the variation in mating success that were ancestrally determined by male-male competition through forced copulations have been largely transferred to female control. Thus, the violence and risk inherent to mate choice in Tooth-billed Bowerbirds has been eliminated in the rest of the family. I assert that male bowerbirds have not been evolutionarily negotiated down toward some intersexual compromise. Rather, males have completely lost control over fertilization as a result of the evolution of new female mating preferences that have successfully established female sexual autonomy. Male reproductive variance is nearly entirely determined by female sexual choice, though males still contribute indirectly through competitive destruction and plundering of each others bowers and collections (Borgia and Mueller 1992; Pruett-Jones and Pruett-Jones 1994) and young marauding males also try to force copulations at other male's bowers (Uy et al. 2000). The fact that bowers continue to evolve and radiate in structure in ways that maintain female autonomy with ongoing evolution in display behavior (Borgia and Presgraves 1998) further demonstrates the persistence of the power of female sexual autonomy in the aesthetic evolution of bowers.

Refining Borgia's (1995) model, I propose explicitly that bowers have evolved through the indirect benefit to females of reducing male sexual coercion through forced copulations. If females pay a indirect fitness costs from forced copulations from non-preferred males because their male offspring will be less attractive, then variations in mating preferences will evolve which preserve the capacity to choose a mate autonomously without interference from sexual coercion. Beginning with an open male display arena with a few displayed objects, like the Tooth-billed Catbird female preferences for any ornamental structures that incidentally protected them from male attack would evolve because such preferences would provide additional leverage against male coercion. Thus, I hypothesize that bowers are a component of the male extended phenotype that has evolved (perhaps twice) by female mating preference because of the indirect benefits that bowers provide to female sexual autonomy over male coercion.

This hypothesis is confirmed by Patricelli's observations that female Satin Bowerbirds exhibit greater tolerance of risky, aggressive display behavior from more attractive males. It is only the indirect risks of sexual violence, not the direct risks of forced copulation, that are minimized by male attractiveness. If females experience threat from the direct harm of sexual attack, they should experience uniform discomfort in any sexual threatening situation because the amount of physical harm from attack should be unrelated to the attractiveness of the male. Or, if physical vigor is an indicator of male quality and vigorous attack is associated with greater harm to the female, then a female should experience greater discomfort with higher quality males. (But for a skeptical discussion of the prediction that vigor evolves as indicator of quality, see Prum 2012). In bowerbirds, however, females are more comfortable with the sexual risk experienced from more desirable males. I predict that this is because female bower preferences evolve through the indirect benefits of choice, and the indirect costs of coercion. If the costs of coercion are unattractive offspring, then females should get more and more comfortable with risk from sexually attractive mates, precisely as shown in Satin Bowerbird (Patricelli et al. 2003).

Lastly, Borgia noted that the protection of females from sexual threat by bower architecture is associated with the explosive diversification of bower ornamentation. By selecting on various types of architecture–avenue or maypole bowers–females have expanded their autonomy over fertilization and male efforts to subvert female mating preferences. The assertion of female sexual autonomy facilitates female capacity to inspect individual males and their gathered materials at intimate distances without loss of sexual control. The result has been a further, amazing differentiation in the breadth and diversity of bowerbird aesthetics. In this way, sexual autonomy facilitates the macroevolution of greater aesthetic diversity (Prum 1997).

11.8 Conclusions

In order to understand the complex interactions between mate choice, sexual coercion, and sexual conflict, it is necessary to recognize the concepts of sexual agency and sexual autonomy in evolutionary biology. The evolution of individual sexual agency through mate choice gives rise to the phenomenon of sexual autonomy. Sexual autonomy evolves by the indirect costs of sexual coercion. This mode of sexual selection can result in the evolution of female resistance to sexual coercion or in the evolution of aesthetic mating preferences with the ancillary contributions to sexual autonomy. Thus, the aesthetic norms established by female preference create a novel source of selection, which reinforces the autonomy of mate choice. In this manner, sexual autonomy has self-organizing, autocatalytic, or self-reinforcing capacities, which likely have broad implications for the evolutionary interactions between mate choice and sexual coercion in organisms that possess sexual agency.

The concepts of sexual agency and sexual autonomy were implicit in Darwin's *Descent of Man*. Darwin may not have articulated them directly out of concern for the radical social implications of his theory. However, sexual agency and sexual autonomy constituted major targets of early criticisms of the mechanism of evolution by mate choice. In particular, the conclusion that female animals were cognitively incapable of sexual agency or sexual autonomy became main elements of the rejection of sexual selection that continued for nearly a century.

Despite substantial interest in developing sexual selection theory that incorporates female sexual agency (e.g. Gowaty 1997b), it is not surprising that a modern Darwinian concept of sexual agency and autonomy have not emerged from contemporary literature dominated by adaptive views of sexual selection. As a consequence of the ubiquitous assumption of natural selection on mating preferences, which is inherent in most research on good genes and direct benefit mechanisms, adaptation is thought to provide a complete account of the evolutionary consequences of female mating preferences. There has been no need to recognize any evolutionary agency at the level of individual mate choice, because all these consequences are already accounted for by natural selection on mating preferences. But the recognition of the possibility of arbitrary mate choice, through either the LK null sexual selection mechanism or sensory/cognitive biases independent of natural selection, requires that we identify and conceptualize a new "evolutionary player" within sexual selection theory. This new player is the independent agency of mate choice.

I have provided two examples of the evolution of sexual autonomy in action. One concerns the evolution of physical, anatomical resistance to enhance sexual autonomy in waterfowl, and the other concerns the evolution of aesthetic bower preferences to further sexual autonomy in polygynous bowerbirds.

Despite pervasive, violent sexual coercion, many species of waterfowl have coevolved complex vaginal morphologies that limit the fertilization success of forced copulations. These structures have most likely evolved through sexual selection for mechanisms of resistance to forced copulation to preserve and expand female sexual autonomy. Thus, sexually antagonistic coevolution by sexual conflict in waterfowl is notably distinct from the earlier examples of the phenomenon because female counter measures do not function in limiting direct harm to females. These resistance anatomies are evolving not by direct natural selection to limit female harm, but by indirect sexual selection to prevent sexually unattractive male offspring that result from coercion.

The narrow sense sexual conflict paradigm creates several intellectual impediments to understanding sexual conflict in species with complex cognitive abilities, distinct mating preferences, mechanisms of resistance, display traits, and mechanisms of coercion (Brennan and Prum 2012). Genital coevolution in waterfowl is distinct from sexually antagonistic genital coevolution in bedbugs (Cimicidae)

(Carayon 1966; Stutt and Siva-Jothy 2001; Arnqvist and Rowe 2005), because resistance to sexual coercion in bedbugs functions to prevent direct fitness losses and perhaps as a barrier to be overcome by the 'best,' most coercive males. In contrast, the derived genital morphologies of female ducks evolve in order to reassert the opportunity for their own autonomous mate choices. Thus, recognition of the concept of sexual autonomy in evolutionary biology facilitates analysis of the origin of resistance through the indirect costs of coercion. It is the prior existence of mating preferences in waterfowl that create the additional source of sexual selection when sexual autonomy is constrained by male coercion.

In bowerbirds, I propose that females have evolved aesthetic preferences for bower architectures that protect them from the indirect costs of sexual assault and forced copulation. Borgia's (1995) formulation of the "threat reduction hypothesis" did not explicitly recognize the role of the indirect costs of sexual coercion. Once recognized, these indirect costs provide an efficient evolutionary mechanism for origin and evolutionary radiation of novel preferences for bower architectures, which assert and maintain female sexual autonomy against male sexual attack.

What do females do with sexual autonomy? An important general evolutionary consequence of female autonomy is the evolutionary elaboration of ornament. In both waterfowl and bowerbirds, females use their sexual autonomy to make aesthetic mate choices. The result is an evolutionary radiation in male intersexual ornament including plumage, song, display behavior, and even extended phenotypic constructions and curated collections of found objects. Sexual autonomy has unexpected macroevolutionary consequences (Prum 1997). Sexual autonomy begets ornamental diversity.

Many workers in sexual selection have expressed skepticism over sexual selection by indirect benefits, usually because of the difficulties of confirming good genes hypotheses (e.g. Kotiaho and Puurtinen 2007). This conclusion can be made because often these researchers have failed to entertain even possibility of an arbitrary LK null sexual selection process and that the evidence of good genes is likely genuinely weak (Prum 2010, 2012). However, examples of the indirect costs of sexual coercion provide strong support for the indirect benefits of mate choice. The sexually antagonistic convolution in waterfowl genitalia provides an excellent example of the indirect benefits of mate choice. Because these genital structures function not in limiting direct harm, but by reducing unwanted fertilizations they can only evolve if there is an evolutionary advantage to mating with a specific, preferred individual. Preferred males must provide indirect fitness benefits to females in order for these complex vaginal structures to have evolved. In the case of bowerbird bowers, the evidence is somewhat less strong given that these structures could have evolved entirely through adaptive sexual selection on direct benefits. However, it is very hard to explain the female protective function of the bower as a direct benefit given that female bowerbirds show greater comfort with more attractive males who would likely provide the same or greater direct harm if they were to attempt a sexual attack. Thus, recognizing the evolution of female sexual autonomy provides a whole new avenue for exploring the evolution by indirect benefits of mate choice.

Acknowledgments I would like to thank Patricia Brennan who brought her interest in duck research to my lab, and thereby inspired this new understanding of the importance of sexual autonomy and its evolution. I would like to thank Thierry Hoquet of the Department of Philosophy of the University of Lyon 3 for organizing and inviting me to participate in this symposium. I am grateful to the BIOSEX project (sponsored by the French *Agence Nationale de la Recherche*), and to the research program "L'Humain en devenir" (sponsored by the University of Paris West) for their financial support to attend the symposium in Paris. The manuscript was greatly improved by reviews from Gerry Borgia, Patricia Brennan, and Gail Patricelli. However, I remain solely responsible for any errors, confusions, or omissions in the paper.

References

Adler M (2009) Sexual conflict in waterfowl: why do females resist extra-pair copulation? Behav Ecol 21:182–192

Alonzo SH (2008) Female mate choice copying affects sexual selection in wild populations of the ocellated wrasse. Animal Behav 75:1715–1723

Arnqvist G, Rowe L (2005) Sexual conflict. Princeton University Press, Princeton

Bailey NW, Moore AJ (2012) Runaway sexual selection without genetic correlations: social environments and flexible mate choice initiate and enhance the Fisher process. Evolution 66:2674–2684

Borgia G (1985) Bower quality, number of decorations and mating success of male satin bowerbirds (*Ptilonorhynchus violaceus*): an experimental analysis. Animal Behav 33:266–271

Borgia G (1995) Why do bowerbirds build bowers? Am Sci 83:542–547

Borgia G, Mueller U (1992) Bower destruction, decoration stealing, and female choice in the spotted bowerbird *Chlamydera maculata*. Emu 92:11–18

Borgia G, Presgraves DC (1998) Coevolution of elaborated male display traits in the spotted bowerbird: an experimental test of the threat reduction hypothesis. Animal Behav 56:1121–1128

Borgia G, Pruett-Jones SG, Pruett-Jones MA (1985) The evolution of bower-building and the assessment of male quality. Zeitschrift für Tierpsychology 67:225–236

Brennan PLR, Prum RO (2012) The limits of sexual conflict in the narrow sense: new insights from waterfowl biology. Proc Royal Soc Lond B 367:2324–2338

Brennan PLR, Prum RO, Mccracken KG, Sorenson MD, Wilson RE, Birkhead TR (2007) Coevolution of male and female genital morphology in waterfowl. PLoS ONE 2:e418

Brennan PLR, Clark CJ, Prum RO (2010) Explosive eversion and functional morphology of the duck penis supports sexual conflict in waterfowl genitalia. Proc Royal Soc Lond B 277:1309–1314

Burns JT, Cheng KM, Mckinney F (1980) Forced copulation in captive mallards. I. Fertilization of eggs. Auk 97:875–879

Carayon J (1966) Traumatic insemination and paragenital system. In: Usinger RL (ed) Monograph of the Cimidae (Hemiptera, Heteroptera). Entomological Society of America, Maryland

Chapman T (2006) Evolutionary conflicts of interest between males and females. Curr Biol 16:R744–R754

Clutton-Brock TH, Parker GA (1995) Sexual coercion in animal societies. Animal Behav 49:1345–1365

Cronin H (1991) The Ant and the Peacock. Cambridge University Press, Cambridge

Darwin C (1871) The descent of man, and selection in relation to sex. John Murray, London

Dawkins R (1999) The Extended Phenotype. Oxford University Press, Oxford

Diamond JM (1982) Evolution of bowerbirds' bowers: animal origins of the aesthetic sense. Nature 297:99–102

Dunn PO, Afton AD, Gloutney ML, Alisauskas RT (1999) Forced copulation results in few extra-pair fertilizations in Ross' and Lesser Snow Geese. Animal Behav 57:1071–1081

Eberhard WG (1996) Female control: sexual selection by cryptic female choice. Princeton University Press, Princeton
Eberhard WG (2002) The function of female resistance behavior: intromission by coercion vs. female cooperation in sepsid flies (Diptera: sepsidae). Rev Biol Trop 50:485–505
Eberhard WG, Cordero C (2003) Sexual conflict and female choice. Trends Ecol Evol 18:439–440
Evarts S, Williams CJ (1987) Multiple paternity in a wild population of mallards. Auk 104:597–602
Fisher RA (1915) The evolution of sexual preference. Eugen Rev 7:184–191
Fisher RA (1930) The genetical theory of natural selection. Clarendon, Oxford
Frith CB, Frith DW (2001) Nesting biology of the spotted catbird, *Ailuruedus melanotis*, a monogamous bowerbird (*Ptilonorhynchidae*), in Australian wet tropical upland rainforests. Aust J Zool 49:279–310
Frith C, Frith D (2004) The bowerbirds. Oxford University Press, Oxford
Fuller RC, Houle D, Travis J (2005) Sensory bias as an explanation for the evolution of mating preferences. Am Nat 166:437–446
Gavrilets S, Arnqvist G, Friberg U (2001) The evolution of female mate choice by sexual conflict. Proc Royal Soc Lond B 268:531–539
Gilliard ET (1969) Birds of paradise and bowerbirds. Natural History Press, New York
Gowaty PA (ed) (1997a) Feminism and evolutionary biology. Chapman and Hall, New York
Gowaty PA (1997b) Sexual dialectics, sexual selection, and variation in reproductive behavior. In: Gowaty PA (ed) Feminism and evolutionary biology. Chapman and Hall, New York
Hrdy SB (1981) The woman who never evolved. Harvard University Press, Cambridge
Johnson KP (2000) The evolution of courtship display repertoire size in dabbling ducks (Anatini). J Evol Biol 13:634–644
Kirkpatrick M (1982) Sexual selection and the evolution of female choice. Evolution 82:1–12
Kokko H (2005) Treat 'em mean keep 'em -sometimes keen: evolution of female preferences for dominant and coercive males. Evol Ecol 19:123–135
Kotiaho JS, Puurtinen M (2007) Mate choice for indirect genetic benefits: scrutiny of the current paradigm. Funct Ecol 21:638–644
Kusmierski R, Borgia G, Crozier RH, Chan BHY (1993) Molecular information on bower phylogeny and the evolution of exaggerated characteristics. J Evol Biol 6:737–752
Kusmierski R, Borgia G, Uy JA, Corzier RH (1997) Labile evolution of display traits in bowerbirds indicate reduced effects of phylogenetic constraint. Proc Royal Soc Lond B 264:307–313
Lande R (1981) Models of speciation by sexual selection on polygenic traits. Proc Natl Acad Sci U S A 78:3721–3725
Madden JR, Balmford A (2004) Spotted bowerbirds *Chlamydera maculata* do not prefer rare or costly bower decorations. Behav Ecol Sociobiol 55:589–595
Marshall AJ (1954) Bower-birds: their displays and breeding cycles. Clarendon, Oxford
Mckinney F, Derrickson SR, Mineau P (1983) Forced copulation in waterfowl. Behavior 86:250–294
Mivart SG (1871) Review of Darwin's *'Descent of man'*. Q Rev 131:47–90
Muller MN, Wrangham RW (eds) (2009) Sexual coercion in primates and humans. Harvard University Press, Cambridge
Patricelli GL, Uy JA, Walsh G, Borgia G (2002) Male displays adjusted to female's response. Nature 415:279–230
Patricelli GL, Uy JA, Borgia G (2003) Multiple male traits interact: attractive bower decorations facilitate attractive behavioural displays in satin bowerbirds. Proc Royal Soc Lond B 270:2389–2395
Patricelli GL, Uy JA, Borgia G (2004a) Female signals enhance the efficiency of mate assessment in satin bowerbirds (*Ptilonorhynchus violaceus*). Behav Ecol 15:297–304
Patricelli GL, Uy JA, Borgia G (2004b) Female signals enhance the efficiency of mate choice assessment in satin bowerbirds (*Ptilonorhynchus violaceus*). Behav Ecol 15:297–304
Peters JL, Brewer GL, Bowe LM (2003) Extrapair paternity and breeding synchrony in gadwalls (*Anas strepera*) in North Dakota. Auk 120:883–888

Pruett-Jones S, Pruett-Jones M (1994) Sexual competition and courtship disruptions: why do male bowerbirds destroy each other's bowers? Animal Behav 47:607–620
Prum RO (1997) Phylogenetic tests of alternative intersexual selection mechanisms: macroevolution of male traits in a polygynous clade (Aves: Pipridae). Am Nat 149:668–692
Prum RO (2010) The Lande-Kirkpatrick mechanism is the null model of evolution by intersexual selection: implications for meaning, honesty, and design in intersexual signals. Evolution 64:3085–3100
Prum RO (2012) Aesthetic evolution by mate choice: Darwin's really dangerous idea. Phil Transact Royal Soc London B 367:2253–2265
Ryan MJ, Akre KL, Kirkpatrick M (2009) Cognitive mate choice. In: Dukas R, Ratcliffe J (eds) Cognitive ecology. University of Chicago Press, Chicago
Smuts B (1985) Sex and friendship in baboons. Aldine, New York
Smuts B, Smuts RW (1993) Male aggression and sexual coercion of females in nonhuman primates and other mammals: evidence and theoretical implications. Adv Study Behav 22:1–63
Soanes C, Stevenson A (eds) (2005) Oxford English Dictionary. Oxford University Press, Oxford
Stutt AD, Siva-Jothy MD (2001) Traumatic insemination and sexual conflict in the bed bug *Cimex lectularius*. Proc Nat Acad Sci U S A 98:5683–5687
Uy JA, Borgia G (2000) Sexual selection drives rapid divergence in bowerbird display traits. Evolution 54:273–278
Uy JA, Patricelli GL, Borgia G (2000) Dynamic mate-searching tactic allows female satin bowerbirds *Ptilonorhynchus violaceus* to reduce searching. Proc Royal Soc Lond B 267:251–256
Uy JA, Patricelli GL, Borgia G (2001) Complex mate searching in the satin bowerbird *Ptilonorhynchus violaceus*. Am Nat 158:530–542
Wallace AR (1889) Darwinism. MacMillian, London
Wallace AR (1895) Natural selection and tropical nature. Macmillan, London

Chapter 12
The Riddle of Attractiveness: Looking for an 'Aesthetic Sense' Within the Hedonic Mind of the Beholders

Michel Kreutzer and Verena Aebischer

> *Beauty, my dear Sir, is not so much a quality of the object beheld, as an effect in him who beholds it.*
> Spinoza, 1674 (1901), Letter to Hugo Boxel

Abstract Darwin conceived the theory of sexual selection in order to explain beauty in animal Kingdom. He hypothesised that most of the male ornaments had been developed to correspond to a female 'sense of beauty'. His successors developed a theory of mate choice in which the aesthetic sense was left out. The male sexual ornaments were considered as salient cues that evolved because they are indicators of males' fitness, which stimulate the female to mate. As a consequence "good genes" would spread to future generations. Such a perspective left no place for the males' appearance and displays as a source of pleasure for females. More recently, authors have considered that male traits might evolve because they make discrimination, stimulus recognition, memorability and learning easier. The winner is the most attractive not necessarily the 'strongest' male. Moreover, male traits might be favoured because they happen to fit an already existing bias in the female sensory system. Such a sensory exploitation determines the direction of a "runaway process".

Today, the "aesthetic sense" is back, the neurosciences study the chemistry and circuitry that support pleasure in the brains of humans and animals; social psychology and animal cognition focus on emotions, categorisation and prototype used for mate choice. Animals and humans in order to make a decision, have to evaluate both

M. Kreutzer (✉)
Laboratoire d'Ethologie, Cognition et Développement (EA 3456), University Paris Ouest, Nanterre—La Défense, 92000 Nanterre, France
e-mail: Michel.Kreutzer@u-paris10.fr

V. Aebischer
Laboratoire Parisien de Psychologie Sociale (EA 4386), University Paris Ouest, Nanterre—La Défense, 92000 Nanterre, France
e-mail: Verena.Aebischer@u-paris10.fr

T. Hoquet (ed.), *Current Perspectives on Sexual Selection,* History, Philosophy and Theory of the Life Sciences 9, DOI 10.1007/978-94-017-9585-2_12

the sensation and the goal directed action. For this a salient hedonic value has to be built by the mind. Here are the processes involved in the 'aesthetic judgement'.

Keywords Animal aesthetics · Attractiveness in humans · Brain reward circuitry and aesthetics · Evolution of beauty · Hedonic mind

12.1 Introduction

"The sense of beauty" is a basic process in the theory of sexual selection that Darwin introduced in the second part of the ninteenth century. Since its publication in 1871 it has met widespread criticism, scepticism and recast of the theory. The idea that female animals are capable of aesthetic choice was not well received among evolutionists. As shown in Sect. 2 and 3, over the past 150 years, many controversies and reformulations have taken place about the function of aesthetic choice. In behavioural studies, until today, much more research has been conducted on the male anatomy and displays supposed to 'stimulate' or to 'charm' the females than studies on female emotions or 'feelings' in front of good looking males. There might be two reasons to that: Either because male behaviour is easier to study than female subjectivity or because authors found it more interesting to study males who are traditionally said to be more 'active' in mate choice. Interestingly, as Sect. 4 illustrates, aesthetic choice has also been absent in the investigation of mate choice in humans explored by social scientists, obsessed as they were by the fitness model of their natural science colleagues. However, as evidenced in the last paragraphs of Sect. 3 and 4 aesthetic choice based on the sense of beauty is coming back with numerous studies on the brain and mind of the receiver, the 'choosy' female. Section 5 illustrates that the sense of beauty or aesthetic sense is much more a process in the mind of the individuals who perceive a signal than a property of the signal per se. Recent work in social psychology, animal cognition and neurosciences reserves an important place to the study of what makes a partner attractive in the eyes of the beholder. The expression "aesthetic judgement" has been used by numerous authors in the context of pair formation and mate choice, and aesthetic choices are explored in humans and animals. Social psychology and animal cognition, for instance, take especially in consideration emotions, categorisation and prototypes related to mate choice. Section 6 is on 'hedonic reward', and 'pleasure', which are prerequisites to "aesthetic judgment" in animals and humans especially when choosing sexual partners. When testing the 'sense of beauty', the neurosciences study the hedonic circuitry that supports the pleasure in the brain of humans and animals opening up new avenues for the understanding of "aesthetic choices".

12.2 A Theory for the Origin of Natural Beauty

12.2.1 To Whom Beauty is Addressed?

When considering nature and especially living species, humans may experience the beauty of some of them. From our admiration for flowers and ornaments in animals or vocalisations emerges the question: to whom this beauty is addressed?

Darwin (1859) had this question in mind when he observed plants and animals. To him the existence of beauty in low animals like coral or sea anemones is the direct result of the structure and the chemical nature of their tissues. Their aesthetic properties fulfil no function at all. They may have something to do with what Portman (1952) called the "unaddressed appearance". Beauty also exists to attract animals; this is the case when it serves the purpose of plants that need animals for their pollination. Then beauty is addressed to pollinating animals who are rewarded with nectar and odour. There is coevolution because beautiful things on one side fit the sensibility of the receiver on the other side.

However, Darwin was most aware of beauty addressed by males to females of their species. As Hoquet (2009) underlines, if, for Darwin, beauty is not useful then it does not serve natural selection. For instance, the evolution of the male bowerbirds' elaborate courtship displays and complex nest decoration presented a problem to Charles Darwin's theory of natural selection, because in no way did such behaviour help the bird to survive in its environment. Both appeared to be colourful and complex elaborations produced by and for an aesthetic sensibility. How could these elaborations during the course of evolution be otherwise explained than by the animals' "taste for the beautiful", their "sense of beauty" or "aesthetic faculty"?

The theory of evolution was first centred on natural selection, which explains variation among species. Natural selection is based on differential survival that depends on the interactions between individuals and their environment. It is a theory of "adaptation", where only traits with life preserving value could evolve. Its crucial point is the survival of the fittest. Darwin was not convinced, contrary to Wallace (1858), co-discoverer of the theory of natural selection, that the 'struggle for life' was the right explanation for all the transformations in animals and especially for their beauty. When he established the concept of natural selection in his 'origin of species' (1859), sexual selection was only slightly mentioned. It left the question of the causation of beauty unsolved. For that reason, Darwin, later (1871), wrote another book, devoted mostly to sexual selection, a theory to explain beauty in animal kingdom.

12.2.2 Sexual Selection and the Female Aesthetic Sense

Sexual selection is a strategy that is complementary and different from natural selection, as it makes bodies, most often the males' bodies, more beautiful and at the same time enhances the aesthetic sense of the receivers, most often the females' aesthetic sense. Milam (2010) summarised Darwin's ideas and the discussion they raised in the following way:

In sexual selection two mechanisms:

> … explained why males and females differed in appearance and behaviour: one was female choice, the other was male-male competition. In female choice, females compared the mating displays of reproductively mature males and chose the most appealing male with which to mate. In male-male competition, males fought to determine which male would have access to the reproductively available female(s) in the area. Both mechanisms, Darwin argued, would result in exaggerated male traits. Over evolutionary time, female choice would lead to aesthetically pleasing male traits (long tails and brightly coloured plumage), and male-male competition to armour or weapons (bony plates or antlers)…. In this way '… female choice presupposed both a sense of aesthetic appreciation and an ability to choose rationally based on this aesthetic sensibility'. (Milam 2010, pp. 1–2)

Sexual selection provided Darwin with an answer to the question: How was it that females and males of the same species differed so significantly in their appearance and behaviour, especially with regard to mating display? Sexual selection does not depend on struggle for life, as natural selection does, but on competition between males for the possession of the females; the result is few or no offspring for the loser. Most of the time fighting between males was replaced by male competition for female attention through courtship displays. Females, on the other hand, actively use their attention and their 'taste for the beautiful' to select the most attractive partners.

Darwin (1871) was very clear on these points, when referring to the male features he said:

> … these organs will have been perfected through sexual selection, that is by the advantage acquired by certain males over their rivals. (Darwin 1871, I, p. 257)

And he said that in birds:

> … those males, which are best able by their various charms to please or excite the female, are under ordinary circumstances accepted. (Darwin 1871, II, p. 124)

The continuity between animals and humans is presented in the following way:

> Sense of Beauty.—This sense has been declared to be peculiar to man. But when we behold male birds elaborately displaying their plumes and splendid colours before the females, whilst other birds not thus decorated make no such display, it is impossible to doubt that the females admire the beauty of their male partners. (Darwin 1871, I, p. 63)

> Why certain bright colours and certain sounds should excite pleasure, when in harmony, cannot, I presume, be explained any more than why certain flavours and scents are agreeable; but assuredly the same colours and the same sounds are admired by us and by many of the lower animals. (Darwin 1871, I, p. 64)

> No doubt the perceptive powers of man and the lower animals are so constituted that brilliant colours and certain forms, as well as harmonious and rhythmical sounds, give pleasure and are called beautiful…. (Darwin 1871, II, p. 353)

Darwin also developed a modern point of view on the coevolution between signal, attractiveness, and brain capacities:

> Everyone who admits the principle of evolution, and yet feels great difficulty in admitting that female mammals, birds, reptiles, and fish, could have acquired the high standard of taste which is implied by the beauty of the males, and which generally coincides with our own standard, should reflect that in each member of the vertebrate series the nerve-cells of the brain are the direct offshoots of those possessed by the common progenitor of the whole

> group. It thus becomes intelligible that the brain and mental faculties should be capable under similar conditions of nearly the same course of development, and consequently of performing nearly the same functions. (Darwin 1871, II, p. 401)

> He who admits the principle of sexual selection will be led to the remarkable conclusion that the cerebral system not only regulates most of the existing functions of the body, but has indirectly influenced the progressive development of various bodily structures and of certain mental qualities. …/… musical organs, both vocal and instrumental, bright colours, stripes and marks, and ornamental appendages, have all been indirectly gained by one sex or the other, through the influence of love and jealousy, through the appreciation of the beautiful in sound, colour or form, and through the exertion of a choice; and these powers of the mind manifestly depend on the development of the cerebral system. (Darwin 1871, II, p. 402)

So it appears that male beauty has been developed for the purpose of being appreciated by the corresponding female sense of beauty. Females will go for the most beautiful males, whose beauty provides the most pleasure. In Darwin's theory of sexual selection females do not appreciate beauty as a signal of fitness; their 'aesthetic judgment' is their only guide. His theory is a theory of the female psychology, not of evolutionary "ultimate" causes.

12.3 Fall and Rise of the Idea of Aesthetic Choices

12.3.1 Who has the Capacity to Appreciate Beauty?

For Darwin different species share the same sense of beauty, which is the reason why humans find bright colours and songs beautiful in birds. As a reverse Watanabe (2012) demonstrated how some visual and auditory dimensions of human art productions have a reinforcing property for non-human animals. Thus Darwin believed in the anatomical, physiological, intellectual and behavioural continuity of humans with other animals; he applied the theory of natural selection and sexual selection in a similar way to both humans and animals. This was far from convincing to his naturalist colleagues, then as it is today. Wallace (1864) preferred to draw a clear line, with humans on one side and animals on the other, and contrary to Darwin, argued that animals did not possess the capacity to reason, therefore they could not choose (see Slotten 2004 for details). Wallace believed that female animals were not capable of evaluating the aesthetic appeal of males' courtship displays and of choosing a mate based on such effects. Romanes (1881) was of a different opinion; he agreed with Darwin's point of view that animals are able to make decisions and to feel the stimuli that determine their choices. Romanes attributed aesthetic 'emotions of the beautiful' to animals exhibiting colourful secondary sexual features and extended such capacities even to the arthropods. Huxley (1914), a famous defender of evolutionary theories, did not agree with the idea that females evaluated the aesthetic of the males. His idea was that of emotional excitement. Looking at the displays of the great crested grebe he noted that both sexes participated in courtship dances; sexual selection could, therefore, not be involved in male beauty.

12.3.2 Genetic Fitness is Declared as the Eternal Winner, the Fall of Beauty

During the course of the twentieth century the importance of the female aesthetic sense fell in disrepute and with it the theory of sexual selection. They were replaced by a theory of mate choice that some authors included in the natural selection processes. One may find a complete analysis of this in Milam (2010). The rise of Mendelian genetics focused evolutionary biology on the genetic change in a single population over time with natural selection depending on the total number and quality of an individual's offspring. Sexual selection was thus eclipsed and evolution redefined as a process of natural selection merely changing gene frequencies along generations (Haldane 1932). The gradual accumulation of micro modifications could drive large modifications and directional evolution over the time.

With the emergence of a synthesis constituting a new evolutionary theory, called neo-Darwinism, biologists investigated evolution as a process of speciation (Mayr 1942). Instead of sexual selection and aesthetic choices, authors began to investigate what happened if some females in a population only preferred to mate with a specific kind of males. Thus, female choice could drive the creation of a reproductively isolated population and potentially lead to speciation. Fisher (1930) provided an explanation for the evolution of secondary sex ornaments by mate choice. If females have a sexual preference for a particular kind of male trait, so he argued, this confers a selective advantage. The trait will became a salient cue and will spread to future generations in a "runaway process". In no way does this theory imply that females perceive beauty, they only "appreciate" secondary sexual ornaments, which, when associated with a slight survival advantage, could became an indicator of fitness.

For most of these authors, the history of mate choice is distinct from the history of sexual selection, the latter being only considered as the Darwinian explanation of beauty in animal kingdom. From their perspective, mind and aesthetic sensibilities ought to be considered as discontinuous elements demarcating animals from humans, thus sharing Wallace's ideas about the differences between humans and animals.

The interest for female sensibility to male traits gained momentum with the discovery of the "rare male effect" Bösiger (1974), Petit (1958). Both authors' intent was to explain why females preferred to mate with male types that were the least frequently found in a population. One held the view that females act to maintain genetic diversity, whereas the other interpreted it in terms of female sensitivity to males' stimulatory courtship behaviours.

12.4 Signal Design: Sexual Selection is Back, but Where is Beauty?

Darwin's theory was forcefully resisted by scientists for over a century, in part because the active choosing of mates seemed to grant too much power to females, who were thought to remain passive in the mating process. It was also resisted because it seemed to blur the animal-human boundary. That those females should go for the beautiful, and for the most beautiful male, because of their sense of beauty, stirred up controversy among communities of biologists who preferred to subsume sexual selection under natural selection. They would rather explain female preferences for males' courtship display in terms of reaction to signals advertising fitness. In other words, what would appear to be a choice of "beauty" should ultimately be understood as a preference for utility.

12.4.1 Selective Female and Male Honest Signalling

Bateman (1948) and Trivers (1972) opened a new line of interest in sexual selection by providing arguments for female choice. Bateman demonstrated that in Drosophila the reproductive success during lifetime is constrained for both male and female; for a male by the number of mates an individual is able to obtain and for the female by the number of births or hatchings. The female does not have as many opportunities to reproduce as the male does. Because of her greater reproductive refractoriness she is supposed to be more attentive to the success of her reproduction by showing greater selectivity when choosing her partners than her male counterparts. Trivers, in a similar way, developed the "theory of parental investment" in which he compared male and female cost to reproduce: gametes, pregnancies, broodings, taking care of the young, etc. According to his theory, the sex that exhibits more costly parental investments (most often the female in birds and mammals) will be the chooser. The sex that exhibits less investment will compete for mating opportunities; as a consequence it is the females that will most often choose the males in higher vertebrates.

The term 'choice' in animal behaviour was used in "behavioural ecology", during the sixties and seventies, to describe the ultimate causes that are causes considered efficient from the evolutionary point of view to explain natural and sexual selection. That does not mean that animals are aware of their determinisms. Game theory (Maynard Smith 1989) applied to animal behaviour has considered individuals as rational agents who try to enhance benefits and to minimise costs. From such a perspective it was hypothesised that the choosy sex will mate with individuals who possess traits that signify overall genetic quality. Females are, therefore, seeking good genes in order to gain an evolutionary advantage for their offspring. Thus the theory of sexual selection is not looking for a female 'taste for the beautiful' but explains that females are looking for indicators of males' fitness. Anderson (1994, p. 22) summarises this as follows:

> There is now much evidence that females often choose their mate, and that such choice favours conspicuous male traits. The exact ways in which female choice selects for such traits are still debated, and so are the ways in which female preferences evolve, which remain a main controversial issue in the theory of sexual selection.

Among the potentially salient traits for females' choice, bilateral symmetry and its deviations (fluctuating asymmetry) as well as pigmentation were examined. Hamilton and Zuk (1982) hypothesised that sexual ornaments are indicators of disease resistance to parasites, and also an index for good health and fertility. In birds, plumage colour, pigment and ornaments symmetry support this hypothesis (Swaddle 1996; Møller 1992, 1996; Hill et al. 1999). Because symmetry is said to reveal disease resistance and the ability of a genotype to undergo stable development of a phenotype under given environmental conditions, this feature was also tested positively in humans (Tovée et al. 2000). Many authors designated as beauty such symmetry and conspicuous ornamentations or colours.

The theory of the handicap principle (Zahavi 1975, 1977) explains how female preferences for males' signals have evolved. The handicap principle describes how evolution may lead to honest signalling. It suggests that to avoid animals' cheating for advantages over conspecifics about their state, capacity or intention, reliable signals must be costly to the signaller. They indicate to the receiver the sexual quality of the emitter; because inferior quality signallers cannot afford to produce such extravagant signals. Males exhibiting the most costly features must also be the most vigorous ones. Sexual ornaments such as bird songs, peacock tails, courtship displays, or bowerbirds' bower, are costly signals, and therefore salient indicators of the male quality. However, the theory does not include a description of the females' feeling when looking at these ornaments.
As Anderson (1994, p. 17) explains:

> The idea that conspicuous male display, colours feather plume, and other secondary sex ornaments evolve through female choice met much early scepticism. Darwin seemed to assume similar sense of beauty in other higher vertebrates as in man. But this assumption, right or wrong, is not necessary for female choice: discrimination among males in relation to size, shape colour or others should suffice.

12.4.2 Sensory Bias, Receiver Psychology and the Design of Animal Signals

For Ryan (1990) a new male trait might be favoured because it happens to fit an already existing bias in the female sensory system. Such a sensory exploitation determines the direction of a "runaway process". Males might evolve traits that exploit pre-existing sensory biases of the female. The theory states that the sensory bias evolves in a non-mating context; for instance the foraging ecology of a species may lead to high sensitivity to certain colours. This bias might favour the evolution of male ornaments with such colours and the female preferences for these traits. Females leave males with poorly developed courtship signals for males that provide more effective stimulation. Ryan concludes (1990, p. 186):

> Such an approach can only increase our understanding of the fascinating process of sexual selection by showing how properties of the receivers exert selection on male traits, and how they are responsible for some of the most bizarre morphologies and behaviours in the animal kingdom. It certainly will inform us about how evolution operates and thus can contribute to hypotheses of the evolution of female preferences.

Female preference may have evolved for reasons not related to fitness advantages for males with the most far-reaching signals. Other factors, such as an initial sensory bias, or selection for species recognition, might help explain mate choice. Some male traits may have evolved simply because they make it easier for females to find the male. Sexual selection would favour males who most effectively stimulate the recipients, that is, with intense, persistent, or otherwise conspicuous signals. These ideas were particularly well developed in the "receiver psychology" approach of Guilford and Dawkins (1991, 1993). They underline the role played by sensory systems of the receiver's brain in constraining the design of the signals emitted by the sender. The receiver's capacity of signal detection (reaction time and detection against background noise), discrimination, stimulus recognition, memory and learning, must be taken into account. In addition Rowe (1999) mentioned the advantage of the multimodal components of a signal, because redundancy facilitates its detection.

Receiver psychology only takes into account some aspects of the theory of signal transmission, detection and recognition, but not cognition, emotion and feeling of the receiver. As in the studies of the male emitter, there is no place for "a taste for the beautiful" in the receiver psychology. The receiver is considered like an automaton.

12.4.3 Beauty is Back

The female "sense of beauty" has been brought back with Burley and Symanski (1998) in their seminal paper: "A taste for the beautiful": Latent aesthetic mate preferences for white crests in two species of Australian Grassfinches. By experimental means they demonstrated that two avian species from a lineage devoid of crested species have mate preferences for opposite sex conspecifics wearing artificial white crests. Other colours of crests that have been studied were not preferred. One may interpret such results as a confirmation of the sensory bias hypothesis. But the fact that in the lineage there was no crest at all does not fit very well with sensory bias For Burley and Symanski these results give "powerful evidence for highly structured aesthetic mate preferences in estrilidae finches and suggest that the preference for such a "structure" is influenced by the central nervous system". In conclusion they hypothesise that aesthetic preferences are a potent force in the early evolution of sexually selected traits, and that "indicator" traits evolve secondarily from traits initially favoured by aesthetic preferences. From this a new question arises: sender or receiver, who is the first during the course of evolution? It may well be the receiver and her need for hedonic rewards.

12.5 From Sexual Selection to Seduction in Humans

The breakthrough in applying sexual selection to humans came in the late 1970s and 1980s in the form of theoretical advances initiated by Buss and colleagues in the fields of psychology and anthropology (Buss 1989, 1994; Buss and Barnes 1986) and what was to become evolutionary psychology[1]. Evolutionary psychologists

[1] In his 1989 study on "Sex differences in human mate preferences: Evolutionary Hypotheses Tested in 37 Cultures," Buss asked 10,047 people in 37 different cultures located in 33 countries to

have found sexual selection theories to be most helpful to account for widespread gendered behaviours among humans. Theories in the field of sexual selection seemed to shed light on human sex differences and hence those pertaining to "human mating" and reproduction.

12.5.1 The Fitness Model Applied to Humans

Interestingly, like his counterparts in biology, Buss developed a series of hypotheses, all of them related to the fitness model. And from the beginning evolutionary psychology was entangled in social and stereotypical conceptions of masculinity and femininity, with Buss wanting to verify a few of the most obvious evolutionary predictions about sex differences in mating preferences; for example, whether men desire youth and physical attractiveness in a mate and whether women desire status and economic security. He found support for greater male than female use of resource display and for greater female use of enhancing physical appearance. Most sexually selected traits or behaviours in humans were explained as proxies for fitness. Thus, he contented (Buss 1998; cited by Haufe 2007, p. 116) that women had a preference for a "reliable man willing to commit to her" and that "the resources, aid, and protection" which those men provided caused her to have "children who survived and thrived". Buss offered no data on reproductive success of women either with a preference for reliable men or without a preference for reliable men. In place of the necessary data, Buss offers the "discovery" that women "place a premium on a man's social status, his ambition and industriousness, and his older age—qualities known to be linked with resource acquisition," (Buss 1998; cited by Haufe 2007, p. 117). The evolutionary approach generates only "after the fact" explanations, relying on models from ethology and experiments conducted with animals combined with observations and findings in psychology, psychiatry, behavioural genetics and neurobiology. There is no demonstration of differences in reproductive success in humans resulting from the valuable resource or the preference for reliable men.

The female preference for reliable, resource-giving men is supposed to have evolved sometime during the Pleistocene. Indeed, as underlined by some of the most influential proponents of evolutionary psychology, Cosmides and Tooby (1997) "our modern skulls house a stone age mind". Behaviour in the present is generated by information-processing mechanisms that exist because they solved adaptive problems in the past, which our hunter-gatherer ancestors faced during our species' evolutionary history. For this reason, evolutionary psychology is past-oriented. However, to Cosmides, contrary to Buss, because these mechanisms solved problems efficiently in the past does not mean that they necessarily generate adaptive behavior in the present (also see Symons (1989) and Tooby and Cosmides (1990).

---

provide information about features which according to Buss have been shown to be theoretically important to human mating preferences.

12.5.2 Sex Roles Without Aesthetics

A different explanation for sex differences in behaviour has been suggested by social structural theory (Eagly and Wood 1999). Rather than to attribute sex differences in contemporary society to sex-typed evolved mechanisms, social structuralists maintain that because men and women tend to occupy different social roles, they become psychologically different in ways that adjust them to these roles. Physical sex differences, in interaction with social and ecological conditions, influence the roles held by men and women because certain activities are more efficiently accomplished by one sex. The benefits of this greater efficiency can be realise when women and men are allied in cooperative relationships and establish a division of labour. The particular character of the activities that each sex performs then determines its placement in the social structure (Wood and Eagly 2000).

Evolutionary theory values and validates as a natural given stereotypical conceptions of masculinity (philandering, social status and strength) and feminity (passivity, valuing of their body and youth). Social structuralists focus on the sexes' efficient cooperation in life tasks given women's capacity for gestation and lactation and men's greater speed and physical strength (Wood and Eagly 2000). Both theories have in common that they have never ever considered the role aesthetics could play in mate choice.

12.5.3 Seduction in Humans

A more complete picture of mate choice is offered by Moore (1985), who has observed nonverbal facial expressions and gestures, exhibited by human females that are commonly labelled "flirting" behaviours. Moore has identified 52 nonverbal solicitations such as giggling, laughing, smiling, head tossing, hair flipping, caressing ones arm or leg. Indeed, men seem to have little chances to seduce a woman if she has not taken notice of them before, and if she has not signalled interest via nonverbal cues. Glancing behaviours are important in signalling interest, glancing at and then away from the male. Glancing behaviour appears to be a significant part of the female role (Cary 1978). Males are generally hesitant to approach without some indication of interest, and repeated eye contact seems to demonstrate a woman's interest (Crook 1972a, b). Women reject suitors by failing to recognise their presence through eye contact. Moore observed, then, head tossing, with the head flipped backwards so that the face was briefly tilted upwards often combined with pushing her fingers through her hair or running her palm along the surface of her hair, but also pouting, eyebrow flashing and, above all, smiling. Smiling is among the most prevalent behaviours and consists of the corners of the mouth being turned upward, in partial or sometimes full exposure of the teeth. The "coy" smile combines a half-smile with a downward gaze or very brief eye contact. These expressions and gestures appear to function as attractants and advertisers of female interest.

According to Moore, nonverbal solicitation is only one of the first steps in the sequence of behaviours beginning with mate attraction and culminating with mate selection. Women who signal often are also those who are most often approached by men. They can elicit a high number of male approaches, allowing them to choose from a number of available men. Or they may direct solicitations at a particular male. Behaviours such as nodding, leaning close to the man, smiling and laughing are in higher frequencies after the man has made contact with the woman. Conversation is initiated and the participants appear highly animated. Women, while talking to men appear excited, laughing, smiling, and gesticulating frequently. In the sexual arousal phase, touching gestures are exchanged.

Not only does the woman first signal her interest in a potential candidate, the actual choice and final decision is also hers and rests on what the man says to the woman in addition to his behaviour towards her and others. However, little is known as to which men are chosen for further interaction, which are rejected, and on what grounds. More investigation on that level of female choice could open up new vistas as to aesthetical considerations involved in that choice.

Another example of female choice and signal sending was offered by Zanna and Pack (1975). Young women were asked to characterise themselves to a male partner who was either attractive or unattractive and whose ideal of a woman conformed either very closely to the traditional female stereotype or to its opposite. When the male partner was attractive and favoured the traditional female stereotype, the young women portrayed themselves as significantly more conventional and traditional than when, a few weeks before, they had been asked to make an objective description of themselves. When he was attractive, but favoured a modern type of woman, they portrayed themselves as more modern, unconventional and independent than before. However, when the partner was unattractive, they did not change their self-description. Nobody knows how the story would have evolved, had these young women really met the attractive or the unattractive partner. However, the experiment shows quite clearly that these young women have first appreciated the partner's attractiveness, before signalling that there could be some commonalities and mutual understanding between them.

12.6 The Mind of the Beholder

Receivers are not simple signal analysers passively waiting to be stimulated. The receptors experience emotions, build representations of their relationships with peers and may feel social pleasure in the presence of some of them. These aspects must be taken into account if we want to understand the aesthetic value that females attribute to some males. However, the study of animals' subjectivity is more complicated than the study of their behaviour. Anthropomorphism, which consists of transposing human emotions or rationality onto animal actions, tended to make animals look more like humans. Most researchers, therefore, prefer to choose the paradigm of parsimony. In doing so, they seem, however, to neglect that the receiver psychology is much more complex than that of an automaton, even a sophisticated one.

12.6.1 Cognition, Categorisation and Prototype

The world is not just a source of isolated stimuli, which act independently from each other, as behaviourists first thought. A substantial amount of studies on animal cognition show that birds and mammals have representations of their world as categories of objects and events. Herrnstein (1984) was one of the first to demonstrate natural concept discrimination in pigeons. Subjects were trained to classify photographs that exemplify categories as trees, individual persons, fish and others ... As stated by Roitblat (1985, p. 306):

> ...these results support the hypothesis that pigeons form abstract representations of the discriminated concepts...

Cognition in animals (Thompson 1995) got further understanding when the theory of prototype (Rosch and Mervis 1975) was extended to them. Similar objects are members of the same category. The categories may have clear or fuzzy boundaries depending on their properties, graduated colour, or not, like an animal. But the important point is that some members are better exemplars than others. For instance, for most people some dogs are better than others to represent the category dog. These central exemplars are prototypes.

Categorisation in animals is an important step towards "aesthetic values" in animals, because if some males are better than others to be chosen as sexual partners, they can be considered to be prototypes. Moreover, if emotions and pleasure are experienced at the sight of these prototypes, then we are at the heart of the category "beautiful partners". The "sense for beauty" could work.

Surprisingly, when it comes to humans, attractiveness based on physical characteristics has captivated researchers' interest mainly from a male perspective. In an overview of recent research, Swami and Furnham (2006), both social psychologists, showed that there are physical characteristics that are attractive across cultures. In most cultures men will rate women with a 0.7 waist to hip ratio WHR as more attractive than a woman with a higher WHR. For faces averageness, symmetry, and sexual dimorphism (masculinity in males, femininity in females) are also preferred across cultures (for reviews, see Fink and Penton-Voak 2002; Thornhill and Gangestad 1999).

Attractiveness is most salient in people's reactions to faces (Olson and Marshuetz 2005). Recent studies have uncovered some of the factors involved in the perception of attractiveness that seem to be universal. When people were asked to rate the attractiveness of computer-composite images of faces (female and male) they perceived attractiveness increases the more different faces of the same sex were averaged and went into each image (composites created by averaging 4 vs. 8 vs. 16 vs. 32 faces). The more images were used, the more idiosyncrasies of particular faces, which may be unusual, become ironed out. Moving a facial image closer to the average increases its attractiveness (Langlois and Roggman 1990). This "beauty-in-averageness effect" only works, however, when the features of the faces are average of the group to which a face belongs (Potter and Corneille 2008).

The beauty in averageness effect is often theoretically explained as reflecting a biological predisposition to interpret prototypes as salient cues for mate choices (Symons 1979).

Referring to the honest signalling theory Thornhill and Gangestad (1993) consider that if facial composites made by combining individual faces are judged to be more attractive than the majority of individual faces it is because the composites possess both symmetry and averageness of features. Facial averageness may reflect high individual protein heterozygosis and thus an array of proteins to which parasites must adapt. As we know heterozygote are better than homozygote in defence against parasites. Facial averageness reflects, therefore, resistance to parasites whereas symmetry certifies overall phenotypic quality and developmental health (Thornhill and Gangestad 1993).

A rather different explanation has been suggested by Zajonc (1980). In humans particular cerebral areas are related to appreciation of beauty (Kawabata and Zeki 2004). The basis for the relation of attractiveness and averageness might be our preference for things that are familiar to us. It is intimately linked to affective reactions, which are often the very first reactions of the organism and are dominant reaction for many species. This form of preference without inference may contribute to perceived attractiveness. In other words, emotions may come to play without our being aware of it. The closeness to the faces people may have seen creates a preference based on a comforting sense of ease. Such prototypes are attractive because they are easy on the mind (Langlois and Roggman 1990; Rhodes and Tremewan 1996; Winkielman et al. 2006). The pleasing prototypes respond to principles of economy.

12.6.2 What is Attractive in Humans?

People who are judged to be physically attractive make generally more favourable impressions on others than do people with lesser looks. There is an important list of qualities that attractive people are supposedly blessed with. They are assumed to have more agreeable personalities, to be more sociable, healthy, intelligent, better students, and teachers than less attractive people (Eagly et al. 1991). And both men and women desire as romantic partners the most attractive women or men they are able to win (Walster et al. 1966). But what makes a person attractive? Some general characteristics are universally considered to be attractive such as symmetry and averageness of faces and a narrow waist-to-hip ratio (Marcus and Miller 2003). However, there are also some specific features of the face and of the body that seem to attract one or both sexes.

According to Jones and Hill (1993), there is more to facial attractiveness than averageness. Although Langlois and her colleagues found that composite faces are more attractive than most of the faces that go into making the composites, few individual faces are consistently rated more attractive than any composite (Alley and Cunningham 1991). Cunningham (1986) showed that photographs of female faces rated attractive in the United States have unusually large eyes, high cheekbones, thin cheeks, and small noses, chins, and jaws. Appealing female faces have a more neotenous "babyface" appearance combining features such as large eyes and a small nose (Jones 1995; Perrett et al. 1998). The ideal male face is closer to the average male face. However, male faces undergo a more thorough remodeling during adolescence than female faces, with a great expansion of the nose, mid-face, brows, chin, and jaw, which reduces the apparent prominence of the eyes and cheekbones (Jones and Hill 1993).

In one recent study, Li and Kenrick (2006, p. 479) found that young psychology male and female undergraduates who had or were considering having casual sex were physically attracted to the person and thought it would feel good. But the male students, far more than the female students, thought that it would allow them to get a sense of their value in the mate market.

Why does physical attractiveness lead to attraction? Aesthetic appeal seems to be desirable, and leads to positive affect (Kenrick et al. 1993). People like to look not only at people, but also at things that they find visually appealing. Even infants show a preference for attractive compared to unattractive people (Langlois et al. 1991). Research also suggests that when men see photos of very attractive women, a particular part of the brain is more strongly activated than when they concentrate on photos of average-looking women (Aharon et al. 2001). The part of these men's brains that is activated in response to beautiful faces is also activated in response to rewarding behaviours such as money and drugs (Sanderson 2010, p. 416).

12.6.3 Emotions in Humans and Animals

Darwin was particularly aware of what animals felt, because in his theory of evolution it was a proof of the continuity between animals and humans, which was the subject of his book, *The expression of the emotions in man and animals* (1872). The following two passages show his awareness that animal like humans experience emotion and pleasure.

> The sexes of many animals incessantly call for each other during the breeding-season; and in not a few cases, the male endeavours thus to charm or excite the female. This, indeed, seems to have been the primeval use and means of development of the voice…. Thus the use of the vocal organs will have become associated with the anticipation of the strongest pleasure which animals are capable of feeling. Animals which live in society often call to each other when separated, and evidently feel much joy at meeting. (Darwin 1872, p. 84)

> When male animals utter sounds in order to please the females, they would naturally employ those which are sweet to the ears of the species; and it appears that the same sounds are often pleasing to widely different animals, owing to the similarity of their nervous systems, as we ourselves perceive in the singing of birds and even in the chirping of certain tree-frogs giving us pleasure. On the other hand, sounds produced in order to strike terror into an enemy, would naturally be harsh or displeasing. (Darwin 1872, p. 91)

Later in a paragraph on "Pleasure, joy, affection" Darwin comments the pleasure and satisfaction that animals experience by referring to an observation made by the French psychiatrist Duchenne:

> Dr Duchenne—and I cannot quote a better authority—informs me that he kept a very tame monkey in his house for a year; and when he gave it during meal-times some choice delicacy, he observed that the corners of its mouth were slightly raised; thus an expression of satisfaction, partaking of the nature of an incipient smile, and resembling that often seen on the face of man, could be plainly perceived in this animal. (Darwin 1872, p. 132)

More recently Bekoff (2007, p. 14) referred to the importance of emotions to understand animals' life as follows:

> Of course there are differences among species. We would expect variations based on social, ecological, and physical factors. However there are compelling similarities despite sometimes extreme differences.... The brain of mice, dogs, elephants, and human differ greatly in size, but all of these species display joy and empathy.

Communication and sending a signal can be considered as a means to manipulate the mental state of the receiver. Music is a good candidate for such a job in humans. Miller published an article, in 2000, reaffirming Darwin's suggestion that human music is manipulative and was shaped by sexual selection to function as a courtship display:

> The vocalizations and gestures do not appear to be telling another individual about the world in the same way as we refer to objects, events and ideas when talking to another individual. Monkeys and apes probably simply do not appreciate that other individuals lack the knowledge and intentions that they themselves possess. Rather than being referential, theirs calls and gestures are manipulative: they are trying to generate some form of desired behaviour in other individual. The multimodal communication, the use of rhythm and melody, enhance synchronisation and the share of emotional state. (Miller 2000, p. 121)

Miller insisted that the function of such aesthetic selection criteria is to enhance selective mate preference in order to improve the outcome of sexual recombination that maintains genetic diversity, promotes speciation, and facilitates evolutionary search through optimal outbreeding.

Mithen (2006, p. 96) looking for "The origins of music in the singing Neanderthals", title of his book, went a step further in the understanding of the feeling of the receiver. Music affects emotion and mood:

> Mood is slightly different from an emotion; the former is a prolonged feeling that lasts over minutes, hours or even days, while the latter may be a very short feeling. The success of music therapy further demonstrates how music can be used both to express and to arouse a wide range of emotions, and also lead to substantial improvements in mental and physical health.

The idea that music importantly modifies the inner state of the listener—receiver has gained support over the past couple of decades since we now know a lot more about centres and circuitry, on one side, and social or behavioural situations implicating receiver's satisfaction, on the other side.

12.7 The Hedonic Receiver

12.7.1 Reward, Dopamine and the Mesocorticolimbic Circuitry

Olds and Milner (1954) introduced a seminal research by demonstrating that rats produced positive reinforcement until exhaustion by electrical stimulation of their septal area and other regions of their brain, especially their lateral hypothalamus (LH). Demonstration was also produced in humans that LH is a key centre of the brain reward function. Indeed Bishop et al. (1963) using intracranial self-stimulation in man obtained results similar to those of the rat experiments. Electrode stimulation evoked desire to stimulate again and strong sexual arousal, while never producing orgasm. What it did was to make humans press the button more.

A considerable body of literature followed these discoveries on the neurobiology of reward, based largely on studies of addiction or substance abuse. The circuitry involved in addiction is most often described as the 'dopamine mesocorticolimbic system'. Relevant circuitry in such studies has included dopamine cell bodies in the ventral tegmental area (VTA) that project to nucleus accumbens (NAcc, especially the core shell) that project via ventral pallidum to thalamus. There is a broad thalamic projection to prefrontal and cingulate cortex. The cortex completes the loop by projecting to VTA. Moreover the VTA projects directly to cortex and amygdala in addition to the nucleus accumbens.

In spite of the fact that addictive drugs act on the brain reward systems, one may consider that the brain had evolved not simply to respond to drugs but to natural rewards, such as food and sex. Appropriate responses to natural rewards were evolutionary important for survival, reproduction, and fitness Kelley and Berridge (2002). Many studies have effectively demonstrated the close correlation between dopamine release and behaviours that are sustained by drive and motivation (as taking food, water search and looking for a sexual partner). Today authors are in accordance when considering that neural circuits using dopamine confer to given stimuli particularities, which elicit them to be relevant to sensory, emotional or affective systems. Wise (2004) gave evidence of the role of such a system in learning. Today for many neuroscientists the primary role of the mesocorticolimbic dopamine circuitry in reward is to facilitate arousal, attention, motivation, memory consolidation, goal directed behaviour, and decision making (Schultz 2006).

Whatever the species, the pleasure of the brain is a central theme in cognitive neurosciences. Animals and humans in order to make a decision, have to evaluate both the sensation and the goal directed action. For this a salient hedonic value has to be built by the brain circuitry (Dickinson and Balleine 2009). Thus the reward processes in humans and animals have led to comparative studies. For Berridge and Kringelbach (2008) a rewarding stimulus activates many brain systems at the same time. There are three types of reward components: (1) liking: the hedonic impact of reward; (2) wanting: the motivation to obtain a reward and (3) representation: the cognitive processes about the rewards. The hedonic brain mechanisms being similar in humans and other animals, the way is open for a comparative "affective neuroscience of pleasure".

Based on these considerations one may conclude that the decision of the 'female beholder' is as much related to the pleasure she experiences as to the characteristics of the signals she receives. Search of sensations that are pleasing for themselves is sufficient a motivation to trigger search behaviour in conjunction with pleasant stimuli, even in the absence of physiological needs. Animals are probably encouraged to simply looking for hedonic rewards. Objects and events that are able to provide such rewards have obviously an incentive value. During the evolution of cognition, pleasentness has allowed the construction of mental categories in which particular objects or events are regarded as being "good". We are here quite close to the Darwinian idea, that the animal brain like the human brain has built a sense of beauty (Kreutzer 2012). Moreover studies on addiction have demonstrated that a "runaway process" can transform the search of satisfaction that was initially related to the pleasure provided by useful objects or events into a search of pleasure for pleasure. The receiver's hedonism can drive to dependence.

12.7.2 Life, Social life and Pleasure

Berridge and Kringelbach (2008, p. 459) considerably enlarge the vision that rewards should only mediate sensory pleasure such as food and sex, mainly because:

> … social interactions with conspecifics are important to propagation of genes in all social animals such as humans … (thus particular social activities) … are also likely to be part of the repertoire of fundamental pleasures. Social pleasures in animals other than humans might be conceived as essentially similar to basis sensory pleasures or conceivably even in some nonhuman species as something more abstract. Social pleasure includes sensory visual features such as faces, touch features of grooming and caress, as well as in humans more abstract and cognitive features of social rewards.

Several studies have clearly demonstrated the link between pleasure and social life, particularly in the context of play, singing, pair formation and attachment (Panksepp1998 for a review; Panksepp 2005). A close association between opiates and play in juvenile rats was discovered by Ikemoto and Panksepp (1999). Low doses of morphine increased play, and opioids blockade with naloxone reduced play. From this one may conclude that opioids may enhance the pleasure and rewards associated with playing. Singing in songbirds, should be also a source of pleasure. Male zebra finches may use 'directed songs' for courtship and 'undirected songs' when singing alone. Hara et al. (2007), Huang and Hessler (2008) demonstrated that in VTA (ventral tegmental area) the neurons are more strongly activated during directed songs (courtship) than during undirected songs. Thus social context provides modulation to forebrain rewards' circuitry. Dopaminergic neurons of these birds work in a similar way as those of the mammalian (VTA) in the rewards circuitry. Moreover, such results support the idea that social encounters can trigger the same pathways as addictive drugs.

Insel (2003) studied the prairie and pine voles who form partner preferences and pair bound after mating contrary to montane and meadow voles who generally do not form such preferences. The neurobiology for pair bound is correlated with the mesolimbic dopamine activation of a particular kind of receptors that are necessary and sufficient for the development of partner preference. The processes involved in social attachment are close to those observed in an addictive process. Aragona et al. (2003) also showed in prairie voles that the administration of haloperidol directly into the NAcc (Nucleus accumbens, core shell) blocked partner preferences induced by mating. But, on the contrary, administration of apomorphine into the NAcc induced partner preference even in the absence of mating. These use of pharmacological techniques thus manipulated both the dopaminergic circuitry and the pair bound formation with its correlated attachment.

These results suggest that widely distinct groups of animals may experience similar positive emotional states while undergoing different types of social interactions. From this we can easily conceive that learning, displays, vocalisations, and partners' perception are impossible to imagine without hedonic processes. Therefore, animals should definitively be regarded as close to humans and not simply as complex mechanistic automatons. As Balcombe said (2006, p. 22):

> Feeling good is a powerful motivator that steers animals towards behaviour that keep them alive and help them reproduce ... survival and pleasure are mutually compatible.

12.7.3 Are Love and Beauty Abstractions or Rooted in Basic Brain Circuitry?

Animal studies and human imaging have contributed to the understanding of the psychobiology of attachment and social bounding. When comparing the neuronal circuitry of maternal love, romantic love, and long-term attachment, neurobiology (Stein 2009) shows that they both overlap and differ. In these circuits, molecules, which have been demonstrated to play a role in the psychobiology of attachment and social bounding, include dopamine, serotonin, opioids, and vasopressin. Relevant circuitry has included the anterior cingulate, medial insula, striatum, and ventral tegmental area (VTA). Both maternal and romantic loves involve: anterior cingulate, medial insula, caudate nucleus, and the VTA. Thus sophisticated human behaviour is rooted in mammalian biology. Complicated concepts and processes such as love, reward, beauty, addiction, emotion or even rejection in love are, therefore, embodied in more basic and overlapping structures. During the course of evolution these functions were rooted in existing neuro circuitry and neurochemistry (Fisher 2004; Fisher et al. 2005, 2010).

From Bartels and Zeki (2004) studies comparing maternal and romantic loves emerge three points: (1) both involve a unique and overlapping set of areas, as well as areas that are specific to each; (2) the activated regions belong to the reward system and are also known to contain a high density of receptors for oxytocin and vasopressin, suggesting that the neuro hormonal control of these strong forms of attachment observed in animals also applied to humans; (3) both forms of attachment suppress activities in regions associated with negative emotions, as well as regions associated with 'minding' and social judgments. This suggests that emotional ties to other persons inhibit not only negative emotions but also affect the network involved in making social judgements about a person. Attachment processes activate a specific pathway of the reward system and at the same time deactivate circuits that are responsible for critical social assessment and for negative emotions.

12.8 Conclusion: What does it mean to Have an "Aesthetic Sense"?

Vallet and Kreutzer 1995; Draganoiu et al. 2002; Suthers et al. 2012 have amply tested the acoustic preferences of the female songbirds and confirmed such sensory bias, sensory trap and honest signalling theories. However, more is needed to understand the female's attraction to a male's songs. When evaluating the signals exhibited by the male, the choosing and choosy female experiences the hedonic impact of reward: that is, emotions, feelings and liking. These are greater when morphology and displays of the male are close to the prototype she has in her mind, and they are necessarily based on an 'aesthetic judgement' depending on her 'sense of beauty' or 'taste for the beautiful'.

The position of Welsch (2004), a convinced advocate that animals, like humans, have an "aesthetic sense" could be summarised in the following way. While Darwin had advocated the existence of an aesthetic sense in some animals, most contemporary evolutionists have reduced aesthetics to mere survival value. They try to unmask aesthetic appreciation as a mere manifestation of fitness. From this (neo-Darwinian and especially sociobiological) perspective, there is simply no space for an aesthetic attitude. Even if one assumes that beauty means fitness in a hidden way, and that this is ultimately the reason why the beautiful is esteemed, one cannot get round the fact that what the female appreciates in the first place is the beautiful as such. The proximate goal and the ultimate goal would not be reached if aesthetic appreciation had not taken place. The aesthetic momentum remains indispensable. But it can only be reached through the perception and estimation of the beauty of the beautiful.

As noted by Welsch (2004), Darwin, contrary to most of his evolutionist successors never used the term "aesthetics judgement" when referring to female choice. Most of the time he mentioned that they had a "taste for the beautiful" or a "sense of beauty", and only rarely that they had an "aesthetic faculty". The concept of "taste", going back to eighteenth century philosophy, held the judgement of beauty to be immediate and disinterested. This means that the pleasure of beauty is not useful. Later, during the nineteenth century, the concept of "aesthetic" replaced the notion of "taste". Darwin not being a philosopher referred to "aesthetic faculty", "taste for the beautiful" and "sense of beauty" as synonyms, without giving a definition or a clear and constant meaning of these concepts.

Today, when authors refer to these Darwinian concepts, they often use the expression "aesthetic judgement". The meaning of "aesthetic judgment" is certainly different from the meaning of "taste", because it implies some underlying general principles that have yet to be discovered. The immediacy and disinterestedness, which encompass the general sense of taste, fit better with "aesthetic attitude", which refers to contemplation and to the work of nature for its own sake. It is also more in line with Darwin's writings that underline the un-usefulness of the males' traits selected by females.

However, "aesthetic attitude" is far from "aesthetics of the object", the objects being the salient cues of the animal's morphology and displays, which fit the psychology of the receiver. In fact most of these authors never use the term aesthetic. It is certainly the concept of "aesthetic experience" that will best correspond to the "conspicuous indicators" that sociobiology and behavioural ecology present as adaptive and revealing the vigour or good genes of the males. Petts (2000) establishes a link between the objects and the feeling they raise, for humans he says:

> Aesthetics experience is a natural felt response … adaptive … revealing value in the world. (Petts 2000, p. 70)

> The aesthetic experience is not a response to having checked that things have worked according to plan; rather there is a feeling that things are just so. This feeling of harmony or rightness that come about in our experience of things is a feature basic to our aesthetically experiencing them. (Petts 2000, p. 65)

As we see, the authors depending on whether they take into account the honest emitter, the hedonic receiver or the adaptive process, may use one or another definition. But, whatever the definition of aesthetic sense, of aesthetic faculty, of aesthetic judgment, of aesthetic attitude or aesthetic experience, they remain useful concepts to explore the continuity within the mind of the beholders, either human or animal.

References

Aharon I, Etcoff N, Ariely D, Chabris CF, O'Connor E, Breiter HC (2001) Beautiful faces have variable reward value: fMRI and behavioral evidence. Neuron 32(3):537–551

Alley TR, Cunningham MR (1991) Average faces are attractive, but very attractive faces are not average. Psychol Sci 2(2):123–125

Anderson M (1994) Sexual selection. Princeton University Press, Princeton

Aragona BJ, Liu Y, Curtis T, Stephan FK, Wang Z (2003) A critical role for Nucleus accumbens dopamine in partner preference formation in male prairie voles. J Neurosci 28(8):3483–3480

Balcombe J (2006) Pleasurable kingdom, animals and the nature of feeling good. Macmillan, New York

Bartels A, Zeki S (2004) The neural correlates of maternal and romantic love. Neuroimage 21:1155–1166

Bateman AJ (1948) Intra-sexual selection in *Drosophila*. Heredity 2:349–368

Bekoff M (2007) The emotional lives of animals. New World Library, California

Berridge KC, Kringelbach ML (2008) Affective neuroscience of pleasure: reward in humans and animals. Psychopharmacology (Berl) 199(3):457–480

Bishop MP, Elder ST, Health RG (1963) Intracranial self-stimulation in man. Science 140:394–396

Bösiger E 1974. The role of sexual selection in the maintenance of the genital heterogeneity of *Drosophila* populations and its genetic basis. In: van Abeelen JFH (ed) Genetic of behaviour. Elsevier, New York
Burley NT, Symanski R (1998) A taste for the beautiful: latent aesthetic mate preferences for white crests in two species of Australian Grassfinches. Am Nat 152(6):792–802
Buss DM (1989) Sex differences in human mate preferences: evolutionary hypotheses tested in 37 cultures. Behav Brain Sci 12(1):1–49
Buss DM (1994) The evolution of desire. Basic Books, New York
Buss DM (1998) The psychology of human mate selection: exploring the complexity of the strategic repertoire. In: Crawford C, Mahwah DKD (eds) Handbook of evolutionary psychology: ideas, issues, and applications. Lawrence Erlbaum Associates, New Jersey, pp 405–429
Buss DM, Barnes M (1986) Preferences in human mate selection. J Pers Soc Psychol 50(3):559–570
Cary MS (1978) The role of gaze in the initiation of conversation. Social Psychol 41(3):269–271
Cosmides L, Tooby J (1997) Evolutionary psychology : a primer. http://www.psych.ucsb.edu/research/cep/primer.html. Accessed 18 May 2012
Crook JH (1972a) Sexual selection, dimorphism, and social organization in primates. In: Campbell B (ed) Sexual selection and the descent of man 1871–1971. Aldine, Chicago, pp 180–230
Crook JH (1972b) The socio-ecology of primates. In: Crook JH (ed) Social behavior in birds and mammals: essays on the social ethology of animals and man. Academic, London
Cunningham MR (1986) Measuring the physical in physical attractiveness: quasi-experiments in the sociobiology of female facial beauty. J Pers Soc Psychol 50(5):925–935
Darwin C (1859) On the origin of species by means of natural selection or the preservation of favoured races in the struggle for life. John Murray, London
Darwin C (1871) The descent of man and selection in relation to sex, vol 1–2. John Murray, London
Darwin C (1872) The expression of the emotions in man and animals. John Murray, London
Dickinson A, Balleine B (2009) Hedonics: the cognitive-motivational interface. In: Kringelbach ML, Berridge KC (eds) Pleasures of the brain. Oxford University Press, Oxford, pp 74–84
Draganoiu T, Nagle L, Kreutzer M (2002) Directional female preference for an exaggerated male trait in canary (*Serinus canaria*) song. Proc Royal Soc Lond 269:2525–2531
Eagly AH, Wood W (1999) The origins of sex differences in human behavior: evolves dispositions versus social roles. Am Psychol 54(6):408–423
Eagly AH, Ashmore RD, Makhijani MG, Longo LC (1991) What is beautiful is good, but…: a meta-analytic review of research on the physical attractiveness stereotype. Psychol Bull 110(1):109–128
Fink B, Penton-Voak I (2002) Evolutionary psychology of facial attractiveness. Curr Directions Psychol Sci 11(5):154–158
Fisher RA (1930) The genetical theory of natural selection. Clarendon, Oxford
Fisher H (2004) Why we love: the nature and chemistry of romantic love. Henry Holt, New York
Fisher H, Aron A, Brown LL (2005) Romantic love : an fMRI study of a neural mechanism for mate choice. J Comp Neurol 493:58–62
Fisher K, Brown LL, Aron A, Strong G, Mashek D (2010) Reward, addiction, and emotion regulation systems associated with rejection in love. J Neurophysiol 104:51–60
Guilford T, Dawkins MS (1991) Receiver psychology and the evolution of animal signals. Anim Behav 42(1):1–14
Guilford T, Dawkins MS (1993) Receiver psychology and the design animal signals. Trends Neurosci 16(11):430–436
Haldane JBS (1932) The causes of evolution. Longmans, Green and Co, London
Hamilton WD, Zuk M (1982) Heritable true fitness and bright birds: a role for parasites? Science 218(4570):384–387
Hara E, Kubikova L, Hessler N, Jarvis E (2007) Role of the midbrain dopaminergic system in modulation of vocal brain activation by social context. J Neurosci 25(11):3406–3416

Haufe C (2007) Sexual selection and mate choice in evolutionary psychology. Biol Philos 23(1):115–128
Herrnstein RJ (1984) Objects, categories and discriminative stimuli. In: Roitblat HL, Bever TG, Terrace (eds) Animal cognition. Erlbaum, Hillsdale, pp 233–261
Hill G, Nolan PM, Stoehr AM (1999) Pairing success relative to male plumage redness and pigment symmetry in the house finch: temporal and geographical constancy. Behav Ecol 10(1):48–53
Hoquet T (2009) Darwin contre Darwin: Comment lire l'origine des espèces? Seuil, Paris
Huang Y, Hessler NA (2008) Social modulation during songbird courtship potentiates midbrain dopaminergic neurons. Plos ONE 3(10):e3281. doi:10. 1371/journal.pone.0003281
Huxley JS (1914) The courtship-habits of the great Crested Grebe (*Podiceps cristatus*). With an addition to the theory of sexual selection. Proc Zool Soc 35:491–562
Ikemoto S, Panksepp J (1999) The role of nucleus accumbens dopamine in motivated behavior: a unifying interpretation with special reference to reward-seeking. Brain Res 31(1):6–41
Insel TR (2003) Is social attachment an addictive disorder? Physiol Behav 79:351–357
Jones D (1995) Sexual selection, physical attractiveness and facial neoteny: cross-cultural evidence and implications. Curr Anthropol 36(5):723–748
Jones D, Hill K (1993) Criteria of facial attractiveness in five populations. Hum Nat 4(3):271–296
Kawabata H, Zeki S (2004) Neural correlates of beauty. J Neurophysiol 91:1699–1705
Kelley AE, Berridge KC (2002) The neuroscience of natural rewards: relevance to addictive drugs. J Neurosci 22(9):3306–3311
Kenrick DT, Montello DR, Gutierres SE, Trost MR (1993) Effects of physical attractiveness on affect and perceptual judgements: when social comparison overrides social reinforcement. Personal Soc Psychol Bull 19(2):195–199
Kreutzer M (2012) La beauté est-elle honnête? Pour la Science 412:63
Langlois JH, Roggman LA (1990) Attractive faces are only average. Psychol Sci 1(2):115–121
Langlois JH, Ritter JM, Roggman LA, Vaughn LS (1991) Facial diversity and infant preferences for attractive faces. Dev Psychol 27(1):79–84
Li NP, Kenrick DT (2006) Sex similarities and differences in preferences for short-term mates: what, whether, and why. J Pers Soc Psychol 90(3):468–489
Marcus DK, Miller RS (2003) Sex differences in judgments of physical attractiveness: a social relations analysis. Personal Soc Psychol Bull 29(3):325–335
Maynard Smith J (1989) Evolutionary genetics, 2nd edn. Oxford University Press, Oxford
Mayr E (1942) Systematics and the origin of species. Columbia University Press, New York
Milam EL (2010) Looking for a few good males. Female choice in evolutionary biology. Johns Hopkins University Press, Baltimore
Miller G (2000) Evolution of human music through sexual selection. In: Wallin NLB, Merker B, Brown S (eds) The origin of music. The MIT Press, Cambridge, pp 329–360
Mithen S (2006) The singing neanderthals, the origins of music, language, mind and body. Harvard University Press, Massachusetts
Møller AP (1992) Female swallow preference for symmetrical male sexual ornements. Nature 357:238–240
Møller AP (1996) Parasitism and developmental instability of hosts: a review. Oikos 75:189–196
Moore MM (1985) Nonverbal courtship patterns in women. Context and consequences. Ethol Sociobiol 6(4):237–247
Olds J, Milner P (1954) Positive reinforcement produced by electrical stimulation of septal area and others regions of rat brain. J Comp Psychol 47:419–427
Olson IR, Marshuetz C (2005) Facial attractiveness is appraised in a glance. Emotion 5(4):498–502
Panksepp J (1998) Affective neuroscience, the foundations of human and animal emotions. Oxford University Press, New York
Panksepp J (2005) Beyong a joke : from animal laugher to human joy. Science 308(5118):62–63
Perrett DI, Lee KJ, Penton-Voak I, Rowland D, Yoshikawa S, Burt DM, Henzik SP, Castles DL, Akamatsu S (1998) Effects of sexual dimorphism on facial attractiveness. Nature 394(6696):884–887

Petit C (1958) Le déterminisme génétique et psycho-physiologique de la compétition sexuelle chez *Drosophila melanogaster*. Bulletin biologique de la France et de la Belgique 92(3):248–329
Petts J (2000) Aesthetic experiences and the revelation of value. J Aesthet Art Critic 58(1):61–71
Portman A (1952) Animal forms and patterns. A study of the appearance of animals. Faber and Faber, London
Potter T, Corneille O (2008) Locating attractiveness in the face space: faces are more attractive when closer to their group prototype. Psychon Bull Rev 15(3):615–622
Rhodes G, Tremewan T (1996) Averageness, exaggeration and facial attractiveness. Psychol Sci 7:105–110
Roitblat HL (1985) Introduction to comparative cognition. W. H. Freeman and Company, New York
Romanes GJ (1881) Animal intelligence. Kegan Paul, Trench, London
Rosch E, Mervis CB (1975) Family resemblances: studies in the internal structure of categories. Cogn Psychol 7:573–605
Rowe C (1999) Receiver psychology and the evolution of multicomponent signals. Animal Behav 58:921–931
Ryan MJ (1990) Sexual selection, sensory systems and sensory exploitation. Oxford Surv Evol Biol 7:157–195
Sanderson CA (2010) Social psychology. Wiley, Hoboken
Schultz W (2006) Behavioral theories and neurophysiology of reward. Annu Rev Psychol 57:87–115
Slotten R (2004) The life of Alfred Wallace, the heretic in Darwin's court. Lewis, Colombia
Spinoza B de (1901) The chiefs works of Benedict de Spinoza, [translated from the Latin, with an introduction by Elwes RHM, vol 2 De Intellectus Emendatione. Ethica. (Selected Letters) Revised edition]. George Bell and Sons, London
Stein DJ (2009) Love and attachment: the psychobiology of social bounding. CNS Spectr 14(5):239–242
Suthers R, Vallet E, Kreutzer M (2012) Bilateral coordination and the motor basis of female preference for sexual signals in canary song. J Exp Biol 215:2950–2959
Swaddle JP (1996) Reproductive success and symmetry in zebra finches. Anim Behav 51:203–210
Swami V, Furnham A (2006) The science of attraction. Psychol 19(6):362–365
Symons D (1979) The evolution of human sexuality. Oxford University Press, New York
Symons D (1989) A critique of Darwinian anthropology. Ethol Sociobiol 10(1–3):131–144
Thompson RKR (1995). Natural and relational concepts in animals. In: Roitblat HL, Meyer J-A (eds) Comparative approaches to cognitive Science. MIT, Cambridge, pp 176–224
Thornhill R, Gangestad SW (1993) Human facial beauty: averageness, symmetry and parasite resistance. Hum Nat 4:237–269
Thornhill R, Gangestad SW (1999) Facial attractiveness. Trends Cogn Sci 3(12):452–460
Tooby J, Cosmides L (1990) The past explains the present: emotional adaptations and the structure of ancestral environments. Ethol Sociobiol 11(4–5):375–424
Tovée MJ, Tasker K, Benson PJ (2000) Is symmetry a visual cue to attractiveness in human female body? Evol Hum Behav 21:191–200
Trivers R (1972) Parental investment and sexual selection. In: Campbell BC (ed) Sexual selection and the descent of man, 1871–1971. Aldine, Chicago
Vallet E, Kreutzer M (1995) Female canaries are sexually responsive to special song phrases. Anim Behav 49:1603–1610
Wallace AR (1858) On the tendency of varieties to depart indefinitely from the original type; instability of varieties supposed to prove the permanent distinctiveness of species. J Proc Linnaean Soc Lond (Zool) 3:53–62
Wallace AR (1864) The origin of human races and the antiquity of man deduced from the theory of 'Natural selection'. J Anthropol Soc Lond 2:clxvi
Walster E, Aronson V, Abrahams D, Rottman L (1966) Importance of physical attractiveness in dating behavior. J Pers Soc Psychol 4(5):508–516
Watanabe S (2012) Animal aesthetics from the perspective of comparative cognition. In: Watanabe S, Kuczaj S (eds) Emotions of animals and humans: comparative perspective. Springer, Tokyo

Welsch W (2004) Animal aesthetics. Contemporary Aesthetics, online journal. http://www.contempaesthetics.org/newvolume/pages/article.php?articleID=243. Accessed 30 April 2012
Winkielman P, Halberstadt J, Fazendeiro T, Catty S (2006) Prototypes are attractive because they are easy on the mind. Psychol Sci 17(9):799–806
Wise RA (2004) Dopamine, learning and motivation. Nat Rev Neurosci 5:483–494
Wood W, Eagly AH (2000) Once again, the origins of sex differences. Am Psychol 55(9):1062–1063
Zahavi A (1975) Mate selection—a selection for a handicap. J Theor Biol 53:205–214
Zahavi A (1977) The cost of honesty (further remarks on the handicap principle). J Theor Biol 67:603–605
Zajonc RB (1980) Feeling and thinking: preferences need no inferences. Am Psychol 35(2):151–175
Zanna M, Pack S (1975) On the self-fulfilling nature of apparent sex differences in behavior. J Exp Soc Psychol 11(6):583–591

Chapter 13
Aesthetics and Reinforcement: A Behavioural Approach to Aesthetics

Shigeru Watanabe

Abstract Aesthetics has three aspects, the cognitive aspect, the hedonic aspect, and creation. In this chapter I focus on the second and third aspects. In some cases, human art, in either the visual or the auditory dimension, has a reinforcing property for non-human animals, suggesting that human art has hedonic value for them. The process of art creation has its own reinforcing property, and artists create art for its own sake; in other words, art has functional autonomy. Observation of primates in the laboratory and bowerbirds and songbirds in nature suggests that their art-like behaviour has functional autonomy, although this requires further experimental study. Their art-like products have, however, no reinforcing property for their conspecifics. Finally, I review evolutionary theories of aesthetics. Honest signals of the quality of the message sender and the physical constraints of materials must have played a role in the evolution of our aesthetics, but almost anything can be art if a considerable population of our species agree to accept it as an art. In other words, beauty is the verbal expression of our preferences.

Keywords Sensory reinforcement · Experimental aesthetics · Sexual selection · Animal art · Visual discrimination

13.1 Introduction

Beauty is a subjective experience or feeling, and aesthetics in the academic sense is a system of explanation of this feeling. Even though our subjective feeling is a private event and beauty is a phenomenon dependent upon the individual, we share this feeling with other persons. In this sense, beauty is based in inter-subjective understanding. Beauty in one society, however, may be different from that in another society. Also, beauty in one era differs from that in another era. In this sense, beauty is a socially constructed idea or institution. However, despite within-species differences in the sense of beauty, we have general preferences for certain types of beauty.

S. Watanabe (✉)
Department of Psychology, Keio University, Tokyo, Japan
e-mail: swat@flet.keio.ac.jp

T. Hoquet (ed.), *Current Perspectives on Sexual Selection,* History, Philosophy and Theory of the Life Sciences 9, DOI 10.1007/978-94-017-9585-2_13

In fact, preference tests across ten countries revealed a common pictorial preference for landscapes with trees and open space (Wypijewski 1997). Thus, *Homo sapiens* may have "*universal aesthetics*" just as we have "*universal grammar*" or "*universal moral.*"

Aesthetics has several aspects. One is the cognitive aspect. The empirical study of beauty was initiated by Gustav Fechner (1876), the founder of experimental aesthetics. He proposed three empirical methods of experimental aesthetics: the method of choice, the method of production, and the method of use. In the first method, an application of his famous psychophysical measurement methods to the perception of beauty, subjects were asked to compare stimuli with respect to pleasantness. Specifically, he asked visitors to a Dresden museum to compare two version of Holbein's "Madonna of Burgomaster Meyer". For Fechner, beauty was perception (see Watanabe 2012, for more details on the cognitive aspect of beauty).

The second aspect of aesthetics is the hedonic aspect. We feel pleasure when we see a beautiful picture or listen to beautiful music. In the framework of behaviourism, Berlyne (1971) proposed a new experimental aesthetics in which he introduced four methods of investigation: verbal judgment, psychophysics, statistical analysis, and measurement of exploratory behaviour. The last method is the most important. Because it measures behaviour, we can apply this method not only to humans but also to non-human animals. An organism explores one stimulus longer than another stimulus because the former is more reinforcing. According to Berlyne, beauty is sensory reinforcement; hence, aesthetics is the study of reinforcers. The first topic of this chapter is therefore "*beauty as sensory reinforcement.*"

The third aspect of aesthetics is creation or production. Sophisticated motor skills are required to produce art, but artistic behaviour must be maintained by self-reinforcement. Artists create art for its own sake, not only for money or social admiration. In other words, art has functional autonomy. Further, artists must know the "goal" of their productive behaviour; in other words, they know what their product should be. Therefore cognitive and reinforcing aspects, in addition to motor skills, are prerequisites for the creation of art. Do animals emit art-like behaviour without other reinforcement? Do they produce art spontaneously too? There are two approaches to the study of animal art. One is the experimental study of art-like behaviour in the laboratory, and the other is the exploration of human-art-like behaviour by animals in nature. I will discuss these studies in the third and fourth sections.

One last topic matters: it is the origin of aesthetics. One extreme view is that aesthetics is a uniquely human trait. In this view, our aesthetics is the result of cultural evolution rather than biological evolution. But if we look around the animal kingdom, we can find many "beautiful" animals and a variety of "beautiful" products by animals, for example, birdsong and the nests of many species, including invertebrates. If we accept that some animals also have their "aesthetics", we can trace the phylogeny of aesthetics and examine phylogenetic contingency. Even though aesthetics is based on a human standard, we can find a lot of beauty in non-human animals. If beauty exists in non-human animals, beauty in the human sense should have evolutionary origins. One traditional evolutionary approach, so-called evolutionary aesthetics (see Voland and Grammer 2003; Grammer et al. 2003; Thornhill 1998), explains aesthetics as the result of such natural selection. According to this theory, the origin of human aesthetics is in part an innate affiliation to plant and animal habitats (Wilson 1983). Hence, a beautiful stimulus is beautiful because it signals a good environment for us. Our ancestors had to find protective and safe habitats, select suitable food, and avoid dangerous animals. This evolutionary history may have resulted in preferences for particular landscapes. Landscape preference tests have revealed that humans prefer a "savanna-like" environment where our ancestors lived (Balling and Falk 1982). Aesthetic judgment is fast, just like moral judgment. Hence, Kaplan (1992) argued that quick, automatic decisions in environmental choice resulted in our environmental aesthetics.

Seeking high glucose and fat was adaptive for our ancestors, but such a tendency results in obesity in modern humans. Cheesecake is delicious because eating glucose and fat was adaptive during human evolution; thus our taste for cheesecake is a by-product of natural selection in the past. Pinker has argued that pleasure caused by art is something like a pleasure caused by cheesecake (1997). He hypothesizes that music in particular is a by-product of language. As I described above, both humans and songbirds have complex auditory communication and also show a reinforcing property of music. Their preference for particular musical stimuli might be a by-product of the evolution of a complex vocal communication system. This theory is interesting and plausible for understanding our preference for music, but it is difficult to identify the original adaptive value of visual beauty.

Another theory of beauty based on natural selection is the camouflage theory (Thayer 1909; Cott 1940). The theory tries to explain the beautiful appearance of animals by natural selection. The first scientist who proposed the camouflage theory was Wallace (1989). Darwin argued for sexual selection as an explanation of male decoration, but Wallace wished to explain their appearance only by natural selection. He examined the color patterns of butterflies and found that cryptic colors could be camouflage in some situations. According to Thayer (1909), who was a developer of military camouflage, every pattern and coloration in animals are camouflage to conceal them from predators or prey. Even male peacock feathers, a well-known example of sexual selection, can be camouflage. The camouflage theory argued that one principle, camouflage, is sufficient to explain every aspect of the "beautiful" appearance of animals. The pink feathers of the flamingo are also camouflage at sunrise and sunset. Sometimes cryptic features of animals have the function of concealing. Disruptive camouflage, which is a rather cryptic feature, conceals the outline of the animal. Several principles of adaptive coloration were identified by Cott (1940).

In a Darwinian context, there is a concern for the adaptive value of beauty. This raises the problem of honest signalling and sensory bias (see Kreutzer and Aebischer, this volume).

13.2 Aesthetics as Sensory Reinforcement

13.2.1 Theory of Reinforcement

Reinforcement is the key concept of this chapter, so I will describe the history of reinforcement theory briefly. Animals behave to get pleasure and avoid pain or distress, and they modify their behaviour by their experience. We humans behave in a similar way. Jeremy Bentham (1789) described human nature as a slave with two masters, pleasure and pain. This utilitarianism was, however, a speculative theory, not an empirical theory. Spencer (1870) claimed that a correlation exists between the feeling of pleasure and adaptive action. This was the first biological theory of reinforcement, but it too was a speculative, rather than empirical, theory.

The first empirical theory of reinforcement or pleasure, *the law of effect*, was proposed by Edward Thorndike (1911). He placed a cat in a so-called puzzle box, a box with a latch. The animal tried to escape from the box and succeeded after several trials and errors. With repetition of the test, the time to escape shortened; that is, the cat learned how to escape from the box. Thorndike noted that actions which accompany or are closely followed by satisfaction will be connected with the situation. Satisfaction was defined as a state which "the animal does nothing to avoid, often doing such things as attain and preserve it." An essential part of Thorndike's theory is the association between action and pleasure and his operational definition of satisfaction (pleasure).

Finally, B. F. Skinner (1950) provided an operational definition of reinforcement. Any event can be a reinforcer when the frequency of the behaviour preceding the reinforcement is increased. In other words, a reinforcer can be identified only a posteriori, not a priori, but with this definition psychologists no longer needed to trouble themselves about the biological nature of reinforcement.

13.2.2 Sensory Reinforcement

According to Skinner's theory, food, water, sex, and any other event can be a reinforcer. In fact, many studies have demonstrated that sensory stimulation has a reinforcing effect (Kish 1966). There are several ways to measure a reinforcing property. The simplest method is measuring the amount of consumption of the reinforcer, for example, measuring the amount of consumption of beer or wine to measure the reinforcing property of ethanol. Similarly, monkeys have been found to spend long periods of time manipulating complex puzzles without other reinforcement (Harlow 1950); therefore the puzzle has a reinforcing property for the monkeys. The method is simple and straightforward, but it has a problem in satiation of the reinforcer. One modification of this measurement method is the choice method, in which the subject has a choice between two or more events. For example, using a T-maze, Montgomery (1954) showed that rats chose an arm connected with a checkerboard maze where they could explore. Bengalese finches in a cage with perches stayed longer at a perch in front of a mirror than at one in front of a frosted mirror (Watanabe 2002).

Measuring operant behaviour that results in presentation of a sensory reinforcer is a clear demonstration of the reinforcing property. The classic example is Berlyne et al.'s (1964) experiment. A rat pressed a lever to light a lamp for 1 s while a yoked control rat that received the same amount of lighting with the same timing without lever pressing did not increase the number of lever presses. The increment of the operant clearly depended on the contingency between operant (lever press) and reinforcer (lighting). Monkeys also showed preference for particular fractal images over other fractal images as measured by choice and also by gazing time (Takebayashi and Funahashi 2009).

Complexity is an important aspect of sensory stimuli and constitutes a dimension of the intensity of the reinforcing value. Barnes and Baron (1961) compared three types of patterns, cross, circle, and rectangle, as well as a combination of the three, and found the cross and the combination caused a higher responding rate of bar-pressing in mice. Rensch (1957, 1958) compared preference for visual patterns in several species and reported that Capuchin monkeys, meerkats, and crows preferred regular patterns to irregular ones. Later, Anderson et al. (2005) confirmed similar preferences in monkeys.

Behavioural theories of sensory reinforcement have skipped analyses of the subjective experience of "pleasure." There is, however, some correlation between behavioural measures and verbal reports in humans. Berlyne (1972) showed line drawings to human subjects and performed a factor analysis of verbal reports and looking time. He found correlations of 0. 82 between scores on complexity-uncertainty (curiosity) and looking time, and 0. 40 between scores on hedonic value and looking time. We are not able to obtain subjective reports from animals, but the staying time of animals should reflect the two aspects of curiosity and hedonic value.

13.2.3 Reinforcing Property of Aesthetics

Biologically relevant or natural visual stimuli have reinforcing value (for example, an image of a conspecific for a Java sparrow, Watanabe 2002, and for macaques, Fujita et al. 1995, and Schwartz et al. 1980). Several studies have demonstrated the reinforcing effect of non-biologically relevant visual stimuli in primates. Wilson and Goldman-Rakic (1994) measured the gaze of rhesus monkeys as they viewed faces, colored pictures obtained from magazines, and colored patterns, and found that the monkeys spent more time looking at the faces and the pictures. Humphrey (1972) trained monkeys to press a button to see a Walt Disney film and compared the reinforcing properties of a repeated display of the same film versus a continuous display of a filmed story. The monkey preferred the continuous story to the looped film. There is, however, no experimental work on the reinforcing properties of paintings in animals.

I measured the reinforcing property of paintings by Mondrian and Kandinsky (Watanabe 2013). Both artists are classified as abstract painters, but their styles are different. Mondrian is a minimalist who demonstrates elemental aspects of pictures, such as line or colour, whereas Kandinsky is an expressionist who conveys his inner images or consciousness. I selected 10 different Mondrian paintings and 10 different paintings by Kandinsky and loaded them on two iPods. One iPod displayed Kandinsky and the other Mondrian in random series, and the two iPods were placed at the ends of two chambers of an apparatus. I examined staying time of mice. Only one of 22 mice showed a statistically significant preference for Kandinsky, but no others showed a consistently longer staying time at a particular artist. Thus, the reinforcing effect of paintings was very rare in mice.

Then I applied a pharmacological paradigm to analyze the discriminative stimulus property of paintings for mice. First, preference between the two types of paintings was measured by staying time at the paintings; then the mice were injected with morphine and restricted in a compartment with one particular type of painting, for example, Kandinsky's. The next day, they were injected with saline and restricted in another compartment with the other type of painting, Mondrian's. After such pharmacological training (a conditioned place preference procedure), the mice stayed longer at the compartment with the paintings associated with the morphine injection. Therefore, they had the ability to discriminate Kandinsky from Mondrian but did not have a preference between them.

Mammals are originally nocturnal animals, and rodents are typical non-visual animals. As a diurnal species, humans are rather exceptional mammals. On the other hand, most birds are diurnal and have highly developed visual cognition. Ikkatai and Watanabe (2010) examined the reinforcing property of paintings in a songbird, the Java sparrow. The procedure was basically similar to the mice experiment described above. In a long experimental chamber, two computer monitors displayed two of three different styles of paintings, Japanese, impressionist, or cubist, and another monitor displayed grey scale patterns. We found considerable individual differences, but 5 of 7 birds preferred cubists to impressionists, 3 preferred Japanese to cubists, 2 preferred cubists to Japanese and 6 did not show a differential preference between the impressionist and Japanese categories. This is an interesting observation, because Japanese paintings influenced the impressionists. This experiment did not clarify the mechanisms of reward, but did demonstrate differential behaviour directed at different styles of paintings.

The birds were also trained to discriminate Japanese paintings from impressionist paintings or cubists from impressionists by conventional operant discrimination with a food reward. To obtain food, the birds had to hop to one perch when a particular type of painting was displayed on a computer monitor but not hop to the perch when the other type of painting was displayed. Four of 5 birds successfully learned the task. Thus, both mice and songbirds demonstrated a discriminative stimulus property in their painting discrimination, but a reinforcing property was observed only in birds. Although procedural and stimulus differences must be considered, the two species nevertheless differed in preference.

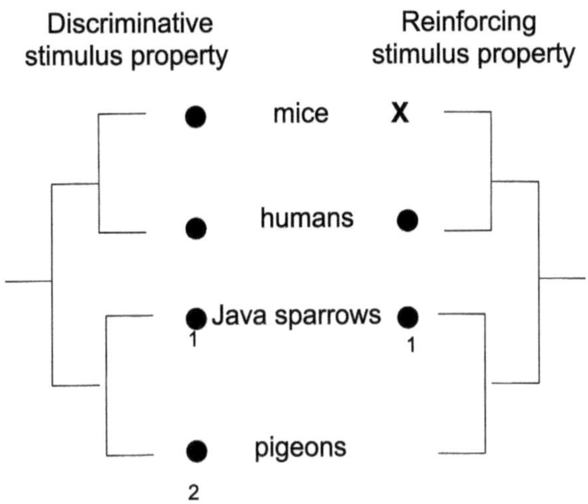

Fig. 13.1 Discriminative and reinforcing properties of paintings. 1: Ikkatai and Watanabe (2010), 2: Watanabe et al. (1995)

I have shown discrimination of painting styles in pigeons, in addition to mice and songbirds (Watanabe et al. 1995; Watanabe 2001), as well as discrimination in pigeons of good and bad paintings by children (Watanabe 2010). The pigeons used local cues to discriminate painting style but used global cues to discriminate their quality, good or bad (Watanabe 2011). Although available data are limited, these results suggest a reinforcing property of complex visual stimuli in animals with developed visual cognition. Figure 13.1 shows a summary of comparative studies of reinforcing and discriminative stimulus properties of paintings for non-human animals.

Hearing music often causes a pleasurable experience in humans (i. e., it has reinforcing properties for us). In other words, music is a sensory reinforcement. It is possible that the reinforcing property of music is human-specific (McDermott and Hauer 2004). Most published results have failed to demonstrate any reinforcing effect of music in animals, even in primates (chimpanzees, Howell et al. 2003; gorillas, Wells et al. 2006; the common marmoset and the cotton-top, McDermott and Hauser 2004, 2007). But Sugimoto et al. (2010) trained an infant chimpanzee to pull a string to hear consonant and dissonant musical stimuli played on the piano or marimba, and the chimpanzee preferred the consonant version.

We trained rats on a choice of levers associated with different styles of music, Bach and Stravinsky (Otsuka et al. 2009), but the rats did not show a strong preference for either style. There are few reports on musical reinforcement in birds. We applied the concurrent chain schedule procedure to pigeons to measure preference for Bach or Stravinsky (Watanabe et al. 2009) and found no clear reinforcing effects of music for pigeons. However, we demonstrated that Java sparrows show a preference for musical style, Bach, Vivaldi, to Schoenfeld and Carter, (Watanabe and Nemoto 1998). Thus, a musical stimulus can have a reinforcing property for Java sparrows. Although a recent study reported that chicks showed a preference for consonant music (Chiandetti and Vallortigara 2011), hens (McAdie et al.

1993) and pigeons (Watanabe et al. 2009) showed no musical preference. Thus, it is reasonable to hypothesize that there are no reinforcing properties of music for non-songbirds. The reinforcing effects of music have not been thoroughly examined, and, at present, humans and songbirds (Java sparrows) are exceptional species in showing preferences for particular types of music (Gess 2007). We have discovered that musical stimuli do not have a reinforcing property for goldfish, but the fish are able to discriminate them (Shinozuka et al. 2013). Figure 13.2 provides a summary of comparative studies of reinforcing and discriminative stimulus properties of music in non-human animals.

These results suggest that music has a reinforcing property for species with complex vocal communication that must be learned through experience. It is likely that the reinforcing property of particular types of auditory stimuli promote their vocal learning.

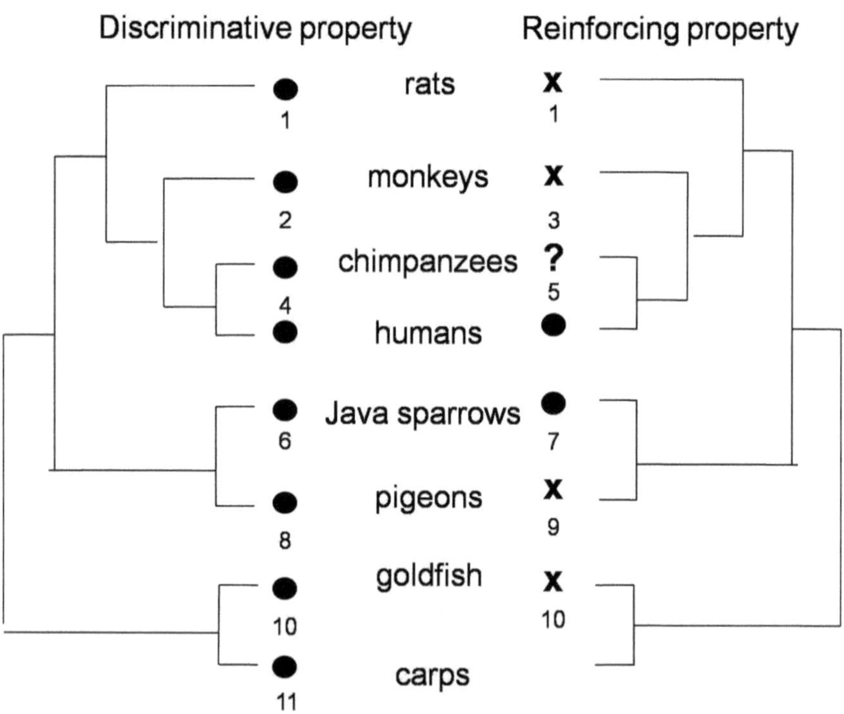

Fig. 13.2 Discriminative and reinforcing properties of music. 1: Ostuka et al (2009), 2: Okaichi and Okaichi (2001), 3: McDermott and Hauser (2004, 2007), 4: Poli and Previde (1991), 5: Howell et al. (2003); Wells et al. (2006); Sugimoto et al. (2010), 6: Watanabe and Sato (1999); Watanabe et al. (2005), 7: Watanabe and Nemoto (1998), 8: Porter and Neuringer (1984), 9: Watanabe et al.(1995), 10: McAdie et al. (1993) but see Chiandetti and Vallortigara (2011), 11: Shinozuka et al. (2013), 12: Chase (2001)

13.3 Art-Like Behaviour in the Laboratory

As I have already pointed out, some animals have a beautiful appearance, but these morphological features are the result of phylogenetic contingency and not the result of ontogenetic contingency or individual experience. Here we discuss the creation of art-like objects by animals with a particular focus on the role of reinforcement in art production. We first discuss art-like behaviour in the laboratory and then art-like behaviour in nature.

13.3.1 Motor Skills

Because art is a product of human behaviour or culture, the first approach is teaching animals human art. Although there have been some studies of the drawing ability of non-human animals (see Zeller 2007), most animal paintings are made for exhibition rather than for scientific research. In fact, there have been many animal "artists" on web sites. The most popular animal art is paintings by chimpanzees (see Lenain 1997). The oldest reports of chimpanzee paintings are included in "*The ape and the child*" by Kellogg and Kellogg (1933) and "*Infant ape and human child*" by Kohts (1935). Morris (1962) documented 32 cases of drawings and paintings by primates. According to Morris, chimpanzee art has symmetrical coverage, rhythmical variations, and beautiful color contrasts. Thus, the pictures have a kind of style and are not random scribbling. Saito et al. (2010) compared chimpanzees' paintings with those of human infants and concluded that chimpanzees did not imitate model drawings. They observed that the chimpanzees drew some marks that resembled the models but they did not complete partial facial outlines.

According to Morris (1962), Congo (a chimpanzee) had a criterion of drawing "completion". If the paper was removed prematurely, he became angry, and after he reached his criterion it was difficult to make him continue. He stopped drawing when he felt it was done and did not add to the drawing even if encouraged to continue. Another interesting observation was the collaboration of the female chimpanzee Kuna with a professional painter (Lenain 1997). At times Kuna responded favorably to the paintings that the artist contributed, and at other times she rubbed out his figures and waited for him to add something. The behaviour suggests that the chimpanzee had a sense of what a completed painting should look like, and that completion is reinforcing.

Scribbling by animals looks like non-representational pictures, in some sense. In 2005, paintings by the chimpanzee Congo came under the hammer at an art auction and were sold for £ 12000. According to Hussain (1965), paintings by chimpanzees have been mistaken for professional art. I once showed a painting by an elephant along with paintings by children to human subjects, and nobody realized that an animal had made one of the paintings. Noll (1966) reported that people judged computer-generated images to be paintings by Mondrian. But there may be special features in human-made non-representational paintings. Hawley-Dolan and Winner

(2011) presented pictures by professional artists, children, and animals to participants and asked their preference. The pictures were correctly or incorrectly labeled as "artist" or "child". The participants preferred professional paintings even when the labels were incorrect. These results suggest that humans have a sense of human paintings even when the paintings are non-representational.

13.3.2 Representation

I once bought a painting by an elephant through an Internet auction: a picture of flowers. Later, I found quite similar elephant paintings on YouTube. The elephants show several fixed patterns of painting and mass-produce their art. Some paintings by elephants look like representational paintings: flowers, other elephants, etc. Representational painting involves mapping or transcription of 3-dimensional objects onto a 2-dimensional canvas. It requires coordination of visual cognition and motor skills. The difficulty lies in deciding whether elephant paintings are really transcriptions of external objects or are rather repertoires of painting behaviour induced by different stimuli. Through training, animals may learn to draw something when object A or a signal by a trainer is presented, while drawing a different something when object B or another signal by the trainer is presented. This type of training is called "conditional discrimination," in which conditional stimulus A or B directs that behaviour a or b should be done. For representational transcription, there should be a general correspondence between many objects and many paintings. In other words, animals have to learn the general rule of "representation" to create representational paintings of novel objects never used during training. Levy (1992) described dolphins who drew shapes with a brush (circles or Ts, etc.) that had also been drawn by a trainer on a different canvas; this demonstrates that they could transcribe the drawing. Dolphins have the ability to mimic human movement without any particular reward (Herman 2002). Thus, dolphins' transcription drawing may be explained in terms of their general mimicking ability.

We often ask children "What is this?" or "What you want to draw?" during their painting behaviour. A direct answer to the question of whether a drawing is representational would be obtained by asking animals about their intention. A chimpanzee who had been taught sign language named one of her drawings "bird" in sign language (as described in Gucwa and Ehmann 1985). Unfortunately, this communicative approach using signing in a face-to-face situation might be contaminated with unconscious signaling by the experimenter.

13.3.3 Functional Autonomy

It is well documented that chimpanzees spontaneously paint without food reinforcement (Boysen et al. 1987; Tanaka et al. 2003). However, most of the painting experiments with primates have been carried out in face-to-face situations, and thus it is impossible to exclude possible social reinforcement. Schiller (1951) described an episode in which his chimpanzee Alpha once tried to draw on a leaf when suitable paper was not available. Such an observation, although anecdotal, supports the notion of self-reinforcement in chimpanzee drawing. In self-reinforcement, behaviour itself, not outcomes of the behaviour, must be a reinforcer. Visual feedback might have reinforcing value, because it has been shown that drawing behaviour on a touch screen decreased when no trace of the drawing appeared on the screen (Tanaka et al. 2003). Interestingly, Morris described how a food reward disturbed, rather than enhanced, painting behaviour of chimpanzees. According to Morris the ape quickly learnt to associate drawing with getting the reward but as soon as this condition had been established the animal took less and less interest in the lines it was drawing (Morris 1962). In other words, even though drawing has a reinforcing effect, a stronger reinforcer, the food, masks the effect. It was, however, also possible to train chimpanzees to trace model lines on a touch screen by food reinforcement (Iversen and Matsuzawa 1997). Tracing by itself may not have enough reinforcing effect, and the chimpanzees may have maintained the tracing only by a food reward.

Another possible explanation of drawing in captive animals is a by-product of captivity. Gucwa and Ehmann (1978) noted that Siri, an Asian elephant, created a lot of drawings, even though her trainer never trained her to do so or rewarded her for such behaviour. Elephants have good motor skills in using sticks or stones, and captive ones often scratch the floor or ground with these materials. That such behaviour occurs spontaneously without training suggests functional autonomy or self-reinforcement, but captive animals usually have a limited environment and often invent new behaviours. Thus, drawing may be a by-product of captivity.

13.3.4 Musical Performance

An interesting example of human-like animal art was observed in a dancing parrot named Snowball (Patel et al. 2009). Complex dance functions as a sexual display in many different species, ranging from insects to fish, birds, and mammals. But in Snowball's case, the dance did not function as an innate sexual display; Snowball was dancing to human music. Patel et al. (2009) experimentally analyzed this parrot's synchronization of his body movements to music. When the tempo changed, the bird spontaneously adjusted his rhythmic movements to fit the new rhythm. Such adjusting is observed in animals that have complex vocal learning skills, such as songbirds, cetaceans, and pinnipeds. Snowball's exact musical history is unknown, but he started rhythmic bobbing movements to music soon after his owner obtained him. Hasegawa et al. (2011) trained budgerigars (relatives of parrots) to peck a key to synchronize with a rhythm given by metronome-like stimuli and found they could match a wide range of tempos. Thus, vocal-mimicking species may have the capability for rhythmic synchronization. Bonobos and chimpanzees do not seem to have such synchronization ability (Kugler and Savage-Rumbaugh 2002). Thus, the ability to synchronize body movements to music seems to have evolved in several evolutionary lines independently.

There are reports of "musical tool" use in animals, just as in animal painting. In a traditional Japanese animal exhibition, four varied tits simultaneously played four different instruments: two different drums, strings, and a bell (Koyama 1999). That type of animal exhibition disappeared around 100 years ago in Japan, but there were many other examples of tool use by tits at that time. These performances were surely the result of training with a food reward, not by self-reinforcement.

13.3.5 *Value of Animal Art for Other Animals*

Humans not only create artistic products but also enjoy them. One essential point of human art is the reinforcing property of artistic products for other members or for society. On the other hand, non-human primates do not enjoy their products, because they often tear the paper after drawing on it, suggesting that the products have no reinforcing value for them. They do not keep them to enjoy them. This is a big difference between animals' art creation behaviour and our own. As I pointed out earlier, human art is socially constructed. Society should reach a consensus about art; this consensus gives artistic products their value within the society. Therefore the products have a reinforcing property for members of the society. Figure 13.3 summarizes the limits of animal art. Animals may be able to discriminate and prefer

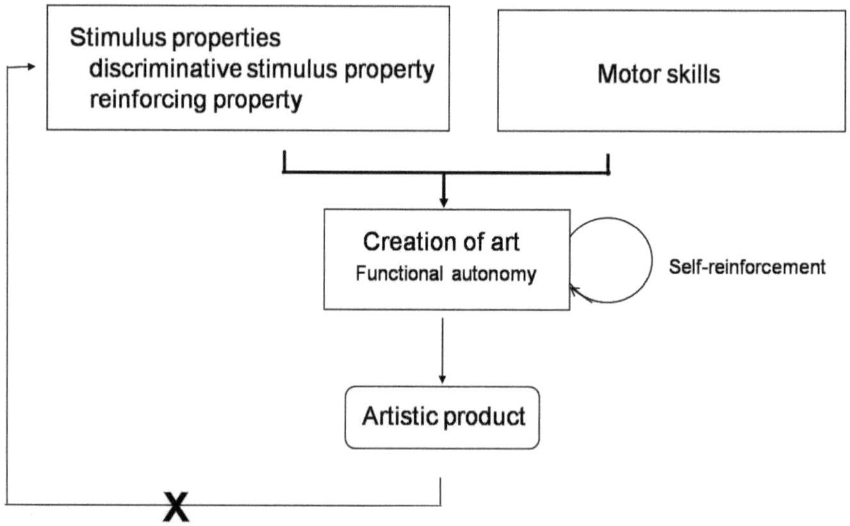

Fig. 13.3 Animal aesthetic. Some animal show discriminative property (cognition) and reinforcing property (pleasure) by art. They probably have self-reinforcement (functional autonomy) of creation of art-like product, but their products do not have reinforcing value to other members

particular sensory stimuli, and they may have the motor skills to make "artistic" products and do so by self-reinforcement without conventional reinforcers such as food reward. They do not, however, enjoy the products.

13.4 Art-Like Behaviour in Nature

The animal art works described above are human-art-like behaviours of animals that take place in artificial settings. But some animals show art-like performance in nature.

13.4.1 Animal Music

The best-documented example of animal music is bird song, but other biologically distant species, including whales, gibbons, and also mice, sing complex songs. To humans, birdsong sounds like music (Rothenberg 2005). In fact, "zoomusicology" is the field of study of animal music from the viewpoint of aesthetics. Recently, musicologist Hollis Taylor and cognitive scientist Dominique Lestel (Taylor and Lestel 2011) analyzed songs of the pied butcherbird in the same way that Western music is analyzed. They found many common features between the songs and the music, for example, variation of theme, inversion, additive and divisive rhythms, accelerandos, crescendos, and decrescendos. Bird song researchers Marler and Peters (1981) also pointed out that birdsongs include improvisation, memorizing and replacing a theme, and systematic transformation. This suggests an appetite for novelty.

Darwin (1781) suggested that musical notes and rhythm was first acquired by the male or female progenitors of mankind for the sake of charming the opposite sex. The main function of birdsong is sexual display, but improvement of song during acquisition is not directly maintained by sexual reinforcement but by a kind of self-reinforcement. Learning of song in songbirds consists of two phases, a sensory phase and a sensori-motor phase. Infants learn the template of songs mostly from their fathers (sensory phase), then young birds start singing a sub-song and gradually shape it into a final crystallized song by matching it to the template that they heard in the sensory phase (sensori-motor phase). The adults sing the finally established (crystallized) song as a sexual display. If songbirds sing songs as a sexual display, male birds do not need to improve their singing after they have secured a mate; however, blackbirds and willow warblers develop their song musically long after mating, and Australian magpies continue to improve their song for a long time. This suggests that they sing for its own sake (Kaplan 2005, 2009). Adult male zebra finches sing crystallized songs for females, but they also sing different variants when they are alone (Stepanek and Doupe 2010). These observations may suggest that singing has functional autonomy in these species.

Behaviours analogous to drumming or other instrumental music are quite rare in animals, the most prominent examples being palm cockatoos (which use sticks to drum on hollow trees, Wood 1984), many species of woodpecker (who seek out resonant trees for display drumming, Dodenhoff et al. 2001; Stark et al. 1998) and various desert rodent species who "drum" with their hind feet on the ground (Randel 1997). Fitch (2005) reported that apes use percussion in nature. But it is not clear that they maintain this musical performance for its own sake.

13.4.2 Animal Architecture

Von Frisch displayed many beautiful products of animals in his interesting book, "*Tiere als Baumeister*" ("*Animal architecture*" 1974). These animal products are impressive because they are constructed in the environment. The best-known example is the bowerbird (see Hansell 2000, 2007; Rotheberg 2005). Male bowerbirds construct complex bowers and decorate them with many colourful materials to attract females (see Madden 2008). Within the bowerbird family, 17 of 20 species build bowers (Frith and Frith 2004). One type is the "maypole", which has a column with decorations, and another type is the "avenue", which is a kind of decorated tunnel. Several ideas have been developed to explain the origin of the bowers. One is the transfer hypothesis, which argues that the original bower was a simple court for a dance display, and eventually the court itself became a display (Gillard 1963). The second is the nest hypothesis, which claims that the bower derived from the nest (Collias and Collias 1984). But females in a tree build the nest, and the bower is constructed by males on the ground. The third is the threat reduction hypothesis (Borgia and Mueller 1992). The avenue and maypole provide the female a place of security when she observes the males violently displaying their dance. But decoration is not necessary to give the female a place for protection.

Birds in one population build and decorate their nests within a particular range of similarity in that population. For example, one group of spotted bowerbird uses solanum berries for decoration, but another group uses reddish pink glass and fruits. Their building is normative and collective. Although bower building has its genetic basis, it is not a genetically fixed action pattern but is flexible in response to learning and experience. Age-dependent progress supports the role of learning, and young birds often visit adult bowers when the owners are absent. Madden (2008) argued that bower building of bowerbirds is considered to be "*cultural behaviour*" in the standard of primate culture. Furthermore, Lestel (2011) pointed out that the phylogenetic basis of western art lies in the practice of the distinctive features found in the species. There have been no empirical studies of originality of the bower and reproductive success, but individual differences in the bowers may suggest that distinctiveness creates attractiveness.

If the decoration is removed experimentally, the chance of copulation is reduced (Borgia 1985); thus females choose males based on the quality of their bowers. Several hypotheses are based on the possibility that the bower is an honest signal. One is the healthy mate hypothesis. Doucet and Montgomery (2003) suggest that good bowers indicate fewer parasites or a good health condition. The second possibility is that the bower is a signal of motor skills (Miller 2000). Male great bowerbirds make courts with grey and white objects that increase in size with distance from the avenue entrance (Endler et al. 2010). This arrangement creates a false perception of size and distance. When the experimenter disrupted their size-distance gradients, the males reconstructed their gradients again, and there was little difference from the original structure that had created a false visual perspective for the audience. This complicated construction requires highly developed motor skills and higher cognitive ability.

The third possibility is that the bower is a signal of social rank. The males steal decoration materials from each other; therefore more decoration means higher social ranking. The berries attract females but also invite rival males who may attack and destroy the bower (Madden 2002), so when additional berries were placed experimentally close to the bower, the male birds removed the berries. Hence, the trade-off results in an optimal amount of ornament, the honest signal of the status of the owner.

The fourth possibility is that the bower acts as a signal of quality of the brain, because complicated behaviour requires much information processing. Madden (2001) took X-rays of the skulls of bowerbirds and reported that bowerbirds had larger brains in comparison to their relative species who do not build bowers. Thus, a complex bower indicates a big brain or sophisticated processing in the brain. Day et al. (2005) examined five brain regions from four species of bowerbirds and found a significant relation between the complexity and size of their cerebellums. Thus, the complexity of bower is a signal of the size of the cerebellum. But these data represent between-species comparisons, and no data about individual-level correlations between the complexity of the bower and volume of the brain have been reported.

13.4.3 Functional Autonomy

As I have already pointed out, human art has functional autonomy, and some researchers have argued that humans have the "desire" or "instinct" for art as a result of natural selection, including sexual selection (Miller 2000; Dutton 2009); others, Driscoll (2006) for example, have claimed that art is the result of cultural selection. Male bowerbirds arrange and rearrange the decorations in their bower (Rogers and Kaplan 2006). Although this suggests that the bowerbirds possess a kind of criterion or standard for their constructions, direct evidence demonstrating functional autonomy of art-like behaviour in nature is not available. However, both song learning and bower construction require much effort and time. Even though the principal function of the song and the construction is sexual display, the long sequence of behaviour during song learning and bower construction may be maintained by self-reinforcement. It is difficult to assume that sexual reinforcement can maintain such long sequences of behaviour because of the long delay between the behaviour and the reinforcement. It is plausible to assume that each sequential behaviour is maintained by self-reinforcement, which may be the basis of higher functional autonomy.

13.5 Conclusion

"Beauty" is an anthropocentric idea. Comparative studies suggest that we verbally describe something as beautiful when we prefer it. Non-human animals also have such preferences. Sometimes human preferences and animal preferences overlap; sometimes they do not. The overlap or convergence stems from common functions of beauty, such as complexity, signalling of health conditions, and common physical bias. Human art behaviour has functional autonomy, art for art's sake. In animals' art-like behaviour, both in the laboratory and in nature, functional autonomy can be observed at a quite basic level, as self-reinforcement. But the reinforcing value of outcomes of the art-like behaviour for conspecifics cannot be observed in animals except for sexual display. The reinforcing value of the art product should be the result of cultural evolution, not natural evolution, and this is a uniquely human characteristic.

Acknowledgements This research was supported by Global COE program at Keio University.

References

Anderson JR, Kuwaharam H, Kuroshimma H, Leighty KA, Fujita K (2005) Are monkeys anesthetists? Rensch (1957) revisited. J Exp Psychol Anim Behav Process 31:71–78

Balling JD, Falk JH (1982) Development of visual preference for natural environments. Environ Behav 14:5–28

Barnes GW, Baron A (1961) Stimulus complexity and sensory reinforcement. J Comp Physiol Psychol 54:466–469

Bentham J (1789) Principles of morals and legislation. Haffner, New York

Berlyne DE (1971) Aesthetics and psychobiology. Appleton-Century-Crofts, New York

Berlyne DE (1972) Ends and means of experimental aesthetics. Can J Psychol 26:303–325

Berlyne EE, Salapatek PH, Gelman RS, Zener SL (1964) Is light increment really rewarding to the rat? J Comp Physiol Psycho 58:148–151

Borgia G (1985) Bower quality, number of decorations and mating success of male satin bowerbirds (*Pilonorhynchus violacus*): an experimental analysis. Anim Behav 33:266–171

Borgia G, Mueller U (1992) Bower destruction, decoration stealing and female choce in the spotted bowerbird *Chlamydera maculata*. Emu 92:11–18

Boysen ST, Berntson GG, Prentice J (1987) Simian scribbles: a reappraisal of drawing in the chimpanzee (*Pan troglodytes*). J Comp Psychol 101:82–89

Chase AR (2001) Music discrimination by carp (*Cyprinus carpio*). Anim Learn Behav 29:336–353

Chiandetti C, Vallortigara G (2011) Chicks like consonant music. Psychol Sci 22:1270–1273
Collias NE, Collias EC (1984) Nest building and bird behaviour. Princeton University, Princeton
Cott H (1940) Adaptive coloration in animals. Methuen, London
Darwin C (1781) The decent of man and selection in relation to sex. John Murry, London
Day LB, Westcott DA, Olster DH (2005) Evolution of bower complexity and cerebellum size in bowerbirds. Brain Behav Evol 66:62–72
Dodenhoff DJ, Stark RD, Johnson EV (2001) Do woodpecker drums encode information for species recognition? Condor 103:143–150
Doucet SM, Montgomery R (2003) Multiple sexual ornaments in satin bowerbirds: ultraviolet plumage and bower signal different aspects of male quality. Behav Ecol 14:503–509
Driscoll C (2006) The bowerbirds and the bees: miller on art, altruism and sexual selection. Philos Psychol 19:507–526
Dutton D (2009) The art instinct: beauty, pleasure and human evolution. Bloomsbury, New York
Endler JA, Endler LC, Doerr NR (2010) Great bowerbirds create theaters with forced perspective when seen by their audience. Curr Biol 20:1679–1684
Fechner von GT (1876) Vorschule der Aesthetik. Breitkopf & Hartel, Leipzig
Fitch WT (2005) The evolution of music in comparative perspective. Ann N Y Acad Sci 1060:1–20
Frith CB, Frith DW (2004) Bower birds. Oxford University, London
Fujita K, Watanabe K (1995) Visual preference for closely related species by Sulawesi macaques. Am J Primatol 37:253–261
Gess A (2007) Birds like music, too. *Science* 317:1864
Gillard ET (1963) The evolution of bowerbirds. Sci Am 209:38–46
Grammer K, Fink B, Moller AP, Thornhill R (2003) Darwinian aesthetics: sexual selection and biology of beauty. Biol Rev 78:385–407
Gucwa D, Ehmann J (1985). To whom it may concern. An investigation of the art of elephants. W. W. Norton & Company: New York
Hansell M (2000) Bird nests and construction behaviour. Cambridge University, New York
Hansell M (2007) Built by animals. Oxford University, New York
Harlow H, Harlow F, Kuenne M, Meyer DR (1950) Learning motivated by a manipulation drive. J Exp Psychol 40:228–234
Hasegawa A, Okanoya K, Hasegawa T, Seki Y (2011) Rhythmic synchronization tapping to an audio–visual metronome in budgerigars. Scientific Reports, 1, doi:10.1038/srep00120
Hawley-Dolan A, Winner E (2011) Seeing the mind behind the art: people can distinguish abstract expressionists paintings from highly similar paintings by children, chimps, monkeys, and elephants. Psychol Sci 22:435–441
Herman LM (2002) Vocal, social, and self-imitation by bottle nosed dolphins. In: Dautenhahn K & Nehaniv CL (eds) Imitation in animals and artifacts. MIT press, Cambridge, pp 63–108
Howell S, Schwandt M, Fritz J, Roeder E, Nelson C (2003) A stereo music system as environment enrichment for captive chimpanzees. Lab Anim 32:31–36
Humphrey NK (1972) 'Interest' and 'pleasure': two determinants of a monkey's visual preferences. Perception 1:395–416
Hussain F (1965) Queques problemes d'esthetiques experimentale. Science de l'Art 2:103–114
Ikkatai Y, Watanabe S (2010) Discriminative and reinforcing properties of paintings in Java sparrows (*Padda oryzivora*). Anim Cogn 14:227–234
Iversen I, Matsuzawa T (1997) Model-guided line drawing in the chimpanzee (*Pan groglodytes*). Jpn Psychol Res 39:154–181
Kaplan S (1992) Environmental preference in a knowledge-seeking, knowledge-using organism. In: Barkow JH, Cosmides L, Tooby J (eds) The adapted mind: evolutionary psychology and the generation of culture. Oxford University, New York, pp 581–591
Kaplan G (2005) The Australian magpie: biology and behaviour of an unusual songbird. University of New South Wales Press, Sydney
Kaplan G (2009) Animal and music: between cultural definitions and sensory evidence. Sign System Studies 37:75–101
Kellogg WN, Kellogg LA (1933) The ape and child. McGraw Hill, New York
Kish GB (1966) Studies of sensory reinforcement. In: Honig WH (ed) Operant behaviour. Appleton-Century-Crofts, NewYork, pp 109–159

Kohts N (1935) Infant ape and human child. Scientific memories of the Museum Darwininun, (Moskow) cited in Morris (1962)
Koyama S (1999) Arts by varied tits (in Japanese). Hosei-University Press, Tokyo
Kugler K, Savage-Rumbaugh S (2002). Rhythmic drumming by Kanzi an adult male bonobo (*Pan paniscus*) at the language Research Center. 25th Meeting of the American Society of Primatologists
Lenain T (1997) Monkey painting. Reakton Books, London
Lestel D (2011) Non-human artistic practices: a challenge to the social science of the future. Soc Sci Inf 50:505–512
Levy BA (1992) Psych aesthetics dolphin project. J Am Art Ther Assoc 9:193–197
Madden JR (2001) Sex, bower and brain. Proc R Soc Lond B 268:269–833-838
Madden J (2002) Bower decorations attract females but provoke other male spotted bowerbirds; bower owners resolve this trade-off. Proc R Soc Lond B 269:1347–1351
Madden JR (2008) Do bowerbirds exhibit cultures? Anim Cogn 11:1–12
Marler P, Peters SS (1981) Sparrows learn adult song and more from memory. Science 213:780–782
McAdie TM, Foster TM, Temple W, Mattews LR (1993) A method for measuring the aversiveness of sounds to domestic hens. Appl Anim Behav Sci 37:223–238
McDermott J, Hauser MD (2004) Are consonant intervals music to their ear? Spontaneous acoustic preferences in non-human primates. Cognition 94:B11–B21
McDermott J, Hauser MD (2007) Nonhuman primates prefer slow tempos but dislike music overall. Cognition 104:654–668
Miller GF (2000) The mating mind. How sexual selection choice shaped the evolution of human nature. Bull Psychol Art 2:20–25
Montgomery KC (1954) The role of the exploratory drive in learning. J Comp Physiol Psychol 47:60–64
Morris D (1962) The biology of art. Methuen, London
Noll MA (1966) Human or machine: a subjective comparison of Piet Mondrian's "composition with line2 (1917) and a computer-generated picture. Psychol Rec 16:1–10
Okaichi Y, Okaichi H (2001) Music discrimination by rats. Jap J Anim Psychol 51:29–34
Otsuka Y, Yanagi J, Watanabe S (2009) Discrimination and reinforcing stimulus properties of music for rats. Behav Process 80:121–127
Patel A, Iversen J, Bregman M, Schulz I (2009) Experimental evidence for synchronization to a musical beat in a nonhuman animal. Curr Biol 19:827–830
Pinker S (1977) How the mind works. Norton, New York
Poli M, Previde EP (1991) Discrimination of musical stimuli by rats (*Rattus norvegicus*). Int J Comp Psychol 5:7–18
Porter D, Neuringer A (1984) Musical discrimination by pigeons. J Exp Psychol Anim Behav Process 10:138–148
Randell JA (1997) Species-specific foot drumming in kangaroo rats: *Dipodomys ingens, D. deserti. D spectabilis*. Anim Behav 54:1167–1175
Rensch B (1957) Asthetische Faktoren bei Farb- und Formbevorzugungen von Affen. Z Tierpsychol 14:71–99
Rensch B (1958). Die Wirksomkeit aesthetischer Factoren bei Wirbeltieren. Z Tierpsychol 15:447–461
Rogers LJ, Kaplan G (2006). Elephants that paint, birds that make music. Cereberum 2:2–14
Rothenberg D (2005) Why birds sing. A journey into the mystery of bird song. Penguin, London
Saito A, Hayashi M, Takeshita H, Matsuzawa T (2010) Drawing behaviour of chimpanzees ad human children: the origin of the representational drawing. Proceedings of the 3rd International Workshop on Kansei, 111–114
Schiller P (1951) Figural preferences in the drawings of a chimpanzee. J Comp Physiol Psychol 44:101–111
Schwartz GG, Rosenblum LA (1980) Novelty, arousal, and nasal marking in the squirrel monkey. Behav Neural Biol 28:116–122
Shinozuka T, Ono H, Watanabe S (2013) Discriminative and reinforcing properties of music for goldfish. Behav Process 99:26–33

Skinner BF (1950) Are the theories of learning necessary? Psychol Rev 57:193–216
Spencer H (1870) The principles of psychology. Appleton-Century-Crofts, New York
Stark RD, Dodenhoff DJ, Johnson EV (1998) A quantitative analysis of woodpecker drumming. Condor 100:350–356
Stepanek L, Doupe AJ (2010) Activity in a cortical-basal ganglia circuit for song is required for social context-dependent vocal variability. J Neurophysiol 104:2474–2486
Sugimoto T, Kobayashi H, Noritomo N, Kiriyama Y, Takeshita H, Nakamura T, Hashiba K (2010) Preference for consonance music over dissonance music by an infant chimpanzee. Primates 51:7–12
Takebayashi M, Funahashi S (2009) Monkeys exhibited preference for biologically non-significant visual stimuli. Psychologia 52:147–161
Tanaka M, Tomonaga M, Matsuzawa T (2003) Finger drawing by infant chimpanzees (*Pan troglodytes*). Anim Cogn 6:245–251
Taylor H, Lestell D (2011) The Australian pied butcherbird and the nature culture continuum. J Interdiscip Music Stud 5:57–83
Thayer G (1909) Concealing coloration in the animal kingdom. Macmillam, New York
Thorndike EL (1911) Animal intelligence. Macmillan, New York
Thornhil R (1998) Darwinian aesthetics. In: Crawford C, Krebs DL (eds) Handbook of evolutionary psychology. Erlbaum: Mahwash, pp 543–572
Tolman EC (1932) Purposive behaviour in animals and men. Appleton-Century-Crofts, New York
Voland E, Grammer K (eds) (2003) Evolutionary aesthetics. Springer, Berlin
von Frisch K (1974) Tieres als Baumeister. Ullstein, Frankfurt a. M.
Wallece AR (1889) Darwinism. Macmillam, London
Watanabe S (2001) Van Gogh, Chagall and pigeons. Anim Cogn 4:147–151
Watanabe S (2002) Preference for mirror images and video image in Java sparrows (*Padda oryzivora*). Behav Process 60:35–39
Watanabe S (2010) Pigeons can discriminate "good" and "bad" paintings by children. Anim Cogn 13:75–85
Watanabe S (2011) Discrimination of painting style and quality: pigeons use different strategies for different tasks. Anim Cogn 14:797–808
Watanabe S (2012) Animal aesthetics from the perspective of comparative cognition. In: Watanabe S, Kuczaj S (eds) Emotions of animals and humans: comparative perspectives. Springer, Tokyo
Watanabe S (2013) Preference for and discrimination of paintings by mice. Plos ONE 8(6):e65335, doi:10.137
Watanabe S, Nemoto M (1998) Reinforcing property of music in Java sparrows (*Padda oryzivora*). Behav Process 43:211–218
Watanabe S, Sato K (1999) Discriminative stimulus properties of music in Java sparrows. Behav Process 47:53–58
Watanabe S, Wakita M, Sakamoto J (1995) Discrimination of monet and picasso in pigeons. J Exp Anal Behav 63:165–174
Watanabe S, Uozumi M, Tanaka K (2005) Discrimination of consonance and dissonance in Java sparrows. Behav Process 70:203–208
Watanabe S, Suzuki T, Yamazaki Y (2009) Reinforcing property of music for non-human animals: analysis with pigeons. Philosophy 121:1–21
Wells DL, Coleman D, Challis MGA (2006) A note on the effect of auditory stimulation on the behaviour and welfare of zoo-housed gorillas. Appl Anim Behav Sci 100:327–332
Wilson EO (1983) Biophilia. Harvard University Press, Cambridge
Wilson, FA, Goldman-Rakic PS (1994) Viewing preferences of rhesus monkeys related to memory for complex pictures, colours and faces. Behav Brain Res 60:79–89
Wood GA (1984) Tool use by the palm cockatoo *Probosciger aterrimus* during display. Corella 8:94–95
Wypijewski J (1997) Painting by numbers; komar and melamid's scientific guide to art. Farrar, Straus & Giroux, New York
Zeller A (2007) "What's in a picture?" A comparison of drawings by ape and children. Semiotica 166:181–214

MIX
Papier aus verantwortungsvollen Quellen
Paper from responsible sources
FSC® C105338

If you have any concerns about our products,
you can contact us on
ProductSafety@springernature.com

In case Publisher is established outside the EU,
the EU authorized representative is:
Springer Nature Customer Service Center GmbH
Europaplatz 3, 69115 Heidelberg, Germany

Printed by Libri Plureos GmbH
in Hamburg, Germany